Genevieve B. Orr   Klaus-Robert Müller   (Eds.)

Neural Networks: Tricks of the Trade

 Springer

Series Editors

Gerhard Goos, Karlsruhe University, Germany
Juris Hartmanis, Cornell University, NY, USA
Jan van Leeuwen, Utrecht University, The Netherlands

Volume Editors

Genevieve B. Orr
Willamette University, Department of Computer Science
Salem, OR 97301, USA
E-mail: gorr@willamette.edu

Klaus-Robert Müller
GMD First (Forschungszentrum Informationstechnik)
Rudower Chaussee 5, D-12489 Berlin, Germany
E-mail: klaus@first.gmd.de

Cataloging-in-Publication data applied for

Die Deutsche Bibliothek - CIP-Einheitsaufnahme

Neural networks : tricks of the trade / Genevieve B. Orr ; Klaus-Robert Müller
(ed.). - Berlin ; Heidelberg ; New York ; Barcelona ; Hong Kong ; London ;
Milan ; Paris ; Singapore ; Tokyo : Springer, 1998
  (Lecture notes in computer science ; Vol. 1524)
  ISBN 3-540-65311-2

CR Subject Classification (1998): F.1, I.2.6, I.5.1, C.1.3

ISSN 0302-9743
ISBN 3-540-65311-2 Springer-Verlag Berlin Heidelberg New York

This work is subject to copyright. All rights are reserved, whether the whole or part of the material is
concerned, specifically the rights of translation, reprinting, re-use of illustrations, recitation, broadcasting,
reproduction on microfilms or in any other way, and storage in data banks. Duplication of this publication
or parts thereof is permitted only under the provisions of the German Copyright Law of September 9, 1965,
in its current version, and permission for use must always be obtained from Springer-Verlag. Violations are
liable for prosecution under the German Copyright Law.

© Springer-Verlag Berlin Heidelberg 1998
Printed in Germany

Typesetting: Camera-ready by author
SPIN 10692833   06/3142 – 5 4 3 2 1 0   Printed on acid-free paper

Table of Contents

Introduction 1

Speeding Learning

Preface 7

1 Efficient BackProp 9
   Yann LeCun, Leon Bottou, Genevieve B. Orr, and Klaus-Robert Müller

Regularization Techniques to Improve Generalization

Preface 51

2 Early Stopping – But When? 55
   Lutz Prechelt

3 A Simple Trick for Estimating the Weight Decay Parameter 71
   Thorsteinn S. Rögnvaldsson

4 Controling the Hyperparameter Search in MacKay's Bayesian Neural
   Network Framework 93
   Tony Plate

5 Adaptive Regularization in Neural Network Modeling 113
   Jan Larsen, Claus Svarer, Lars Nonboe Andersen, and Lars Kai Hansen

6 Large Ensemble Averaging 133
   David Horn, Ury Naftaly, and Nathan Intrator

Improving Network Models and Algorithmic Tricks

Preface 141

7 Square Unit Augmented, Radially Extended, Multilayer Perceptrons . . . 145
   Gary William Flake

8 A Dozen Tricks with Multitask Learning 165
   Rich Caruana

9 Solving the Ill-Conditioning in Neural Network Learning 193
   Patrick van der Smagt and Gerd Hirzinger

10 Centering Neural Network Gradient Factors 207
   Nicol N. Schraudolph

11 Avoiding Roundoff Error in Backpropagating Derivatives 227
   Tony Plate

Representing and Incorporating Prior Knowledge in Neural Network Training

Preface 235

12 Transformation Invariance in Pattern Recognition – Tangent Distance
   and Tangent Propagation 239
   Patrice Y. Simard, Yann A. LeCun, John S. Denker, and Bernard Victorri

13 Combining Neural Networks and Context-Driven Search for On-Line,
   Printed Handwriting Recognition in the Newton 275
   Larry S. Yaeger, Brandyn J. Webb, and Richard F. Lyon

14 Neural Network Classification and Prior Class Probabilities 299
   Steve Lawrence, Ian Burns, Andrew Back, Ah Chung Tsoi, and C. Lee Giles

15 Applying Divide and Conquer to Large Scale Pattern Recognition Tasks 315
   Jürgen Fritsch and Michael Finke

Tricks for Time Series

Preface 343

16 Forecasting the Economy with Neural Nets: A Survey of Challenges
   and Solutions 347
   John Moody

17 How to Train Neural Networks 373
   Ralph Neuneier and Hans Georg Zimmermann

Author Index 425

Subject Index 427

Introduction

It is our belief that researchers and practitioners acquire, through experience and word-of-mouth, techniques and heuristics that help them successfully apply neural networks to difficult real world problems. Often these "tricks" are theoretically well motivated. Sometimes they are the result of trial and error. However, their most common link is that they are usually hidden in people's heads or in the back pages of space-constrained conference papers. As a result newcomers to the field waste much time wondering why their networks train so slowly and perform so poorly.

This book is an outgrowth of a 1996 NIPS workshop called *Tricks of the Trade* whose goal was to begin the process of gathering and documenting these tricks. The interest that the workshop generated motivated us to expand our collection and compile it into this book. Although we have no doubt that there are many tricks we have missed, we hope that what we have included will prove to be useful, particularly to those who are relatively new to the field. Each chapter contains one or more tricks presented by a given author (or authors). We have attempted to group related chapters into sections, though we recognize that the different sections are far from disjoint. Some of the chapters (e.g., 1, 13, 17) contain entire systems of tricks that are far more general than the category they have been placed in.

Before each section we provide the reader with a summary of the tricks contained within, to serve as a quick overview and reference. However, we do not recommend applying tricks before having read the accompanying chapter. Each trick may only work in a particular context that is not fully explained in the summary. This is particularly true for the chapters that present systems where combinations of tricks must be applied together for them to be effective.

Below we give a rough roadmap of the contents of the individual chapters.

Speeding Learning

The book opens with a chapter based on Leon Bottou and Yann LeCun's popular workshop on efficient backpropagation where they present a system of tricks for speeding the minimization process. Included are tricks that are very simple to implement as well as more complex ones, e.g., based on second-order methods. Though many of the readers may recognize some of these tricks, we believe that this chapter provides both a thorough explanation of their theoretical basis and an understanding of the subtle interactions among them.

This chapter provides an ideal introduction for the reader. It starts by discussing fundamental tricks addressing input representation, initialization, target values, choice of learning rates, choice of the nonlinearity, and so on. Subsequently, the authors introduce in great detail tricks for estimation and approximation of the Hessian in neural networks. This provides the basis for a discussion of second-order algorithms, fast training methods like the stochastic Levenberg-Marquardt algorithm, and tricks for learning rate adaptation.

Regularization Techniques to Improve Generalization

Fast minimization is important but only if we can also insure good generalization. We therefore next include a collection of chapters containing a range of approaches for improving generalization. As one might expect, there are no tricks that work well in all situations. However, many examples and discussions are included to help the reader to decide which will work best for their own problem.

Chapter 2 addresses what is one of the most commonly used techniques: early stopping. Here Lutz Prechelt discusses the pitfalls of this seemingly simple technique. He quantifies the tradeoff between generalization and training time for various stopping criteria, which leads to a trick for picking an appropriate criterion.

Using a weight decay penalty term in the cost function is another common method for improving generalization. The difficulty, however, is in finding a good estimate of the weight decay parameter. In Chapter 3, Thorsteinn Rögnvaldsson presents a fast technique for finding a good estimate, surprisingly, by using information measured at the early stopping point. Experimental evidence for its usefulness is given in several applications.

Tony Plate in Chapter 4 treats the penalty terms along the lines of MacKay, i.e. as hyperparameters to be found through iterative search. He presents and compares tricks for making the hyperparameter search in classification networks work in practice by speeding it up and simplifying it. Key to his success is a control of the frequency of the hyperparameter updates and a better strategy in cases where the Hessian becomes out-of-bounds.

In Chapter 5, Jan Larsen et al. present a trick for adapting regularization parameters by using simple gradient descent (with respect to the regularization parameters) on the validation error. The trick is tested on both classification and regression problems.

Averaging over multiple predictors is a well known method for improving generalization. Two questions that arise are how many predictors are "enough" and how the number of predictors affects the stopping criteria for early stopping. In the final chapter of this section, David Horn et al. present solutions to these questions by providing a method for estimating the error of an infinite number of predictors. They then demonstrate this trick for a prediction task.

Improving Network Models and Algorithmic Tricks

In this section we examine tricks that help improve the network model. Even though standard multilayer perceptrons (MLPs) are, in theory, universal approximators, other architectures may provide a more natural fit to a problem. A better fit means that training is faster and that there is a greater likelihood of finding a good and stable solution. For example, radial basis functions (RBFs) are preferred for problems that exhibit local features in a finite region. Of course, which architecture to choose is not always obvious.

In Chapter 7, Gary Flake presents a trick that gives MLPs the power of both an MLP and an RBF so that one does not need to choose between them. This trick is simply to add extra inputs whose values are the square of the regular inputs. Both a theoretical and intuitive explanation are presented along with a number of simulation examples.

Rich Caruana in Chapter 8 shows that performance can be improved on a main task by adding extra outputs to a network that predict related tasks. This technique, known as multi-task learning (MTL), trains these extra outputs in parallel with the main task. This chapter presents multiple examples of what one might use as these extra outputs as well as techniques for implementing MTL effectively. Empirical examples include mortality rankings for pneumonia and road-following in a network learning to steer a vehicle.

Patrick van der Smagt and Gerd Hirzinger consider in Chapter 9 the ill-conditioning of the Hessian in neural network training and propose using what they call a linearly augmented feed-forward network, employing input/output short-cut connections that share the input/hidden weights. This gives rise to better conditioning of the learning problem and, thus, to faster learning, as shown in a simulation example with data from a robot arm.

In Chapter 10, Nicol Schraudolph takes the idea of scaling and centering the inputs even further than Chapter 1 by proposing to center all factors in the neural network gradient: inputs, activities, error signals and hidden unit slopes. He gives experimental evidence for the usefulness of this trick.

In Chapter 11, Tony Plate's short note reports a numerical trick for computing derivatives more accurately with only a small memory overhead.

Representation and Incorporating Prior Knowledge in Neural Network Training

Previous chapters (e.g., Chapter 1) present very general tricks for transforming inputs to improve learning: prior knowledge of the problem is not taken into account explicitly (of course regularization, as discussed in Chapters 2–5, implicitly assumes a prior but on the weight distribution). For complex, difficult problems, however, it is not enough to take a black box approach, no matter how good that black box might be. This section examines how prior knowledge about a problem can be used to greatly improve learning. The questions asked include how to best represent the data, how to make use of this representation for training, and how to take advantage of the invariances that are present. Such issues are key for proper neural network training. They are also at the heart of the tricks pointed out by Patrice Simard et al. in the first chapter of this section. Here, the authors present a particularly interesting perspective on how to incorporate prior knowledge into data. They also give the first review of the tangent distance classification method and related techniques evolving from it such as tangent prop. These methods are applied to the difficult task of optical character recognition (OCR).

In Chapter 13, Larry Yaeger et al. give an overview of the tricks and techniques for on-line handwritten character recognition that were eventually used in Apple Computer's Newton MessagePad ®and eMate®. Anyone who has used these systems knows that their handwriting recognition capability works exceedingly well. Although many of the issues that are discussed in this chapter overlap with those in OCR, including representation and prior knowledge, the solutions are complementary. This chapter also gives a very nice overview of what design choices proved to be efficient as well as how different tricks such as choice of learning rate, over-representation of more difficult patterns, negative training, error emphasis, and so on work together.

Whether it be handwritten character recognition, speech recognition or medical applications, a particularly difficult problem encountered is the unbalanced class prior probabilities that occur, for example, when certain writing styles and subphoneme classes are uncommon or certain illnesses occur less frequently. Chapter 13 briefly discusses this problem in the context of handwriting recognition and presents a heuristic which controls the frequency with which samples are picked for training.

In Chapter 14, Steve Lawrence et al. discuss the issue of unbalanced class prior probabilities in greater depth. They present and compare several different heuristics (prior scaling, probabilistic sampling, post scaling and class membership equalization) one of which is similar to that in Chapter 13. They demonstrate their tricks by solving an ECG classification problem and provide some theoretical explanations.

Many training techniques work well for nets of small to moderate size. However, when problems consist of thousands of classes and millions of examples, not uncommon in applications such as speech recognition, many of these techniques break down. This chapter by Jürgen Fritsch and Michael Finke is devoted to the issue of large scale classification problems and representation design in general. Here the problem of unbalanced class prior probabilities is also tackled.

Although Fritsch and Finke specifically exemplify their design approach for the problem of building a large vocabulary speech recognizer, it becomes clear that these techniques are also applicable to the general construction of an appropriate hierarchical decision tree. A particularly interesting result in this paper is that the structural design for incorporating prior knowledge about speech done by a human speech expert was outperformed by their machine learning technique using an agglomerative clustering algorithm for choosing the structure of the decision tree.

Tricks for Time Series

We close the book with two papers on the subject of time series and economic forecasting. In the first of these chapters, John Moody presents an excellent survey of both the challenges of macroeconomic forecasting and a number of neural network solutions. The survey is followed by a more detailed description of smoothing regularizers, model selection methods (e.g., AIC, effective number of parameters, nonlinear cross-validation), and input selection via sensitivity-based input pruning. Model interpretation and visualization are also discussed.

In the final chapter, Ralph Neuneier and Hans Georg Zimmermann present an impressive integrated system for neural network training of time series and economic forecasting. Every aspect of the system is discussed including input preprocessing, cost functions, handling of outliers, architecture, regularization techniques, and solutions for dealing with the problem of bottom-heavy networks, i.e., the input dimension is large while the output dimension is very small. There is also a thought-provoking discussion of the Observer-Observer dilemma: we want both to create a model based on observed data and, at the same time, to use this model to judge the correctness of new incoming data. Even those people not interested specifically in economic forecasting are encouraged to read this very useful example of how to incorporate prior (system) knowledge into training.

Final Remark

As a final remark, we note that some of the views taken in the chapters contradict each other, e.g., some authors favor one regularization method over another, while other authors make exactly the opposite statement. On the one hand, one can explain these discrepancies by stating that the field is still very active and therefore opposing viewpoints will inevitably exist until more is understood. On the other hand, it may be that both (contradicting) views are correct but on different data sets and in different applications, e.g., an approach that considers noisy time-series needs algorithms with a completely different robustness than in, say, an OCR setting. In this sense, the present book mirrors an active field and a variety of applications with its diversity of views.

August 1998   Jenny & Klaus

Acknowledgments

We would like to thank all authors for their collaboration. Special thanks to Steven Lemm for considerable help with the typesetting. K.-R.M. acknowledges partial financial support from DFG (grant JA 379/51 and JA 379/7) and EU ESPRIT (grant 25387-STORM).

Speeding Learning

Preface

There are those who argue that developing fast algorithms is no longer necessary because computers have become so fast. However, we believe that the complexity of our algorithms and the size of our problems will always expand to consume all cycles available, regardless of the speed of our machines. Thus, there will never come a time when computational efficiency can or should be ignored. Besides, in the quest to find solutions faster, we also often find better and more stable solutions as well. This section is devoted to techniques for making the learning process in backpropagation (BP) faster and more efficient. It contains a single chapter based on a workshop by Leon Bottou and Yann LeCun. While many alternative learning systems have emerged since the time BP was first introduced, BP is still the most widely used learning algorithm. The reason for this is its simplicity, efficiency, and its general effectiveness on a wide range of problems. Even so, there are many pitfalls in applying it, which is where all these tricks enter.

Chapter 1 begins gently by introducing us to a few practical tricks that are very simple to implement. Included are easy to understand qualitative explanations of each. There is a discussion of **stochastic (on-line) vs batch mode learning** where the advantages and disadvantages of both are presented while making it clear that stochastic learning is most often preferred (p. 13). There is a trick that aims at maximizing the per iteration information presented to the network simply by knowing **how best to shuffle the examples** (p. 15). This is followed by an entire set of tricks that must be coordinated together for maximum effectiveness. These include:

- how to **normalize, decorrelate, and scale the inputs** (p. 16)
- how to **choose the sigmoid** (p. 17)
- how to **set target values** (classification) (p. 19)
- how to **initialize the weights** (p. 20)
- how to **pick the learning rates** (p. 20).

Additional issues discussed include the effectiveness of momentum and the choice between radial basis units and sigmoid units (p. 21).

Chapter 1 then introduces us to a little of the theory, providing deeper understanding of some of the preceding tricks. Included are discussions of the effect of learning rates on the speed of learning and of the relationship between the Hessian matrix, the error surface, and the learning rates. Simple examples of linear and multilayer nets are provided to illustrate the theoretical results.

The chapter next enters more difficult territory by giving an overview of second order methods (p. 31). Quickly summarized here, they are

Newton method: generally impractical to use since it requires inverting the full Hessian and works only in batch mode.
conjugate gradient: an $O(N)$ algorithm that doesn't use the Hessian, but requires a line search and so works only in batch mode.
Quasi-Newton, Broyden-Fletcher-Goldfarb-Shanno (BFGS) method: an $O(N^2)$ algorithm that computes an estimate of the inverse Hessian. It requires line search and also only works in batch mode.
Gauss-Newton method: an $O(N^3)$ algorithm that uses the square Jacobi approximation of the Hessian. Mainly used for batch and works only for mean squared error loss functions.
Levenberg Marquardt method: extends the Gauss-Newton method to include a regularization parameter for stability.

Second order methods can greatly speed learning at each iteration but often at an excessive computational cost. However, by replacing the exact Hessian with an approximation of either the full or partial Hessian, the benefits of second order information can still be reaped without incurring as great a computational cost.

The first and most direct method for **estimating the full Hessian** is finite differences which simply requires little more than two backpropagations to compute each row of the Hessian (p. 36). Another is to use the square Jacobian approximation which guarantees a positive semi-definite matrix which may be beneficial for improving stability. If even more simplification is desired, one can just compute the diagonal elements of the Hessian. All of the methods mentioned here are easily implemented using BP.

Unfortunately, for very large networks, many of the classical second order methods do not work well because storing the Hessian is far too expensive and because batch mode, required by most of the methods, is too slow. *On-line* second order methods are needed instead. One such technique presented here is a **stochastic diagonal Levenberg Marquardt method** (p. 41).

If all that is needed is the **product of the Hessian with an arbitrary vector** rather than the Hessian itself, then much time can be saved using a method that computes this entire product directly using only a single backpropagation step (p. 38). Such a technique can be used to compute the largest eigenvalue and associated eigenvector of the Hessian. The inverse of the largest eigenvalue can then be used to obtain a good estimate of the learning rate.

Finally, three useful tricks are presented for **computing the principal eigenvalue** and vector without having to compute the Hessian: the power method, Taylor expansion, and an on-line method (p. 44).

Jenny & Klaus

1
Efficient BackProp

Yann LeCun[1], Leon Bottou[1], Genevieve B. Orr[2], and Klaus-Robert Müller[3]

[1] Image Processing Research Department AT&T Labs - Research, 100 Schulz Drive, Red Bank, NJ 07701-7033, USA
[2] Willamette University, 900 State Street, Salem, OR 97301, USA
[3] GMD FIRST, Rudower Chaussee 5, 12489 Berlin, Germany
{yann,leonb}@research.att.com, gorr@willamette.edu, klaus@first.gmd.de

Abstract. The convergence of back-propagation learning is analyzed so as to explain common phenomenon observed by practitioners. Many undesirable behaviors of backprop can be avoided with tricks that are rarely exposed in serious technical publications. This paper gives some of those tricks, and offers explanations of why they work.
Many authors have suggested that second-order optimization methods are advantageous for neural net training. It is shown that most "classical" second-order methods are impractical for large neural networks. A few methods are proposed that do not have these limitations.

1.1 Introduction

Backpropagation is a very popular neural network learning algorithm because it is conceptually simple, computationally efficient, and because it often works. However, getting it to work well, and sometimes to work at all, can seem more of an art than a science. Designing and training a network using backprop requires making many seemingly arbitrary choices such as the number and types of nodes, layers, learning rates, training and test sets, and so forth. These choices can be critical, yet there is no foolproof recipe for deciding them because they are largely problem and data dependent. However, there are heuristics and some underlying theory that can help guide a practitioner to make better choices.

In the first section below we introduce standard backpropagation and discuss a number of simple heuristics or tricks for improving its performance. We next discuss issues of convergence. We then describe a few "classical" second-order non-linear optimization techniques and show that their application to neural network training is very limited, despite many claims to the contrary in the literature. Finally, we present a few second-order methods that do accelerate learning in certain cases.

1.2 Learning and Generalization

There are several approaches to automatic machine learning, but much of the successful approaches can be categorized as *gradient-based learning methods*. The learning machine, as represented in Figure 1.1, computes a function $M(Z^p, W)$ where Z^p is the p-th input pattern, and W represents the collection of adjustable parameters in the system. A cost function $E^p = C(D^p, M(Z^p, W))$, measures the discrepancy between D^p, the "correct" or desired output for pattern Z^p, and the output produced by the system. The average cost function $E_{train}(W)$ is the average of the errors E^p over a set of input/output pairs called the training set $\{(Z^1, D^1), ...(Z^P, D^P)\}$. In the simplest setting, the learning problem consists in finding the value of W that minimizes $E_{train}(W)$. In practice, the performance of the system on a training set is of little interest. The more relevant measure is the error rate of the system in the field, where it would be used in practice. This performance is estimated by measuring the accuracy on a set of samples disjoint from the training set, called the test set. The most commonly used cost function is the Mean Squared Error:

$$E^p = \frac{1}{2}(D^p - M(Z^p, W))^2, \qquad E_{train} = \frac{1}{P}\sum_{p=1}^{P} E^p$$

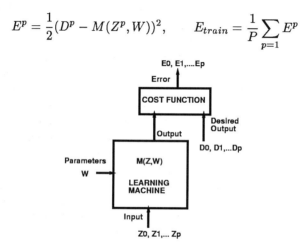

Fig. 1.1. Gradient-based learning machine.

This chapter is focused on strategies for improving the process of minimizing the cost function. However, these strategies must be used in conjunction with methods for maximizing the network's ability to *generalize*, that is, to predict the correct targets for patterns the learning system has not previously seen (e.g. see chapters 2, 3, 4, 5 for more detail).

To understand generalization, let us consider how backpropagation works. We start with a set of samples each of which is an input/output pair of the function to be learned. Since the measurement process is often noisy, there may be errors in the samples. We can imagine that if we collected multiple *sets* of samples then each set would look a little different because of the noise and because of the different points sampled. Each of these data sets would also result in networks with minima that are slightly different from each other and from the true function. In this chapter, we concentrate on improving the process of finding the minimum for the particular set of examples that we are given. Generalization techniques try to correct for the errors introduced into the network as a result of our choice of dataset. Both are important.

Several theoretical efforts have analyzed the process of learning by minimizing the error on a training set (a process sometimes called Empirical Risk Minimization) [40, 41].

Some of those theoretical analyses are based on decomposing the generalization error into two terms: bias and variance (see e.g. [12]). The bias is a measure of how much the network output, averaged over all possible data sets differs from the desired function. The variance is a measure of how much the network output varies between datasets. Early in training, the bias is large because the network output is far from the desired function. The variance is very small because the data has had little influence yet. Late in training, the bias is small because the network has learned the underlying function. However, if trained too long, the network will also have learned the noise specific to that dataset. This is referred to as overtraining. In such a case, the variance will be large because the noise varies between datasets. It can be shown that the minimum total error will occur when the sum of bias and variance are minimal.

There are a number of techniques (e.g. early stopping, regularization) for maximizing the generalization ability of a network when using backprop. Many of these techniques are described in later chapters 2, 3, 5, 4.

The idea of this chapter, therefore, is to present minimization strategies (given a cost function) and the tricks associated with increasing the speed and quality of the minimization. It is however clear that the choice of the model (model selection), the architecture and the cost function is crucial for obtaining a network that generalizes well. So keep in mind that if the wrong model class is used and no proper model selection is done, then even a superb minimization will clearly not help very much. In fact, the existence of overtraining has led several authors to suggest that inaccurate minimization algorithms can be better than good ones.

1.3 Standard Backpropagation

Although the tricks and analyses in this paper are primarily presented in the context of "classical" multi-layer feed-forward neural networks, many of them also apply to most other gradient-based learning methods.

The simplest form of multilayer learning machine trained with gradient-based learning is simply a stack of modules, each of which implements a function $X_n = F_n(W_n, X_{n-1})$, where X_n is a vector representing the output of the module, W_n is the vector of tunable parameters in the module (a subset of W), and X_{n-1} is the module's input vector (as well as the previous module's output vector). The input X_0 to the first module is the input pattern Z^p. If the partial derivative of E^p with respect to X_n is known, then the partial derivatives of E^p with respect to W_n and X_{n-1} can be computed using the backward recurrence

$$\frac{\partial E^p}{\partial W_n} = \frac{\partial F}{\partial W}(W_n, X_{n-1})\frac{\partial E^p}{\partial X_n}$$

$$\frac{\partial E^p}{\partial X_{n-1}} = \frac{\partial F}{\partial X}(W_n, X_{n-1})\frac{\partial E^p}{\partial X_n} \tag{1.1}$$

where $\frac{\partial F}{\partial W}(W_n, X_{n-1})$ is the Jacobian of F with respect to W evaluated at the point (W_n, X_{n-1}), and $\frac{\partial F}{\partial X}(W_n, X_{n-1})$ is the Jacobian of F with respect to X. The Jacobian of a vector function is a matrix containing the partial derivatives of all the outputs with respect to all the inputs. When the above equations are applied to the modules in reverse order, from layer N to layer 1, all the partial derivatives of the cost function with respect to all the parameters can be computed. The way of computing gradients is known as back-propagation.

Traditional multi-layer neural networks are a special case of the above system where the modules are alternated layers of matrix multiplications (the weights) and component-wise sigmoid functions (the units):

$$Y_n = W_n X_{n-1} \tag{1.2}$$
$$X_n = F(Y_n) \tag{1.3}$$

where W_n is a matrix whose number of columns is the dimension of X_{n-1}, and number of rows is the dimension of X_n. F is a vector function that applies a sigmoid function to each component of its input. Y_n is the vector of weighted sums, or *total inputs*, to layer n.

Applying the chain rule to the equation above, the classical backpropagation equations are obtained:

$$\frac{\partial E^p}{\partial y_n^i} = f'(y_n^i)\frac{\partial E^p}{\partial x_n^i} \tag{1.4}$$

$$\frac{\partial E^p}{\partial w_n^{ij}} = x_{n-1}^j \frac{\partial E^p}{\partial y_n^i} \tag{1.5}$$

$$\frac{\partial E^p}{\partial x_{n-1}^k} = \sum_i w_n^{ik} \frac{\partial E^p}{\partial y_n^i}. \tag{1.6}$$

The above equations can also be written in matrix form:

$$\frac{\partial E^p}{\partial Y_n} = F'(Y_n)\frac{\partial E^p}{\partial X_n} \tag{1.7}$$

$$\frac{\partial E^p}{\partial W_n} = X_{n-1}\frac{\partial E^p}{\partial Y_n} \tag{1.8}$$

$$\frac{\partial E^p}{\partial X_{n-1}} = W_n^T \frac{\partial E^p}{\partial Y_n}. \tag{1.9}$$

The simplest learning (minimization) procedure in such a setting is the gradient descent algorithm where W is iteratively adjusted as follows:

$$W(t) = W(t-1) - \eta\frac{\partial E}{\partial W}. \tag{1.10}$$

In the simplest case, η is a scalar constant. More sophisticated procedures use variable η. In other methods η takes the form of a diagonal matrix, or is an estimate of the inverse Hessian matrix of the cost function (second derivative matrix) such as in the Newton and Quasi-Newton methods described later in the chapter. A proper choice of η is important and will be discussed at length later.

1.4 A Few Practical Tricks

Backpropagation can be very slow particularly for multilayered networks where the cost surface is typically non-quadratic, non-convex, and high dimensional with many local minima and/or flat regions. There is no formula to guarantee that (1) the network will converge to a good solution, (2) convergence is swift, or (3) convergence even occurs at all. However, in this section we discuss a number of tricks that can greatly improve the chances of finding a good solution while also decreasing the convergence time often by orders of magnitude. More detailed theoretical justifications will be given in later sections.

1.4.1 Stochastic versus Batch learning.

At each iteration, equation (1.10) requires a complete pass through the entire dataset in order to compute the *average* or true gradient. This is referred to as batch learning since an entire "batch" of data must be considered before weights are updated. Alternatively, one can use stochastic (online) learning where a *single* example $\{Z^t, D^t\}$ is chosen (e.g. randomly) from the training set at each iteration t. An *estimate* of the true gradient is then computed based on the error E^t of that example, and then the weights are updated:

$$W(t+1) = W(t) - \eta \frac{\partial E^t}{\partial W}. \qquad (1.11)$$

Because this estimate of the gradient is noisy, the weights may not move precisely down the gradient at each iteration. As we shall see, this "noise" at each iteration can be advantageous. Stochastic learning is generally the preferred method for basic backpropagation for the following three reasons:

Advantages of Stochastic Learning
1. Stochastic learning is usually *much* faster than batch learning.
2. Stochastic learning also often results in better solutions.
3. Stochastic learning can be used for tracking changes.

Stochastic learning is most often *much* faster than batch learning particularly on large redundant datasets. The reason for this is simple to show. Consider the simple case where a training set of size 1000 is inadvertently composed of 10 identical copies of a set with 100 samples. Averaging the gradient over all 1000 patterns gives the exact same result as computing the gradient based on just the first 100. Thus, batch gradient descent is wasteful because it recomputes the same quantity 10 times before one parameter update. On the other hand, stochastic gradient will see a full epoch as 10 iterations through a 100-long training set. In practice, examples rarely appear more than once in a dataset, but there are usually clusters of patterns that are very similar. For example in phoneme classification, all of the patterns for the phoneme /æ/ will (hopefully) contain much of the same information. It is this redundancy that can make batch learning much slower than on-line.

Stochastic learning also often results in better solutions because of the noise in the updates. Nonlinear networks usually have multiple local minima of differing depths. The goal of training is to locate one of these minima. Batch learning will discover the minimum of whatever basin the weights are initially placed. In stochastic learning, the noise present in the updates can result in the weights jumping into the basin of another, possibly deeper, local minimum. This has been demonstrated in certain simplified cases [15, 30].

Stochastic learning is also useful when the function being modeled is changing over time, a quite common scenario in industrial applications where the data distribution changes gradually over time (e.g. due to wear and tear of the machines). If the learning machine does not detect and follow the change it is impossible to learn the data properly and large generalization errors will result. With batch learning, changes go undetected and we obtain rather bad results since we are likely to average over several rules, whereas on-line learning – if operated properly (see below in section 1.4.7) – will track the changes and yield good approximation results.

Despite the advantages of stochastic learning, there are still reasons why one might consider using batch learning:

Advantages of Batch Learning
1. Conditions of convergence are well understood.
2. Many acceleration techniques (e.g. conjugate gradient) only operate in batch learning.
3. Theoretical analysis of the weight dynamics and convergence rates are simpler.

These advantages stem from the same noise that make stochastic learning advantageous. This noise, which is so critical for finding better local minima also prevents full convergence to the minimum. Instead of converging to the exact minimum, the convergence stalls out due to the weight fluctuations. The size of the fluctuations depend on the degree of noise of the stochastic updates. The variance of the fluctuations around the local minimum is proportional to the learning rate η [28, 27, 6]. So in order to reduce the fluctuations we can either decrease (anneal) the learning rate or have an adaptive batch size. In theory [13, 30, 36, 35] it is shown that the optimal annealing schedule of the learning rate is of the form

$$\eta \sim \frac{c}{t}, \tag{1.12}$$

where t is the number of patterns presented and c is a constant. In practice, this may be too fast (see chapter 13).

Another method to remove noise is to use "mini-batches", that is, start with a small batch size and increase the size as training proceeds. Møller discusses one method for doing this [25] and Orr [31] discusses this for linear problems. However, deciding the rate at which to increase the batch size and which inputs to include in the small batches is as difficult as determining the proper learning rate. Effectively the size of the learning rate in stochastic learning corresponds to the respective size of the mini batch.

Note also that the problem of removing the noise in the data may be less critical than one thinks because of generalization. Overtraining may occur long before the noise regime is even reached.

Another advantage of batch training is that one is able to use second order methods to speed the learning process. Second order methods speed learning by estimating not just the gradient but also the curvature of the cost surface. Given the curvature, one can estimate the approximate location of the actual minimum.

Despite the advantages of batch updates, stochastic learning is still often the preferred method particularly when dealing with very large data sets because it is simply much faster.

1.4.2 Shuffling the Examples

Networks learn the fastest from the most unexpected sample. Therefore, it is advisable to choose a sample at each iteration that is the most unfamiliar to the system. Note, this applies only to stochastic learning since the order of input presentation is irrelevant for batch[1]. Of course, there is no simple way to know which inputs are information rich, however, a very simple trick that crudely implements this idea is to simply choose successive examples that are from *different* classes since training examples belonging to the same class will most likely contain similar information.

Another heuristic for judging how much new information a training example contains is to examine the error between the network output and the target value when this input is presented. A large error indicates that this input has not been learned by the network and so contains a lot of new information. Therefore, it makes sense to present this input more frequently. Of course, by "large" we mean relative to all of the other training examples. As the network trains, these relative errors will change and so should the frequency of presentation for a particular input pattern. A method that modifies the probability of appearance of each pattern is called an *emphasizing scheme*.

> **Choose Examples with Maximum Information Content**
> 1. Shuffle the training set so that successive training examples never (rarely) belong to the same class.
> 2. Present input examples that produce a large error more frequently than examples that produce a small error.

---

[1] The order in which gradients are summed in batch may be affected by roundoff error if there is a significant range of gradient values.

However, one must be careful when perturbing the normal frequencies of input examples because this changes the relative importance that the network places on different examples. This may or may not be desirable. For example, *this technique applied to data containing outliers can be disastrous* because outliers can produce large errors yet should not be presented frequently. On the other hand, this technique can be particularly beneficial for boosting the performance for infrequently occurring inputs, e.g. /z/ in phoneme recognition (see chapter 13, 14).

1.4.3 Normalizing the Inputs

Convergence is usually faster if the average of each input variable over the training set is close to zero. To see this, consider the extreme case where all the inputs are positive. Weights to a particular node in the first weight layer are updated by an amount proportional to δx where δ is the (scalar) error at that node and x is the input vector (see equations (1.5) and (1.10)). When all of the components of an input vector are positive, all of the updates of weights that feed into a node will be the same sign (i.e. sign(δ)). As a result, these weights can only all decrease or all increase *together* for a given input pattern. Thus, if a weight vector must change direction it can only do so by zigzagging which is inefficient and thus very slow.

In the above example, the inputs were all positive. However, in general, any shift of the average input away from zero will bias the updates in a particular direction and thus slow down learning. Therefore, it is good to shift the inputs so that the average over the training set is close to zero. This heuristic should be applied at all layers which means that we want the average of the *outputs* of a node to be close to zero because these outputs are the inputs to the next layer [19], chapter 10. This problem can be addressed by coordinating how the inputs are transformed with the choice of sigmoidal activation function. Here we discuss the input transformation. The discussion of the sigmoid follows.

Convergence is faster not only if the inputs are shifted as described above but also if they are scaled so that all have about the same covariance, C_i, where

$$C_i = \frac{1}{P}\sum_{p=1}^{P}(z_i^p)^2. \qquad (1.13)$$

Here, P is the number of training examples, C_i is the covariance of the ith input variable and z_i^p is the ith component of the pth training example. Scaling speeds learning because it helps to balance out the rate at which the weights connected to the input nodes learn. The value of the covariance should be matched with that of the sigmoid used. For the sigmoid given below, a covariance of 1 is a good choice.

The exception to scaling all covariances to the same value occurs when it is known that some inputs are of less significance than others. In such a case, it can be beneficial to scale the less significant inputs down so that they are "less visible" to the learning process.

> **Transforming the Inputs**
> 1. The average of each input variable over the training set should be close to zero.
> 2. Scale input variables so that their covariances are about the same.
> 3. Input variables should be uncorrelated if possible.

The above two tricks of shifting and scaling the inputs are quite simple to implement. Another trick that is quite effective but more difficult to implement is to decorrelate the inputs. Consider the simple network in Figure 1.2. If inputs are uncorrelated then it is possible to solve for the value of w_1 that minimizes the error without any concern for w_2, and vice versa. In other words, the two variables are independent (the system of equations is diagonal). With correlated inputs, one must solve for both simultaneously which is a much harder problem. Principal component analysis (also known as the Karhunen-Loeve expansion) can be used to remove *linear* correlations in inputs [10].

Inputs that are linearly dependent (the extreme case of correlation) may also produce degeneracies which may slow learning. Consider the case where one input is always twice the other input ($z_2 = 2z_1$). The network output is constant along lines $W_2 = v - (1/2)W_1$, where v is a constant. Thus, the gradient is zero along these directions (see Figure 1.2). Moving along these lines has absolutely no effect on learning. We are trying to solve in 2-D what is effectively only a 1-D problem. Ideally we want to remove one of the inputs which will decrease the size of the network.

Figure 1.3 shows the entire process of transforming inputs. The steps are (1) shift inputs so the mean is zero, (2) decorrelate inputs, and (3) equalize covariances.

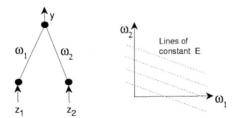

Fig. 1.2. Linearly dependent inputs.

1.4.4 The Sigmoid

Nonlinear activation functions are what give neural networks their nonlinear capabilities. One of the most common forms of activation function is the sigmoid which is a monotonically increasing function that asymptotes at some finite value as $\pm\infty$ is approached. The most common examples are the standard logistic function $f(x) = 1/(1 + e^{-x})$ and hyperbolic tangent $f(x) = \tanh(x)$ shown in

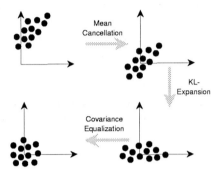

Fig. 1.3. Transformation of inputs.

Figure 1.4. Sigmoids that are symmetric about the origin (e.g. see Figure 1.4b) are preferred for the same reason that inputs should be normalized, namely, because they are more likely to produce outputs (which are *inputs* to the next layer) that are on average close to zero. This is in contrast, say, to the logistic function whose outputs are always positive and so must have a mean that is positive.

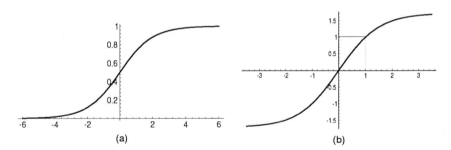

Fig. 1.4. (a) Not recommended: the standard logistic function, $f(x) = 1/(1 + e^{-x})$. (b) Hyperbolic tangent, $f(x) = 1.7159 \tanh\left(\frac{2}{3}x\right)$.

Sigmoids
1. Symmetric sigmoids such as hyperbolic tangent often converge faster than the standard logistic function.
2. A recommended sigmoid [19] is: $f(x) = 1.7159 \tanh\left(\frac{2}{3}x\right)$. Since the tanh function is sometimes computationally expensive, an approximation of it by a ratio of polynomials can be used instead.
3. Sometimes it is helpful to add a small linear term, e.g. $f(x) = \tanh(x) + ax$ so as to avoid flat spots.

The constants in the recommended sigmoid given above have been chosen so that, *when used with transformed inputs* (see previous discussion), the variance of the outputs will also be close to 1 because the effective gain of the sigmoid is roughly 1 over its useful range. In particular, this sigmoid has the properties (a) $f(\pm 1) = \pm 1$, (b) the second derivative is a maximum at $x = 1$, and (c) the effective gain is close to 1.

One of the potential problems with using symmetric sigmoids is that the error surface can be *very* flat near the origin. For this reason it is good to avoid initializing with very small weights. Because of the saturation of the sigmoids, the error surface is also flat far from the origin. Adding a small linear term to the sigmoid can sometimes help avoid the flat regions (see chapter 9).

1.4.5 Choosing Target Values

In classification problems, target values are typically binary (e.g. {-1,+1}). Common wisdom might seem to suggest that the target values be set at the value of the sigmoid's asymptotes. However, this has several drawbacks.

First, instabilities can result. The training process will try to drive the output as close as possible to the target values, which can only be achieved asymptotically. As a result, the weights (output and even hidden) are driven to larger and larger values where the sigmoid derivative is close to zero. The very large weights increase the gradients, however, these gradients are then multiplied by an exponentially small sigmoid derivative (except when a twisting term[2] is added to the sigmoid) producing a weight update close to zero. As a result, the weights may become stuck.

Second, when the outputs saturate, the network gives no indication of confidence level. When an input pattern falls near a decision boundary the output class is uncertain. Ideally this should be reflected in the network by an output value that is in between the two possible target values, i.e. not near either asymptote. However, large weights tend to force all outputs to the tails of the sigmoid regardless of the uncertainty. Thus, the network may predict a wrong class without giving any indication of its low confidence in the result. Large weights that saturate the nodes make it impossible to differentiate between typical and nontypical examples.

A solution to these problems is to set the target values to be within the range of the sigmoid, rather than at the asymptotic values. Care must be taken, however, to insure that the node is not restricted to only the linear part of the sigmoid. Setting the target values to the point of the maximum second derivative on the sigmoid is the best way to take advantage of the nonlinearity without saturating the sigmoid. This is another reason the sigmoid in Figure 1.4b is a good choice. It has maximum second derivative at ± 1 which correspond to the binary target values typical in classification problems.

> **Targets**
> Choose target values at the point of the maximum second derivative on the sigmoid so as to avoid saturating the output units.

---

[2] A twisting term is a small linear term added to the node output, e.g. $f(x) = \tanh(x) + ax$.

1.4.6 Initializing the weights

The starting values of the weights can have a significant effect on the training process. Weights should be chosen randomly but in such a way that the sigmoid is primarily activated in its linear region. If weights are all very large then the sigmoid will saturate resulting in small gradients that make learning slow. If weights are very small then gradients will also be very small. Intermediate weights that range over the sigmoid's linear region have the advantage that (1) the gradients are large enough that learning can proceed and (2) the network will learn the linear part of the mapping before the more difficult nonlinear part.

Achieving this requires coordination between the training set normalization, the choice of sigmoid, and the choice of weight initialization. We start by requiring that the distribution of the outputs of each node have a standard deviation (σ) of approximately 1. This is achieved at the input layer by normalizing the training set as described earlier. To obtain a standard deviation close to 1 at the output of the first hidden layer we just need to use the above recommended sigmoid together with the requirement that the input to the sigmoid also have a standard deviation $\sigma_y = 1$. Assuming the inputs, y_i, to a unit are uncorrelated with variance 1, the standard deviation of the units weighted sum will be

$$\sigma_{y_i} = \left(\sum_j w_{ij}^2 \right)^{1/2}. \tag{1.14}$$

Thus, to insure that the σ_{y_i} are approximately 1 the weights should be randomly drawn from a distribution with mean zero and a standard deviation given by

$$\sigma_w = m^{-1/2} \tag{1.15}$$

where m is the number of inputs to the unit.

Initializing Weights

Assuming that:

1. the training set has been normalized, and
2. the sigmoid from Figure 1.4b has been used

then weights should be randomly drawn from a distribution (e.g. uniform) with mean zero and standard deviation

$$\sigma_w = m^{-1/2} \tag{1.16}$$

where m is the fan-in (the number of connections feeding *into* the node).

1.4.7 Choosing Learning rates

There is at least one well-principled method (described in section 1.9.2) for estimating the ideal learning rate η. Many other schemes (most of them rather empirical) have been proposed in the literature to automatically adjust the learning rate. Most of those schemes decrease the learning rate when the weight vector "oscillates", and increase it when the weight vector follows a relatively steady direction. The main problem with these methods is that they are not appropriate for stochastic gradient or on-line learning because the weight vector fluctuates all the time.

Beyond choosing a single global learning rate, it is clear that picking a different learning rate η_i for each weight can improve the convergence. A well-principled way of doing this, based on computing second derivatives, is described in section 1.9.1. The main philosophy is to make sure that all the weights in the network converge roughly at the same speed.

Depending upon the curvature of the error surface, some weights may require a small learning rate in order to avoid divergence, while others may require a large learning rate in order to converge at a reasonable speed. Because of this, learning rates in the lower layers should generally be larger than in the higher layers (see Figure 1.21). This corrects for the fact that in most neural net architectures, the second derivative of the cost function with respect to weights in the lower layers is generally smaller than that of the higher layers. The rationale for the above heuristics will be discussed in more detail in later sections along with suggestions for how to choose the actual value of the learning rate for the different weights (see section 1.9.1).

If shared weights are used such as in time-delay neural networks (TDNN) [42] or convolutional networks [20], the learning rate should be proportional to the square root of the number of connections sharing that weight, because we know that the gradients are a sum of more-or-less independent terms.

Equalize the Learning Speeds
- give each weight its own learning rate
- learning rates should be proportional to the square root of the number of inputs to the unit
  weights in lower layers should typically be larger than in the higher layers
---

Other tricks for improving the convergence include:

Momentum Momentum

$$\Delta w(t+1) = \eta \frac{\partial E_{t+1}}{\partial w} + \mu \Delta w(t),$$

can increase speed when the cost surface is highly nonspherical because it damps the size of the steps along directions of high curvature thus yielding a larger effective learning rate along the directions of low curvature [43] (μ denotes the strength of the momentum term). It has been claimed that momentum generally helps more in batch mode than in stochastic mode, but no systematic study of this are known to the authors.

Adaptive learning rates Many authors, including Sompolinsky et al. [37], Darken & Moody [9], Sutton [38], Murata et al. [28] have proposed rules for automatically adapting the learning rates (see also [16]). These rules control the speed of convergence by increasing or decreasing the learning rate based on the error.

We assume the following facts for a learning rate adaptation algorithm: (1) the smallest eigenvalue of the Hessian (see Eq.(1.27)) is sufficiently smaller than the second smallest eigenvalue and (2) therefore after a large number of iterations, the parameter vector $w(t)$ will approach the minimum from the direction of the minimum eigenvector of the Hessian (see Eq.(1.27), Figure 1.5). Under these conditions the evolution of the estimated parameter can be thought of as a one-dimensional process and the minimum eigenvector $\mathbf{v}$ can be approximated (for a large number of iterations: see Figure 1.5) by

$$\mathbf{v} = \langle \frac{\partial E}{\partial w} \rangle / \| \langle \frac{\partial E}{\partial w} \rangle \|,$$

where $\| \ \|$ denotes the L^2 norm. Hence we can adopt a projection

$$\xi = \langle \mathbf{v}^T \frac{\partial E}{\partial w} \rangle = \| \langle \frac{\partial E}{\partial w} \rangle \|$$

to the approximated minimum Eigenvector $\mathbf{v}$ as a one dimensional measure of the distance to the minimum. This distance can be used to control the learning rate (for details see [28])

$$w(t+1) = w(t+1) - \eta_t \frac{\partial E_t}{\partial w}, \tag{1.17}$$

$$\mathbf{r}(t+1) = (1-\delta)\mathbf{r}(t) + \delta \frac{\partial E_t}{\partial w}, \quad (0 < \delta < 1) \tag{1.18}$$

$$\eta(t+1) = \eta(t) + \alpha\eta(t)\left(\beta\|\mathbf{r}(t+1)\| - \eta(t)\right), \tag{1.19}$$

where δ controls the leak size of the average, α, β are constants and $\mathbf{r}$ is used as auxiliary variable to calculate the leaky average of the gradient $\frac{\partial E}{\partial w}$.

Note that this set of rules is easy to compute and straightforward to implement. We simply have to keep track of an additional vector in Eq.(1.18): the averaged gradient $\mathbf{r}$. The norm of this vector then controls the size of the learning rate (see Eq.(1.19)). The algorithm follows the simple intuition: far away from the minimum (large distance ξ) it proceeds in big steps and close to the minimum it anneals the learning rate (for theoretical details see [28]).

1.4.8 Radial Basis Functions vs Sigmoid Units

Although most systems use nodes based on dot products and sigmoids, many other types of units (or layers) can be used. A common alternative is the radial basis function (RBF) network (see [7, 26, 5, 32]) In RBF networks, the dot product of the weight and input vector is replaced with a Euclidean distance between

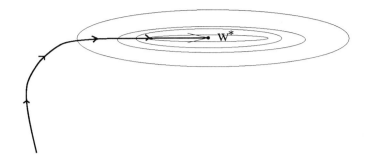

Fig. 1.5. Convergence of the flow. During the final stage of learning the average flow is approximately one dimensional towards the minimum $\mathbf{w}^*$ and it is a good approximation of the minimum eigenvalue direction of the Hessian.

the input and weight and the sigmoid is replaced by an exponential. The output activity is computed, e.g. for one output, as

$$g(x) = \sum_{i=1}^{N} w_i \exp\left(-\frac{1}{2\sigma_i^2}\|x - \nu_i\|^2\right),$$

where ν_i (σ_i) is the mean (standard deviation) of the i-th Gaussian. These units can replace or coexist with the standard units and they are usually trained by combination of gradient descent (for output units) and unsupervised clustering for determining the means and widths of the RBF units.

Unlike sigmoidal units which can cover the entire space, a single RBF unit covers only a small local region of the input space. This can be an advantage because learning can be faster. RBF units may also form a better set of basis functions to model the input space than sigmoid units, although this is highly problem dependent (see chapter 7). On the negative side, the locality property of RBFs may be a disadvantage particularly in high dimensional spaces because may units are needed to cover the spaces. RBFs are more appropriate in (low dimensional) upper layers and sigmoids in (high dimensional) lower layers.

1.5 Convergence of Gradient Descent

1.5.1 A Little Theory

In this section we examine some of the theory behind the tricks presented earlier. We begin in one dimension where the update equation for gradient descent can be written as

$$W(t+1) = W(t) - \eta \frac{dE(W)}{dW}. \tag{1.20}$$

We would like to know how the value of η affects convergence and the learning speed. Figure 1.6 illustrates the learning behavior for several different sizes of η

when the weight W starts out in the vicinity of a local minimum. In one dimension, it is easy to define the optimal learning rate, η_{opt}, as being the learning rate that will move the weight to the minimum, W_{min}, in precisely one step (see Figure 1.6(i)b). If η is smaller than η_{opt} then the stepsize will be smaller and convergence will take multiple timesteps. If η is between η_{opt} and $2\eta_{opt}$ then the weight will oscillate around W_{min} but will eventually converge (Figure 1.6(i)c). If η is more than twice the size of η_{opt} (Figure 1.6(i)d) then the stepsize is so large that the weight ends up farther from W_{min} than before. Divergence results.

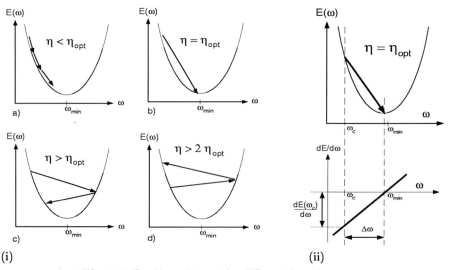

Fig. 1.6. Gradient descent for different learning rates.

What is the optimal value of the learning rate η_{opt}? Let us first consider the case in 1-dimension. Assuming that E can be approximated by a quadratic function, η_{opt} can be derived by first expanding E in a Taylor series about the current weight, W_c:

$$E(W) = E(W_c) + (W - W_c)\frac{dE(W_c)}{dW} + \frac{1}{2}(W - W_c)^2\frac{d^2E(W_c)}{dW^2} + \ldots, \quad (1.21)$$

where we use the shorthand $\frac{dE(W_c)}{dW} \equiv \frac{dE}{dW}|_{W=W_c}$. If E is quadratic the second order derivative is constant and the higher order terms vanish. Differentiating both sides with respect to W then gives

$$\frac{dE(W)}{dW} = \frac{dE(W_c)}{dW} + (W - W_c)\frac{d^2E(W_c)}{dW^2}. \quad (1.22)$$

Setting $W = W_{min}$ and noting that $dE(W_{min})/dW = 0$, we are left after rearranging with

$$W_{min} = W_c - \left(\frac{d^2E(W_c)}{dW^2}\right)^{-1}\frac{dE(W_c)}{dW}. \quad (1.23)$$

Comparing this with the update equation (1.20), we find that we can reach a minimum in one step if

$$\eta_{opt} = \left(\frac{d^2 E(W_c)}{dW^2}\right)^{-1}. \qquad (1.24)$$

Perhaps an easier way to obtain this same result is illustrated in Figure 1.6(ii). The bottom graph plots the gradient of E as a function of W. Since E is quadratic, the gradient is simply a straight line with value zero at the minimum and $\frac{\partial E(W_c)}{\partial W}$ at the current weight W_c. $\partial^2 E/\partial^2 W$ is simply the slope of this line and is computed using the standard slope formula

$$\partial^2 E/\partial^2 W = \frac{\partial E(W_c)/\partial W - 0}{W_c - W_{min}}. \qquad (1.25)$$

Solving for W_{min} then gives equation (1.23).

While the learning rate that gives fastest convergence is η_{opt}, the largest learning rate that can be used without causing divergence is (also see Figure 1.6(i)d)

$$\eta_{max} = 2\eta_{opt}. \qquad (1.26)$$

If E is not exactly quadratic then the higher order terms in equation (1.21) are not precisely zero and (1.23) is only an approximation. In such a case, it may take multiple iterations to locate the minimum even when using η_{opt}, however, convergence can still be quite fast.

In multiple dimensions, determining η_{opt} is a bit more difficult because the right side of (1.24) is a matrix H^{-1} where H is called the Hessian whose components are given by

$$H_{ij} \equiv \frac{\partial^2 E}{\partial W_i \partial W_j} \qquad (1.27)$$

with $1 \leq i, j \leq N$, and N equal to the total number of weights.

H is a measure of the curvature of E. In two dimensions, the lines of constant E for a quadratic cost are oval in shape as shown in Figure 1.7. The eigenvectors of H point in the directions of the major and minor axes. The eigenvalues measure the steepness of E along the corresponding eigendirection.

Example. In the least mean square (LMS) algorithm, we have a single layer linear network with error function

$$E(W) = \frac{1}{2P}\sum_{p=1}^{P} |d^p - \sum_i w_i x_i^p|^2 \qquad (1.28)$$

where P is the number of training vectors. The Hessian in this case turns out the be the same as the covariance matrix of the inputs,

$$H = \frac{1}{P}\sum_p x^p x^{pT}. \qquad (1.29)$$

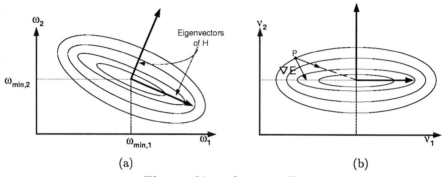

Fig. 1.7. Lines of constant E.

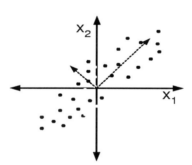

Fig. 1.8. For the LMS algorithm, the eigenvectors and eigenvalues of H measure the spread of the inputs in input space.

Thus, each eigenvalue of H is also a measure of the covariance or spread of the inputs along the corresponding eigendirection as shown in Figure 1.8.

Using a scalar learning rate is problematic in multiple dimensions. We want η to be large so that convergence is fast along the shallow directions of E (small eigenvalues of H), however, if η is too large the weights will diverge along the steep directions (large eigenvalues of H). To see this more specifically, let us again expand E, but this time about a minimum

$$E(W) \approx E(W_{min}) + \frac{1}{2}(W - W_{min})^T H_{(W_{min})}(W - W_{min}). \quad (1.30)$$

Differentiating (1.30) and using the result in the update equation (1.20) gives

$$W(t+1) = W(t) - \eta \frac{\partial E(t)}{\partial W} \quad (1.31)$$
$$= W(t) - \eta H_{(W_{min})}(W(t) - W_{min}). \quad (1.32)$$

Subtracting W_{min} from both sides gives

$$(W(t+1) - W_{min}) = (I - \eta H(W_{min}))(W(t) - W_{min}). \quad (1.33)$$

If the prefactor $(I - \eta H(W_{min}))$ is a matrix transformation that always shrinks a vector (i.e. its eigenvalues all have magnitude less than 1) then the update equation will converge.

How does this help with choosing the learning rates? Ideally we want different learning rates along the different eigendirections. This is simple if the eigendirections are lined up with the coordinate axes of the weights. In such a case, the weights are uncoupled and we can assign each weight its own learning rate based on the corresponding eigenvalue. However, if the weights are coupled then we must first rotate H such that H is diagonal, i.e. the coordinate axes line up with the eigendirections (see Figure 1.7b). This is the purpose of diagonalizing the Hessian discussed earlier.

Let Θ be the rotation matrix such that

$$\Lambda = \Theta H \Theta^T \quad (1.34)$$

where Λ is diagonal and $\Theta^T \Theta = I$. The cost function then can be written as

$$E(W) \approx E(W_{min}) + \frac{1}{2}\left[(W - W_{min})^T \Theta^T\right] \left[\Theta H_{(W_{min})} \Theta^T\right] \left[\Theta(W - W_{min})\right]. \quad (1.35)$$

Making a change of coordinates to $\nu = \Theta(W - W_{min})$ simplifies the above equation to

$$E(\nu) \approx E(0) + \frac{1}{2}\nu^T \Lambda \nu \quad (1.36)$$

and the transformed update equation becomes

$$\nu(t+1) = (I - \eta \Lambda)\nu(t). \quad (1.37)$$

Note that $I - \eta\Lambda$ is diagonal with diagonal components $1 - \eta\lambda_i$. This equation will converge if $|1 - \eta\lambda_i| < 1$, i.e. $\eta < \frac{2}{\lambda_i}$ for all i. If constrained to have a *single* scalar learning rate for all weights then we must require

$$\eta < \frac{2}{\lambda_{max}} \qquad (1.38)$$

in order to avoid divergence, where λ_{max} is the largest eigenvalue of H. For fastest convergence we have

$$\eta_{opt} = \frac{1}{\lambda_{max}}. \qquad (1.39)$$

If λ_{min} is a lot smaller than λ_{max} then convergence will be very slow along the λ_{min} direction. In fact, convergence time is proportional to the condition number $\kappa \equiv \lambda_{max}/\lambda_{min}$ so that it is desirable to have as small an eigenvalue spread as possible.

However, since we have rotated H to be aligned with the coordinate axes, (1.37) consists actually of N independent 1-dimensional equations. Therefore, we can choose a learning rate for each weight independent of the others. We see that the optimal rate for the ith weight ν_i is $\eta_{opt,i} = \frac{1}{\lambda_i}$.

1.5.2 Examples

Linear Network Figure 1.10 displays a set of 100 examples drawn from two Gaussian distributed classes centered at (-0.4,-0.8) and (0.4,0.8). The eigenvalues of the covariance matrix are 0.84 and 0.036. We train a single layer linear network with 2 inputs, 1 output, 2 weights, and 1 bias (see Figure (1.9)) using the LMS algorithm in batch mode. Figure 1.11 displays the weight trajectory and error during learning when using a learning rates of $\eta = 1.5$ and 2.5. Note that the learning rate (see Eq. 1.38) $\eta_{max} = 2/\lambda_{max} = 2/.84 = 2.38$ will cause divergences as is evident for $\eta = 2.5$.

Figure 1.12 shows the same example using stochastic instead of batch mode learning. Here, a learning rate of $\eta = 0.2$ is used. One can see that the trajectory is much noisier than in batch mode since only an estimate of the gradient is used at each iteration. The cost is plotted as a function of epoch. An epoch here is simply defined as 100 input presentations which, for stochastic learning, corresponds to 100 weight updates. In batch, an epoch corresponds to one weight update.

Multilayer Network Figure 1.14 shows the architecture for a very simple multilayer network. It has 1 input, 1 hidden, and 1 output node. There are 2 weights and 2 biases. The activation function is $f(x) = 1.71 \tanh((2/3)x)$. The training set contains 10 examples from each of 2 classes. Both classes are Gaussian distributed with standard deviation 0.4. Class 1 has a mean of -1 and class 2 has a mean of +1. Target values are -1 for class 1 and +1 for class 2. Figure 1.13 shows the stochastic trajectory for the example.

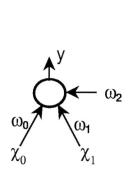

Fig. 1.9. Simple linear network.

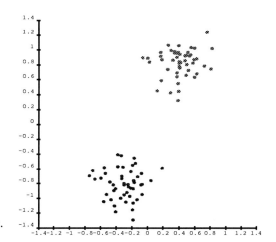

Fig. 1.10. Two classes drawn from gaussian distributions centered at (-0.4,-0.8) and (0.4,0.8).

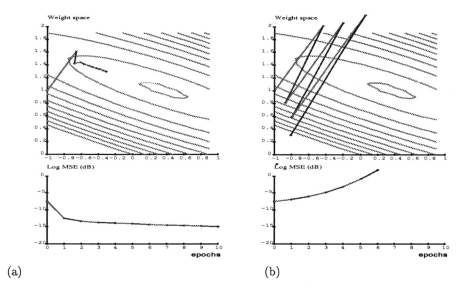

Fig. 1.11. Weight trajectory and error curve during learning for (a) $\eta = 1.5$ and (b) $\eta = 2.5$.

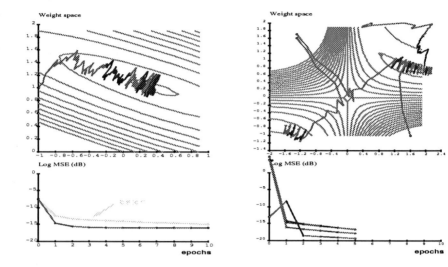

Fig. 1.12. Weight trajectory and error curve during stochastic learning for $\eta = 0.2$.

Fig. 1.13. Weight trajectories and errors for 1-1-1 network trained using stochastic learning.

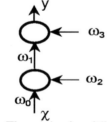

Fig. 1.14. The minimal multilayer network.

1.5.3 Input Transformations and Error Surface Transformations Revisited

We can use the results of the previous section to justify several of the tricks discussed earlier.

> **Subtract the means from the input variables**

The reason for the above trick is that a nonzero mean in the input variables creates a *very large* eigenvalue. This means the condition number will be large, i.e. the cost surface will be steep in some directions and shallow in others so that convergence will be very slow. The solution is to simply preprocess the inputs by subtracting their means.

For a single linear neuron, the eigenvectors of the Hessian (with means subtracted) point along the principal axes of the cloud of training vectors (recall Figure 1.8). Inputs that have a large variation in spread along different directions of the input space will have a large condition number and slow learning. And so we recommend:

> **Normalize the variances of the input variables.**

If the input variables are correlated, this will not make the error surface spherical, but it will possibly reduce its eccentricity.

Correlated input variables usually cause the eigenvectors of H to be rotated away from the coordinate axes (Figure 1.7a versus 1.7b) thus weight updates are not decoupled. Decoupled weights make the "one learning rate per weight" method optimal, thus, we have the following trick:

> **Decorrelate the input variables.**

Now suppose that the input variables of a neuron have been decorrelated, the Hessian for this neuron is then diagonal and its eigenvalues point along the coordinate axes. In such a case the gradient is not the best descent direction as can be seen in Fig 1.7b. At the point P, an arrow shows that gradient does not point towards the minimum. However, if we instead assign each weight its own learning rate (equal the inverse of the corresponding eigenvalue) then the descent direction will be in the direction of the other arrow that points directly towards the minimum:

> **Use a separate learning rate for each weight.**

1.6 Classical second order optimization methods

In the following we will briefly introduce the Newton, conjugate gradient, Gauss-Newton, Levenberg Marquardt and the Quasi-Newton (BFGS) method (see also [11, 34, 3, 5]).

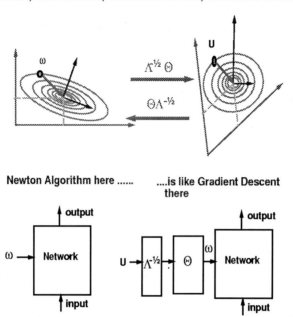

Fig. 1.15. Sketch of the whitening properties of the Newton algorithm.

1.6.1 Newton Algorithm

To get an understanding of the Newton method let us recapitulate the results from section 1.5.1. Assuming a quadratic loss function E (see Eq.(1.21)) as depicted in Figure 1.6(ii), we can compute the weight update along the lines of Eq.(1.21)-(1.23)

$$\Delta w = \eta \left(\frac{\partial^2 E}{\partial w^2}\right)^{-1} \frac{\partial E}{\partial w} = \eta H(w)^{-1} \frac{\partial E}{\partial w}, \tag{1.40}$$

where η must to be chosen in the range $0 < \eta < 1$ since E is in practice not perfectly quadratic. In this equation information about the Hessian H is taken into account. If the error function was quadratic one step would be sufficient to converge.

Usually the energy surface around the minimum is rather ellipsoid, or in the extreme like a taco shell, depending on the conditioning of the Hessian. A whitening transform, well known from signal processing literature [29] can change this ellipsoid shape to a spherical shape through $u = \Theta \Lambda^{1/2} w$ (see Figure 1.15 and Eq.(1.34)). So the inverse Hessian in Eq.(1.40) basically spheres out the error surface locally. The following two approaches can be shown to be equivalent: (a) use the Newton algorithm in an untransformed weight space and (b) do usual gradient descent in a whitened coordinate system (see Figure 1.15) [19].

Summarizing, the Newton algorithm converges in one step if the error function is quadratic and (unlike gradient descent) it is invariant with respect to linear transformations of the input vectors. This means that the convergence time is not affected by shifts, scaling and rotation of input vectors. However

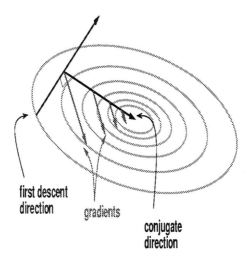

Fig. 1.16. Sketch of conjugate gradient directions in a 2D error surface.

one of the main drawbacks is that an $N \times N$ Hessian matrix must be stored and inverted, which takes $O(N^3)$ per iterations and is therefore impractical for more than a few variables. Since the error function is in general non-quadratic, there is no guarantee of convergence. If the Hessian is not positive definite (if it has some zero or even negative Eigenvalues where the error surface is flat or some directions are curved downward), then the Newton algorithm will diverge, so the Hessian *must be* positive definite. Of course the Hessian matrix of multi-layer networks is in general not positive definite everywhere. For these reasons the Newton algorithm in its original form is not usable for general neural network learning. However it gives good insights for developing more sophisticated algorithms, as discussed in the following.

1.6.2 Conjugate Gradient

There are several important properties in conjugate gradient optimization: (1) it is a $O(N)$ method, (2) it doesn't use the Hessian explicitly, (3) it attempts to find descent directions that try to minimally spoil the result achieved in the previous iterations, (4) it uses a line search, and most importantly, (5) it works only for batch learning.

The third property is shown in Figure 1.16. Assume we pick a descent direction, e.g. the gradient, then we minimize along a line in this direction (line search). Subsequently we should try to find a direction along which the gradient does not change its direction, but merely its length (conjugate direction), because moving along this direction will not spoil the result of the previous iteration. The evolution of the descent directions ρ_k at iteration k is given as

$$\rho_k = -\nabla E(w_k) + \beta_k \rho_{k-1}, \qquad (1.41)$$

where the choice of β_k can be done either according to Fletcher and Reeves [34]

$$\beta_k = \frac{\nabla E(w_k)^T \nabla E(w_k)}{\nabla E(w_{k-1})^T \nabla E(w_{k-1})} \qquad (1.42)$$

or Polak and Ribiere

$$\beta_k = \frac{(\nabla E(w_k) - \nabla E(w_{k-1}))^T \nabla E(w_k)}{\nabla E(w_{k-1})^T \nabla E(w_{k-1})}. \qquad (1.43)$$

Two directions ρ_k and ρ_{k-1} are defined as conjugate if

$$\rho_k^T H \rho_{k-1} = 0,$$

i.e. conjugate directions are orthogonal directions in the space of an identity Hessian matrix (see Figure 1.17). Very important for convergence in both choices

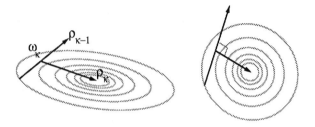

Fig. 1.17. Sketch of conjugate gradient directions in a 2D error surface.

is a good line search procedure. For a perfectly quadratic function with N variables a convergence within N steps can be proved. For non-quadratic functions Polak and Ribiere's choice seems more robust. Conjugate gradient (1.41) can also be viewed as a smart choice for choosing the momentum term known in neural network training. It has been applied with large success in multi-layer network training on problems that are moderate sized with rather low redundancy in the data. Typical applications range from function approximation, robotic control [39], time-series prediction and other real valued problems where high accuracy is wanted. Clearly on large and redundant (classification) problems stochastic backpropagation is faster. Although attempts have been made to define mini-batches [25], the main disadvantage of conjugate gradient methods remains that it is a batch method (partly due to the precision requirements in line search procedure).

1.6.3 Quasi-Newton (BFGS)

The Quasi-Newton (BFGS) method (1) iteratively computes an estimate of the inverse Hessian, (2) is an $O(N^2)$ algorithm, (3) requires line search and (4) it works only for batch learning.

The positive definite estimate of the inverse Hessian is done directly without requiring matrix inversion and by only using gradient information. Algorithmically this can be described as follows: (1) first a positive definite matrix M is chosen, e.g. $M = I$, (2) then the search direction is set to

$$\rho(t) = M(t)\nabla E(w(t)),$$

(3) a line search is performed along ρ, which gives the update for the parameters at time t

$$w(t) = w(t-1) - \eta(t)\rho(t).$$

Finally (4) the estimate of the inverse Hessian is updated. Compared to the Newton algorithm the Quasi-Newton approach only needs gradient information. The most successful Quasi-Newton algorithm is the Broyden-Fletcher-Goldfarb-Shanno (BFGS) method. The update rule for the estimate of the inverse Hessian is

$$M(t) = M(t-1)\left(1 + \frac{\phi^T M \phi}{\delta^T \phi}\right)\frac{\delta \delta^T}{\delta^T \phi} - \left(\frac{\delta \phi^T M + M \phi \delta^T}{\delta^T \phi}\right), \quad (1.44)$$

where some abbreviations have been used for the following $N \times 1$ vectors

$$\phi = \nabla E(w(t)) - \nabla E(w(t-1))$$
$$\delta = w(t) - w(t-1). \quad (1.45)$$

Although, as mentioned above, the complexity is only $O(N^2)$, we are still required to store a $N \times N$ matrix, so the algorithm is only practical for small networks with non-redundant training sets. Recently some variants exist that aim to reduce storage requirements (see e.g. [3]).

1.6.4 Gauss-Newton and Levenberg Marquardt

Gauss-Newton and Levenberg Marquardt algorithm (1) use the square Jacobi approximation, (2) are mainly designed for batch learning, (3) have a complexity of $O(N^3)$ and (4) most important, they work only for mean squared error loss functions. The Gauss-Newton algorithm is like the Newton algorithm, however the Hessian is approximated by the square of the Jacobian (see also section 1.7.2 for a further discussion)

$$\Delta w = \left(\sum_p \frac{\partial f(w, x_p)}{\partial w}^T \frac{\partial f(w, x_p)}{\partial w}\right)^{-1} \nabla E(w). \quad (1.46)$$

The Levenberg Marquardt method is like the Gauss-Newton above, but it has a regularization parameter μ that prevents it from blowing up, if some eigenvalues are small

$$\Delta w = \left(\sum_p \frac{\partial f(w, x_p)}{\partial w}^T \frac{\partial f(w, x_p)}{\partial w} + \mu I\right)^{-1} \nabla E(w), \quad (1.47)$$

where I denotes the unity matrix. The Gauss Newton method is valid for quadratic cost functions however a similar procedure also works with Kullback-Leibler cost and is called Natural Gradient (see e.g. [1, 44, 2]).

1.7 Tricks to compute the Hessian information in multilayer networks

We will now discuss several techniques aimed at computing full or partial Hessian information by (a) finite difference method, (b) square Jacobian approximation (for Gauss-Newton and Levenberg-Marquardt algorithm), (c) computation of the diagonal of the Hessian and (d) by obtaining a product of the Hessian and a vector without computing the Hessian. Other semi-analytical techniques that allow the computation of the full Hessian are omitted because they are rather complicated and also require many forward/backward propagation steps [5, 8].

1.7.1 Finite Difference

We can write the k-th line of the Hessian

$$H^{(k)} = \frac{\partial(\nabla E(w))}{\partial w_k} \sim \frac{\nabla E(w + \delta\phi_k) - \nabla E(w)}{\delta},$$

where $\phi_k = (0, 0, 0, \ldots, 1, \ldots, 0)$ is a vector of zeros and only one 1 at the k-th position. This can be implemented with a simple recipe: (1) compute the total gradient by multiple forward and backward propagation steps. (2) Add δ to the k-th parameter and compute again the gradient, and finally (3) subtract both results and divide by δ. Due to numerical errors in this computation scheme the resulting Hessian might not be perfectly symmetric. In this case it should be *symmetrized* as described below.

1.7.2 Square Jacobian approximation for the Gauss-Newton and Levenberg-Marquardt algorithms

Assuming a mean squared cost function

$$E(w) = \frac{1}{2} \sum_p (d_p - f(w, x_p))^T (d_p - f(w, x_p)) \qquad (1.48)$$

then the gradient is

$$\frac{\partial E(w)}{\partial w} = -\sum_p (d_p - f(w, x_p))^T \frac{\partial f(w, x_p)}{\partial w} \qquad (1.49)$$

and the Hessian follows as

$$H(w) = \sum_p \frac{\partial f(w, x_p)}{\partial w}^T \frac{\partial f(w, x_p)}{\partial w} + \sum_p (d_p - f(w, x_p))^T \frac{\partial^2 f(w, x_p)}{\partial w \partial w}. \qquad (1.50)$$

A simplifying approximation of the Hessian is the square of the Jacobian which is a positive semi-definite matrix of dimension: $N \times O$

$$H(w) \sim \sum_p \frac{\partial f(w, x_p)}{\partial w}^T \frac{\partial f(w, x_p)}{\partial w}, \qquad (1.51)$$

where the second term from Eq.(1.50) was dropped. This is equivalent to assuming that the network is a linear function of the parameters w. Again this is readily implemented for the k-th column of the Jacobian: for all training patterns, (1) we forward propagate, then (2) set the activity of the output units to 0 and only the k-th output to 1, (3) a backpropagation step is taken and the gradient is accumulated.

1.7.3 Backpropagating second derivatives

Let us consider a multi-layer system with some functional blocks with N_i inputs, N_o outputs and N parameters of the form $O = F(W, X)$. Now assume we knew $\partial^2 E/\partial O^2$, which is a $N_o \times N_o$ matrix. Then it is straight forward to compute this matrix

$$\frac{\partial^2 E}{\partial W^2} = \frac{\partial O}{\partial W}^T \frac{\partial^2 E}{\partial O^2} \frac{\partial O}{\partial W} + \frac{\partial E}{\partial O}\frac{\partial^2 O}{\partial W^2}. \tag{1.52}$$

We can drop the second term in Eq.(1.52) and the resulting estimate of the Hessian is positive semi-definite. A further reduction is achieved, if we ignore all but the diagonal terms of $\frac{\partial^2 E}{\partial O^2}$:

$$\frac{\partial^2 E}{\partial w_i^2} = \sum_k \frac{\partial^2 E}{\partial o_k^2}\left(\frac{\partial o_k}{\partial w_i}\right)^2. \tag{1.53}$$

A similar derivation can be done to obtain the N_i times N_i matrix $\partial^2 E/\partial x^2$.

1.7.4 Backpropagating the diagonal Hessian in neural nets

Backpropagation procedures for computing the diagonal Hessian are well known [18, 4, 19]. It is assumed that each layer in the network has the functional form $o_i = f(y_i) = f(\sum_j w_{ij}x_j)$ (see Figure 1.18 for the sigmoidal network). Using the Gauss-Newton approximation (dropping the term that contain $f''(y)$) we obtain:

$$\frac{\partial^2 E}{\partial y_k^2} = \frac{\partial^2 E}{\partial o_k^2}(f'(y_k))^2, \tag{1.54}$$

$$\frac{\partial^2 E}{\partial w_{ki}^2} = \frac{\partial^2 E}{\partial y_k^2}x_i^2 \tag{1.55}$$

and

$$\frac{\partial^2 E}{\partial x_i^2} \sum_k \frac{\partial^2 E}{\partial y_k^2}w_{ki}^2. \tag{1.56}$$

With f being a Gaussian nonlinearity as shown in Figure 1.18 for the RBF networks we obtain

$$\frac{\partial^2 E}{\partial w_{ki}^2} = \frac{\partial^2 E}{\partial y_k^2}(x_i - w_{ki})^2 \tag{1.57}$$

and

$$\frac{\partial^2 E}{\partial x_i^2} = \sum_k \frac{\partial^2 E}{\partial y_k^2}(x_i - w_{ki})^2. \tag{1.58}$$

The cost of computing the diagonal second derivatives by running these equations from the last layer to the first one is essentially the same as the regular backpropation pass used for the gradient, except that the square of the weights are used in the weighted sums. This technique is applied in the "optimal brain damage" pruning procedure (see [21]).

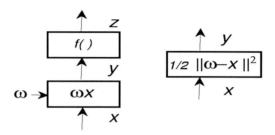

Fig. 1.18. Backpropagating the diagonal Hessian: sigmoids (left) and RBFs (right).

1.7.5 Computing the product of the Hessian and a vector

In many methods that make use of the Hessian, the Hessian is used exclusively in products with a vector. Interestingly, there is a way of computing such products *without* going through the trouble of computing the Hessian itself. The finite difference method can fulfill this task for an arbitrary vector Ψ

$$H\Psi \sim \frac{1}{\alpha}\left(\frac{\partial E}{\partial w}(w + \alpha\Psi) - \frac{\partial E}{\partial w}(w)\right), \tag{1.59}$$

using only two gradient computations (at point w and $w + \alpha\Psi$ respectively), which can be readily computed with backprop (α is a small constant).

This method can be applied to compute the principal eigenvector and eigenvalue of H by the power method. By iterating and setting

$$\Psi(t+1) = \frac{H\Psi(t)}{\|\Psi(t)\|}, \tag{1.60}$$

the vector $\Psi(t)$ will converge to the largest eigenvector of H and $\|\Psi(t)\|$ to the corresponding eigenvalue [23, 14, 10]. See also [33] for an even more accurate method that (1) does not use finite differences and (2) has similar complexity.

1.8 Analysis of the Hessian in multi-layer networks

It is interesting to understand how some of the tricks shown previously influence on the Hessian, i.e. how does the Hessian change with architecture and details of the implementation. Typically, the eigenvalue distribution of the Hessian looks like the one sketched in Figure 1.20: a few small eigenvalues, many medium ones and few very large ones. We will now argue that the *large eigenvalues* will cause the trouble in the training process because [23, 22]

- non-zero mean inputs or neuron states [22] (see also chapter 10)
- wide variations of the second derivatives from layer to layer
- correlation between state variables.

To exemplify this, we show the eigenvalue distribution of a network trained on OCR data in Figure 1.20. Clearly, there is a wide spread of eigenvalues (see Figure 1.19) and we observe that the ratio between e.g. the first and the eleventh eigenvalue is about 8. The long tail of the eigenvalue distribution (see Figure 1.20) is rather painful because the ratio between the largest and smallest eigenvalue gives the conditioning of the learning problem. A large ratio corresponds to a big difference in the axis of the ellipsoidal shaped error function: the larger the ratio, the more we find a taco-shell shaped minima, which are extremely steep towards the small axis and very flat along the long axis.

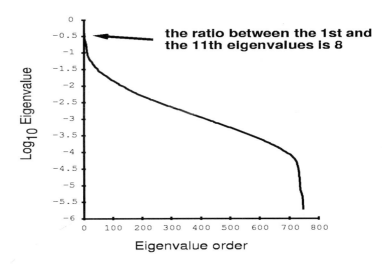

Fig. 1.19. Eigenvalue spectrum in a 4 layer shared weights network ($256 \times 128 \times 64 \times 10$) trained on 320 handwritten digits.

Another general characteristic of the Hessian in multi-layer networks is the spread between layers. In Figure 1.21 we roughly sketch how the shape of the Hessian varies from being rather flat in the first layer to being quite steep in

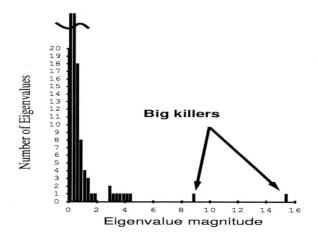

Fig. 1.20. Eigenvalue spectrum in a 4 layer shared weights network ($256 \times 128 \times 64 \times 10$) trained on 320 handwritten digits.

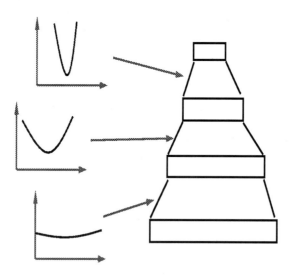

Fig. 1.21. Multilayered architecture: the second derivative is often smaller in lower layers.

the last layer. This affects the learning speed and can provide an ingredient to explain the slow learning in lower layers and the fast (sometime oscillating) learning in the last layer. A trick to compensate this different scale of learning is to use the inverse diagonal Hessian to control the learning rate (see also section 1.6, chapter 17).

1.9 Applying Second Order Methods to Multilayer Networks

Before we concentrate in this section on how to tailor second order techniques for training large networks, let us first repeat some rather pessimistic facts about applying classical second order methods. Techniques using full Hessian information (Gauss -Newton, Levenberg-Marquardt and BFGS) can only apply to very small networks trained in batch mode, however those small networks are not the ones that need speeding up the most. Most second order methods (conjugate gradient, BFGS, ...) require a line-search and can therefore not be used in the stochastic mode. Many of the tricks discussed previously apply only to batch learning. From our experience we know that a carefully tuned stochastic gradient descent is hard to beat on large classification problems. For smaller problems that require accurate real-valued outputs like in function approximation or control problems, we see that conjugate gradient (with Polak-Ribiere Eq.(1.43)) offers the best combination of speed, reliability and simplicity. Several attempts using "mini batches" in applying conjugate gradient to large and redundant problems have been made recently [17, 25, 31]. A variant of conjugate gradient optimization (called scaled CG) seems interesting: here the line search procedure is replaced by a 1D Levenberg Marquardt type algorithm [24].

1.9.1 A stochastic diagonal Levenberg Marquardt method

To obtain a stochastic version of the Levenberg Marquardt algorithm the idea is to compute the diagonal Hessian through a running estimate of the second derivative with respect to each parameter. The instantaneous second derivative can be obtained via backpropagation as shown in the formulas of section 1.7. As soon as we have those running estimates we can use them to compute individual learning rates for each parameter

$$\eta_{ki} = \frac{\epsilon}{\langle \frac{\partial^2 E}{\partial w_{ki}^2} \rangle + \mu}, \tag{1.61}$$

where ϵ denotes the global learning rate, and $\langle \frac{\partial^2 E}{\partial w_{ki}^2} \rangle$ is a running estimate of the diagonal second derivative with respect to w_{ki}. μ is a parameter to prevent η_{ki} from blowing up in case the second derivative is small, i.e. when the optimization moves in flat parts of the error function. The running estimate is computed as

$$\langle \frac{\partial^2 E}{\partial w_{ki}^2} \rangle_{new} = (1-\gamma) \langle \frac{\partial^2 E}{\partial w_{ki}^2} \rangle_{old} + \gamma \frac{\partial^2 E^p}{\partial w_{ki}^2}, \tag{1.62}$$

where γ is a small constant that determines the amount of memory that is being used. The second derivatives can be computed prior to training over e.g. a subset of the training set. Since they change only very slowly they only need to be reestimated every few epochs. Note that the additional cost over regular backpropagation is negligible and convergence is – as a rule of thumb – about three times faster than a carefully tuned stochastic gradient algorithm.

In Figure 1.22 and 1.23 we see the convergence of the stochastic diagonal Levenberg Marquardt method (1.61) for a toy example with two different sets of learning rates. Obviously the experiment shown Figure 1.22 contains fewer fluctuations than in Figure 1.23 due to smaller learning rates.

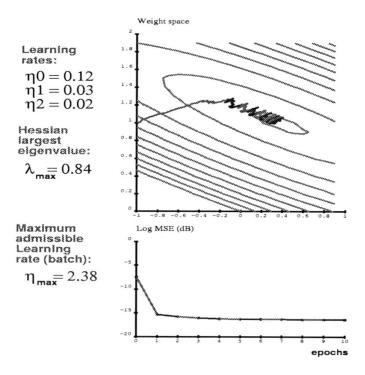

Fig. 1.22. Stochastic diagonal Levenberg-Marquardt algorithm. Data set from 2 Gaussians with 100 examples. The network has one linear unit, 2 inputs and 1 output, i.e. three parameters (2 weights, 1 bias).

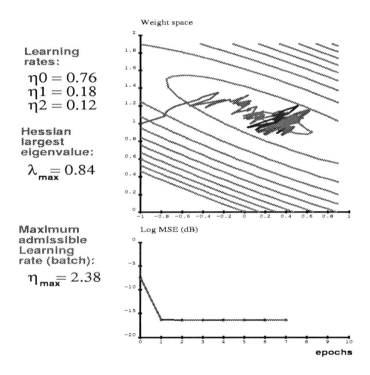

Fig. 1.23. Stochastic diagonal Levenberg-Marquardt algorithm. Data set from 2 Gaussians with 100 examples. The network has one linear unit, 2 inputs and 1 output, i.e. three parameters (2 weights, 1 bias).

1.9.2 Computing the principal Eigenvalue/vector of the Hessian

In the following we give three tricks for computing the principal eigenvalue/Vector of the Hessian without having to compute the Hessian itself. Remember that in section 1.4.7 we also introduced a method to approximate the smallest eigenvector of the Hessian (without having to compute the Hessian) through averaging (see also [28]).

Power Method We repeat the result of our discussion in section 1.7.5: starting from a random initial vector Ψ, the iteration

$$\Psi_{new} = H \frac{\Psi_{old}}{\|\Psi_{old}\|},$$

will eventually converge to the principal eigenvector (or a vector in the principal eigenspace) and $\|\Psi_{old}\|$ will converge to the corresponding eigenvalue [14, 10].

Taylor Expansion Another method makes use of the fact that small perturbations of the gradient also lead to the principal eigenvector of H

$$\Psi_{new} = \frac{1}{\alpha} \left(\frac{\partial E}{\partial w}(w + \alpha \frac{\Psi_{old}}{\|\Psi_{old}\|}) - \frac{\partial E}{\partial w}(w) \right), \tag{1.63}$$

where α is a small constant. One iteration of this procedure requires two forward and two backward propagation steps for each pattern in the training set.

Online Computation of Ψ The following rule makes use of the running average to obtain the largest eigenvalue of the average Hessian very fast

$$\Psi_{new} = (1-\gamma)\Psi + \frac{1}{\alpha} \left(\frac{\partial E^p}{\partial w}(w + \alpha \frac{\Psi_{old}}{\|\Psi_{old}\|}) - \frac{\partial E}{\partial w}(w) \right). \tag{1.64}$$

To summarize, the eigenvalue/vector computations:

1. a random vector is chosen for initialization of Ψ,
2. an input pattern is presented with desired output, a forward and backward propagation, step is performed and the gradients $G(w)$ are stored,
3. $\alpha \frac{\Psi_{old}}{\|\Psi_{old}\|}$ is added to the current weight vector w,
4. a forward and backward propagation step is performed with the perturbed weight vector and the gradients $G(w')$ are stored,
5. the difference $1/\alpha(G(w') - G(w))$ is computed and the running average of the eigenvector is updated,
6. we loop from (2)-(6) until a reasonably stable result is obtained for Ψ,
7. the optimal learning rate is then given as

$$\eta_{opt} = \frac{1}{\|\Psi\|}.$$

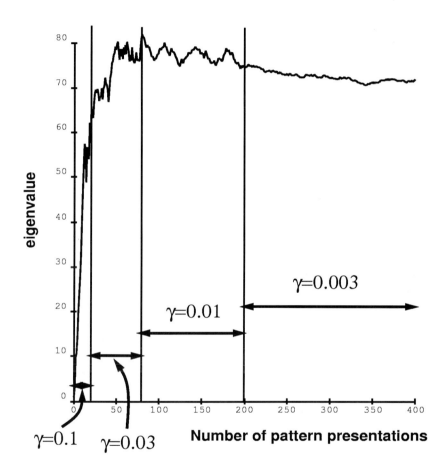

Fig. 1.24. Evolution of the eigenvalue as a function of the number of pattern presentations for a shared weight network with 5 layers, 64638 connections and 1278 free parameters. The training set consists of 1000 handwritten digits.

In Figure 1.24 we see the evolution of the eigenvalue as a function of the number of pattern presentations for a neural network in a handwritten character recognition task. In practice we adapt the leak size of the running average in order to get fewer fluctuations (as also indicated on the figure). In the figure we see that after fewer than 100 pattern presentations the correct order of magnitude for the eigenvalue, i.e the learning rate is reached. From the experiments we also observe that the fluctuations of the average Hessian over training are small.

In Figure 1.25 and 1.26 we start with the same initial conditions, and perform a fixed number of epochs with learning rates computed by multiplying the predicted learning rate by a predefined constant. Choosing constant 1 (i.e. using the predicted optimal rate) always gives residual errors which are very close to the error achieved by the best choice of the constant. In other words, the "predicted optimal rate" is optimal enough.

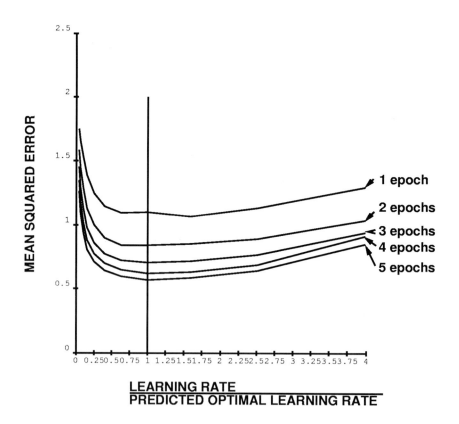

Fig. 1.25. Mean squared error as a function of the ratio between learning rate and predicted optimal learning rate for a fully connected network (784 × 30 × 10). The training set consists of 300 handwritten digits.

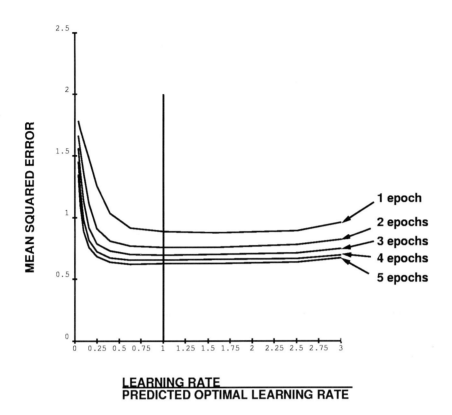

Fig. 1.26. Mean squared error as a function of the ratio between learning rate and predicted optimal learning rate for a shared weight network with 5 layers (1024 × 1568 × 392 × 400 × 100 × 10), 64638 (local) connections and 1278 free parameters (shared weights). The training set consists of 1000 handwritten digits.

1.10 Discussion and Conclusion

According to the recommendations mentioned above, a practitioner facing a multi-layer neural net training problem would go through the following steps:

- shuffle the examples
- center the input variables by subtracting the mean
- normalize the input variable to a standard deviation of 1
- if possible, decorrelate the input variables.
- pick a network with the sigmoid function shown in figure 1.4
- set the target values within the range of the sigmoid, typically +1 and -1.

- initialize the weights to random values as prescribed by 1.16.

The preferred method for training the network should be picked as follows:

- if the training set is large (more than a few hundred samples) and redundant, and if the task is classification, use stochastic gradient with careful tuning, or use the stochastic diagonal Levenberg Marquardt method.
- if the training set is not too large, or if the task is regression, use conjugate gradient.

Classical second-order methods are impractical in almost all useful cases.

The non-linear dynamics of stochastic gradient descent in multi-layer neural networks, particularly as it pertains to generalization, is still far from being well understood. More theoretical work and systematic experimental work is needed.

Acknowledgement Y.L. & L.B. & K.-R. M. gratefully acknowledge mutual exchange grants from DAAD and NSF.

References

1. S. Amari. Neural learning in structured parameter spaces — natural riemannian gradient. In Michael C. Mozer, Michael I. Jordan, and Thomas Petsche, editors, *Advances in Neural Information Processing Systems*, volume 9, page 127. The MIT Press, 1997.
2. S. Amari. Natural gradient works efficiently in learning. *Neural Computation*, 10(2):251–276, 1998.
3. R. Battiti. First- and second-order methods for learning: Between steepest descent and newton's method. *Neural Computation*, 4:141–166, 1992.
4. S. Becker and Y. LeCun. Improving the convergence of backbropagation learning with second oder metho ds. In David Touretzky, Geoffrey Hinton, and Terrence Sejnowski, editors, *Proceedings of the 1988 Connectionist Models Summer School*, pages 29–37. Lawrence Erlbaum Associates, 1989.
5. C. M. Bishop. *Neural Networks for Pattern Recognition*. Clarendon Press, Oxford, 1995.
6. L. Bottou. Online algorithms and stochastic approximations. In David Saad, editor, *Online Learning in Neural Networks (1997 Workshop at the Newton Institute)*, Cambridge, 1998. The Newton Institute Series, Cambridge University Press.
7. D. S. Broomhead and D. Lowe. Multivariable function interpolation and adaptive networks. *Complex Systems*, 2:321–355, 1988.
8. W. L. Buntine and A. S. Weigend. Computing second order derivatives in Feed-Forward networks: A review. *IEEE Transactions on Neural Networks*, 1993. To appear.
9. C. Darken and J. E. Moody. Note on learning rate schedules for stochastic optimization. In R. P. Lippmann, J. E. Moody, and D. S. Touretzky, editors, *Advances in Neural Information Processing Systems*, volume 3, pages 832–838. Morgan Kaufmann, San Mateo, CA, 1991.
10. K. I. Diamantaras and S. Y. Kung. *Principal Component Neural Networks*. Wiley, New York, 1996.
11. R. Fletcher. *Practical Methods of Optimization*, chapter 8.7 : Polynomial time algorithms, pages 183–188. John Wiley & Sons, New York, second edition, 1987.

12. S. Geman, E. Bienenstock, and R. Doursat. Neural networks and the bias/variance dilemma. *Neural Computation*, 4(1):1–58, 1992.
13. L. Goldstein. Mean square optimality in the continuous time Robbins Monro procedure. Technical Report DRB-306, Dept. of Mathematics, University of Southern California, LA, 1987.
14. G. H. Golub and C. F. Van Loan. *Matrix Computations*, 2nd ed. Johns Hopkins University Press, Baltimore, 1989.
15. T.M. Heskes and B. Kappen. On-line learning processes in artificial neural networks. In J. G. Tayler, editor, *Mathematical Approaches to Neural Networks*, volume 51, pages 199–233. Elsevier, Amsterdam, 1993.
16. Robert A. Jacobs. Increased rates of convergence through learning rate adaptation. *Neural Networks*, 1:295–307, 1988.
17. A. H. Kramer and A. Sangiovanni-Vincentelli. Efficient parallel learning algorithms for neural networks. In D. S. Touretzky, editor, *Advances in Neural Information Processing Systems. Proceedings of the 1988 Conference*, pages 40–48, San Mateo, CA, 1989. Morgan Kaufmann.
18. Y. LeCun. *Modeles connexionnistes de l'apprentissage (connectionist learning models)*. PhD thesis, Universitè P. et M. Curie (Paris VI), 1987.
19. Y. LeCun. Generalization and network design strategies. In R. Pfeifer, Z. Schreter, F. Fogelman, and L. Steels, editors, *Connectionism in Perspective*, Amsterdam, 1989. Elsevier. Proceedings of the International Conference Connectionism in Perspective, University of Zürich, 10. – 13. October 1988.
20. Y. LeCun, B. Boser, J. S. Denker, D. Henderson, R. E. Howard, W. Hubbard, and L. D. Jackel. Handwritten digit recognition with a backpropagation network. In D. S. Touretsky, editor, *Advances in Neural Information Processing Systems, vol. 2*, San Mateo, CA, 1990. Morgan Kaufman.
21. Y. LeCun, J.S. Denker, and S.A. Solla. Optimal brain damage. In D. S. Touretsky, editor, *Advances in Neural Information Processing Systems, vol. 2*, pages 598–605, 1990.
22. Y. LeCun, I. Kanter, and S. A. Solla. Second order properties of error surfaces. In *Advances in Neural Information Processing Systems, vol. 3*, San Mateo, CA, 1991. Morgan Kaufmann.
23. Y. LeCun, P. Y. Simard, and B. Pearlmutter. Automatic learning rate maximization by on-line estimation of the hessian's eigenvectors. In Giles, Hanson, and Cowan, editors, *Advances in Neural Information Processing Systems, vol. 5*, San Mateo, CA, 1993. Morgan Kaufmann.
24. M. Møller. A scaled conjugate gradient algorithm for fast supervised learning. *Neural Networks*, 6:525–533, 1993.
25. M. Møller. Supervised learning on large redundant training sets. *International Journal of Neural Systems*, 4(1):15–25, 1993.
26. J. E. Moody and C. J. Darken. Fast learning in networks of locally-tuned processing units. *Neural Computation*, 1:281–294, 1989.
27. N. Murata. *(in Japanese)*. PhD thesis, University of Tokyo, 1992.
28. N. Murata, K.-R. Müller, A. Ziehe, and S. Amari. Adaptive on-line learning in changing environments. In Michael C. Mozer, Michael I. Jordan, and Thomas Petsche, editors, *Advances in Neural Information Processing Systems*, volume 9, page 599. The MIT Press, 1997.
29. A.V. Oppenheim and R.W. Schafer. *Digital Signal Processing*. Prentice Hall, Englewood Cliffs, 1975.
30. G. B. Orr. *Dynamics and Algorithms for Stochastic learning*. PhD thesis, Oregon Graduate Institute, 1995.

31. G.B. Orr. Removing noise in on-line search using adaptive batch sizes. In Michael C. Mozer, Michael I. Jordan, and Thomas Petsche, editors, *Advances in Neural Information Processing Systems*, volume 9, page 232. The MIT Press, 1997.
32. M.J.L. Orr. Regularization in the selection of radial basis function centers. *Neural Computation*, 7(3):606–623, 1995.
33. B.A. Pearlmutter. Fast exact multiplication by the hessian. *Neural Computation*, 6:147–160, 1994.
34. W. H. Press, B. P. Flannery, S. A. Teukolsky, and W. T. Vetterling. *Numerical Recipies in C: The art of Scientific Programming*. Cambridge University Press, Cambridge, England, 1988.
35. D. Saad, editor. *Online Learning in Neural Networks (1997 Workshop at the Newton Institute)*. The Newton Institute Series, Cambridge University Press, Cambridge, 1998.
36. D. Saad and S. A. Solla. Exact solution for on-line learning in multilayer neural networks. *Physical Review Letters*, 74:4337–4340, 1995.
37. H. Sompolinsky, N. Barkai, and H.S. Seung. On-line learning of dichotomies: algorithms and learning curves. In J-H. Oh, C. Kwon, and S. Cho, editors, *Neural Networks: The Statistical Mechanics Perspective*, pages 105–130. Singapore: World Scientific, 1995.
38. R.S. Sutton. Adapting bias by gradient descent: An incremental version of delta-bar-delta. In William Swartout, editor, *Proceedings of the 10th National Conference on Artificial Intelligence*, pages 171–176, San Jose, CA, July 1992. MIT Press.
39. P. van der Smagt. Minimisation methods for training feed-forward networks. *Neural Networks*, 7(1):1–11, 1994.
40. V. Vapnik. *The Nature of Statistical Learning Theory*. Springer Verlag, New York, 1995.
41. V. Vapnik. *Statistical Learning Theory* Wiley, New York, 1998.
42. A. Waibel, T. Hanazawa, G. Hinton, K. Shikano, and K. J. Lang. Phoneme recognition using time-delay neural networks. *IEEE Transactions on Acoustics, Speech, and Signal Processing*, ASSP-37:328–339, 1989.
43. W. Wiegerinck, A. Komoda, and T. Heskes. Stochastic dynamics of learning with momentum in neural networks. *Journal of Physics A*, 27:4425–4437, 1994.
44. H.H. Yang and S. Amari. The efficiency and the robustness of natural gradient descent learning rule. In Michael I. Jordan, Michael J. Kearns, and Sara A. Solla, editors, *Advances in Neural Information Processing Systems*, volume 10. The MIT Press, 1998.

Regularization Techniques to Improve Generalization

Preface

Good tricks for regularization are extremely important for improving the generalization ability of neural networks. The first and most commonly used trick is **early stopping**, which was originally described in [11]. In its simplest version, the trick is as follows:

> *Take an independent validation set, e.g. take out a part of the training set, and monitor the error on this set during training. The error on the training set will decrease, whereas the error on the validation set will first decrease and then increase. The early stopping point occurs where the error on the validation set is the lowest. It is here that the network weights provide the best generalization.*

As Lutz Prechelt points out in chapter 2, the above picture is highly idealized. In practice, the shape of the error curve on the validation set is more likely very ragged with multiple minima. Choosing the "best" early stopping point then involves a trade-off between (1) improvement of generalization and (2) speed of learning. If speed is not an issue then, clearly, the safest strategy is to train all the way until the minimum error on the training set is found, while monitoring the location of the lowest error rate on the validation set. Of course, this can take a prohibitive amount of computing time. This chapter presents less costly strategies employing a number of different stopping criteria, e.g. when the ratio between the *generalization loss* and the *progress* exceeds a given threshold (see p. 59). A large simulation study using various benchmark problems is used in the discussion and analysis of the differences (with respect to e.g. robustness, effectiveness, training time, ...) between these proposed **stopping criteria** (see p. 62ff.). So far theoretical studies [12, 1, 6] have not studied this trade-off.

Weight decay is also a commonly used technique for controlling capacity in neural networks. Early stopping is considered to be fast, but it is not well defined (keep in mind the pitfalls mentioned in chapter 2). On the other hand, weight decay regularizers [5, 2] are well understood, but finding a suitable parameter λ to control the strength of the weight decay term can be tediously time consuming. Thorsteinn Rögnvaldsson proposes a simple trick for **estimating** λ by making use of the *best of both worlds* (see p. 77): simply compute the gradient at the early stopping solution $\boldsymbol{W}_{es}$ and divide it by the norm of $\boldsymbol{W}_{es}$,

$$\hat{\lambda} = \|\nabla E(\boldsymbol{W}_{es})\| / \|2\boldsymbol{W}_{es}\|.$$

Other penalties are also possible. The trick is speedy, since we neither have to do a complete training nor a scan of the whole λ parameter space, and the accuracy of the determined $\hat{\lambda}$ is good, as seen from some interesting simulations.

Tony Plate in chapter 4 treats the penalty factors for the weights (hyperparameters) along the **Bayesian framework** of MacKay [8] and Neal [9]. There are two levels in searching for the best network. The inner loop is a minimization of the training error keeping the hyperparameters fixed, whereas the outer loop searches the hyperparameter space with the goal of maximizing the evidence of having generated the data. This whole procedure is rather slow and computationally expensive, since, in theory, the inner search needs to converge (to a local minimum) at each outer loop search step. When applied to classification networks using the cross-entropy error function the outer-loop search can be unstable with the hyperparameter values oscillating wildly or going to inappropriate extremes. To make this Bayesian framework work better in practice, Tony Plate proposes a number of tricks that speed and simplify the **hyperparameter search** strategies (see p. 98). In particular, his search strategies center around the questions: (1) how often (when) should the hyperparameters be updated (see p. 98) and (2) what should be done if the Hessian is out-of-bounds (see p. 99ff.). To discuss the effects of the choices made in (1) and (2), Tony Plate uses simulations based on artificial examples and concludes with a concise set of rules for making the hyperparameter framework work better.

In chapter 5, Jan Larsen et al. formulate an iterative gradient descent scheme for **adapting** their **regularization parameters** (note, different regularizers can be used for input/hidden and hidden/output weights). The trick is simple: perform gradient descent on the validation set errors with respect to the regularization parameters, and iteratively use the results for updating the estimate of the regularization parameters (see p. 118). This method holds for a variety of penalty terms (e.g. weight decay). The computational overhead is negligible for computing the gradients, however, an inverse Hessian has to be estimated. If second order methods are used for training, then the inverse Hessian may already be available, so there is little additional effort. Otherwise obtaining full Hessian information is rather tedious and limits the approach to smaller applications (see discussion in chapter 1). Nevertheless approximations of the Hessian (e.g. diagonal) could also be used to limit the computation time. Jan Larsen, et al., demonstrate the applicability of their trick on classification (vowel data) and regression (time-series prediction) problems.

Averaging over multiple predictors is a well known method for improving generalization (see e.g. [10, 3, 7, 13]). David Horn et al. raises two questions in ensemble training: (1) how many predictors are "enough" and (2) how does the number of predictors affect the stopping criteria for early stopping (see p. 136). They present solutions for answering these questions by providing a method for estimating the error of an infinite number of predictors and they demonstrate the usefulness of their trick for the sunspot prediction task. Additional theoretical reasoning is given to explain their success in terms of variance minimization within the ensemble.

Jenny & Klaus

References

1. S. Amari, N. Murata, K.-R. Müller, M. Finke, and H. H. Yang. Asymptotic statistical theory of overtraining and cross-validation. *IEEE Transactions on Neural Networks*, 8(5):985–996, September 1997.
2. C. M. Bishop. *Neural Networks for Pattern Recognition*. Clarendon Press, Oxford, 1995.
3. L. Breiman. Bagging predictors. *Machine Learning*, 26(2):123–140, 1996.
4. J. D. Cowan, G. Tesauro, and J. Alspector, editors. *Advances in Neural Information Processing Systems 6*, San Mateo, CA, 1994. Morgan Kaufman Publishers Inc.
5. F. Girosi, M. Jones, and T. Poggio. Regularization theory and neural networks architectures. *Neural Computation*, 7(2):219–269, 1995.
6. M. Kearns. A bound on the error of cross validation using the approximation and estimation rates, with consequences for the training-test split. *Neural Computation*, 9(5):1143–1161, 1997.
7. W. P. Lincoln and J. Skrzypek. Synergy of clustering multiple back propagation networks. In D. S. Touretzky, editor, *Advances in Neural Information Processing Systems 2*, pages 650–657, San Mateo, CA, 1990. Morgan Kaufmann.
8. D. J. C. McKay. A practical Bayesian framework for backpropagation networks. *Neural Computation*, 4:448–472, 1992.
9. R. M. Neal. *Bayesian Learning for Neural Networks*. Number 118 in Lecture Notes in Statistics. Springer, New York, 1996.
10. M. P. Perrone. *Improving Regression Estimation: Averaging Methods for Variance Reduction with Extensions to General Convex Measure Optimization*. PhD thesis, Brown University, May 1993.
11. D. C. Plaut, S. J. Nowlan, and Geoffrey E. Hinton. Experiments on learning by back-propagation. Technical Report Computer Science Dept. Tech. Report, Pittsburgh, PA, 1986.
12. C. Wang, S. S. Venkatesh, and J. S. Judd. Optimal stopping and effective machine complexity in learning. In *[4]*, 1994.
13. D. H. Wolpert. Stacked generalization. *Neural Networks*, 5(2):241–259, 1992.

2
Early Stopping – But When?

Lutz Prechelt

Fakultät für Informatik; Universität Karlsruhe
D-76128 Karlsruhe; Germany
prechelt@ira.uka.de
http://www.ipd.ira.uka.de/~prechelt/

Abstract. Validation can be used to detect when overfitting starts during supervised training of a neural network; training is then stopped before convergence to avoid the overfitting ("early stopping"). The exact criterion used for validation-based early stopping, however, is usually chosen in an ad-hoc fashion or training is stopped interactively. This trick describes how to select a stopping criterion in a systematic fashion; it is a trick for either speeding learning procedures or improving generalization, whichever is more important in the particular situation. An empirical investigation on multi-layer perceptrons shows that there exists a tradeoff between training time and generalization: From the given mix of 1296 training runs using different 12 problems and 24 different network architectures I conclude slower stopping criteria allow for small improvements in generalization (here: about 4% on average), but cost *much* more training time (here: about factor 4 longer on average).

2.1 Early stopping is not quite as simple

2.1.1 Why early stopping?

When training a neural network, one is usually interested in obtaining a network with optimal generalization performance. However, all standard neural network architectures such as the fully connected multi-layer perceptron are prone to overfitting [10]: While the network *seems* to get better and better, i.e., the error on the training set decreases, at some point during training it actually begins to get worse again, i.e., the error on unseen examples increases. The idealized expectation is that during training the generalization error of the network evolves as shown in Figure 2.1. Typically the generalization error is estimated by a validation error, i.e., the average error on a *validation set*, a fixed set of examples not from the training set.

There are basically two ways to fight overfitting: reducing the number of dimensions of the parameter space or reducing the effective size of each dimension. Techniques for reducing the number of parameters are greedy constructive learning [7], pruning [5, 12, 14], or weight sharing [18]. Techniques for reducing the size of each parameter dimension are regularization, such as weight decay

[13] and others [25], or early stopping [17]. See also [8, 20] for an overview and [9] for an experimental comparison.

Early stopping is widely used because it is simple to understand and implement and has been reported to be superior to regularization methods in many cases, e.g. in [9].

2.1.2 The basic early stopping technique

In most introductory papers on supervised neural network training one can find a diagram similar to the one shown in Figure 2.1. It is claimed to show the evolution over time of the per-example error on the training set and on a validation set not used for training (the *training error curve* and the *validation error curve*). Given this behavior, it is clear how to do early stopping using validation:

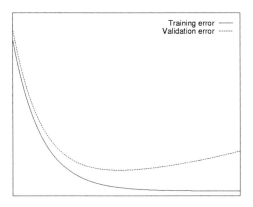

Fig. 2.1. Idealized training and validation error curves. Vertical: errors; horizontal: time

1. Split the training data into a training set and a validation set, e.g. in a 2-to-1 proportion.
2. Train only on the training set and evaluate the per-example error on the validation set once in a while, e.g. after every fifth epoch.
3. Stop training as soon as the error on the validation set is higher than it was the last time it was checked.
4. Use the weights the network had in that previous step as the result of the training run.

This approach uses the validation set to anticipate the behavior in real use (or on a test set), assuming that the error on both will be similar: The validation error is used as an estimate of the generalization error.

2.1.3 The uglyness of reality

However, for real neural network training the validation set error does not evolve as smoothly as shown in Figure 2.1, but looks more like in Figure 2.2. See Sec-

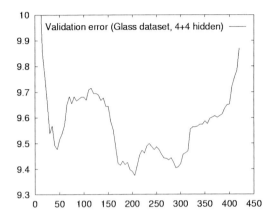

Fig. 2.2. A real validation error curve. Vertical: validation set error; horizontal: time (in training epochs).

tion 2.4 for a rough explanation of this behavior. As we see, the validation error can still go further down after it has begun to increase — plus in a realistic setting we do never know the exact generalization error but estimate it by the validation set error instead. There is no obvious rule for deciding when the minimum of the generalization error is obtained. Real validation error curves almost always have more than one local minimum. The above curve exhibits as many as 16 local minima before severe overfitting begins at about epoch 400. Of these local minima, 4 are the global minimum up to where they occur. The optimal stopping point in this example would be epoch 205. Note that stopping in epoch 400 compared to stopping shortly after the first "deep" local minimum at epoch 45 trades an about sevenfold increase of learning time for an improvement of validation set error by 1.1% (by finding the minimum at epoch 205). If representative data is used, the validation error is an unbiased estimate of the actual network performance; so we expect a 1.1% decrease of the generalization error in this case. Nevertheless, overfitting might sometimes go undetected because the validation set is finite and thus not perfectly representative of the problem.

Unfortunately, the above or any other validation error curve is not *typical* in the sense that all curves share the same qualitative behavior. Other curves might never reach a better minimum than the first, or than, say, the third; the mountains and valleys in the curve can be of very different width, height, and shape. The only thing all curves seem to have in common is that the differences between the first and the following local minima are not huge.

As we see, choosing a stopping criterion predominantly involves a tradeoff between training time and generalization error. However, some stopping criteria may typically find better tradeoffs that others. This leads to the question of which criterion to use with cross validation to decide when to stop training. This is why we need the present trick: To tell us how to *really* do early stopping.

2.2 How to do early stopping best

What we need is a predicate that tells us when to stop training. We call such a predicate a *stopping criterion*. Among all possible stopping criteria we are searching for those which yield the lowest generalization error and also for those with the best "price-performance ratio", i.e., that require the least training for a given generalization error or that (on average) result in the lowest generalization error for a certain training time.

2.2.1 Some classes of stopping criteria

There are a number of plausible stopping criteria and this work considers three classes of them. To formally describe the criteria, we need some definitions first. Let E be the objective function (error function) of the training algorithm, for example the squared error. Then $E_{tr}(t)$, the training set error (for short: training error), is the average error per example over the training set, measured after epoch t. $E_{va}(t)$, the validation error, is the corresponding error on the validation set and is used by the stopping criterion. $E_{te}(t)$, the test error, is the corresponding error on the test set; it is not known to the training algorithm but estimates the generalization error and thus benchmarks the quality of the network resulting from training. In real life, the generalization error is usually unknown and only the validation error can be used to estimate it.

The value $E_{opt}(t)$ is defined to be the lowest validation set error obtained in epochs up to t:

$$E_{opt}(t) := \min_{t' \leq t} E_{va}(t')$$

Now we define the *generalization* at epoch t to be the relative increase of the validation error over the minimum-so-far (in percent):

$$GL(t) = 100 \cdot \left(\frac{E_{va}(t)}{E_{opt}(t)} - 1 \right)$$

High generalization loss is one obvious candidate reason to stop training, because it directly indicates overfitting. This leads us to the **first class of stopping criteria: stop as soon as the generalization loss exceeds a certain threshold.** We define the class GL_α as

$$GL_\alpha : \text{ stop after first epoch } t \text{ with } GL(t) > \alpha$$

However, we might want to suppress stopping if the training is still progressing very rapidly. The reasoning behind this approach is that when the training error still decreases quickly, generalization losses have higher chance to be "repaired"; we assume that often overfitting does not begin until the error decreases only slowly. To formalize this notion we define a *training strip of length k* to be a sequence of k epochs numbered $n+1\ldots n+k$ where n is divisible by k. The training *progress* (in per thousand) measured after such a training strip is then

$$P_k(t) := 1000 \cdot \left(\frac{\sum_{t'=t-k+1}^{t} E_{tr}(t')}{k \cdot \min_{t'=t-k+1}^{t} E_{tr}(t')} - 1 \right)$$

that is, "how much was the average training error during the strip larger than the minimum training error during the strip?" Note that this progress measure is high for unstable phases of training, where the training set error goes up instead of down. This is intended, because many training algorithms sometimes produce such "jitter" by taking inappropriately large steps in weight space. The progress measure is, however, guaranteed to approach zero in the long run unless the training is globally unstable (e.g. oscillating).

Now we can define the **second class of stopping criteria: use the quotient of generalization loss and progress.**

$$PQ_\alpha : \text{stop after first end-of-strip epoch } t \text{ with } \frac{GL(t)}{P_k(t)} > \alpha$$

In the following we will always assume strips of length 5 and measure the validation error only at the end of each strip.

A completely different kind of stopping criterion relies only on the sign of the changes in the generalization error. We define the **third class of stopping criteria: stop when the generalization error increased in s successive strips.**

$$UP_s : \text{stop after epoch } t \text{ iff } UP_{s-1} \text{ stops after epoch } t-k \text{ and}$$
$$E_{va}(t) > E_{va}(t-k)$$
$$UP_1 : \text{stop after first end-of-strip epoch } t \text{ with } E_{va}(t) > E_{va}(t-k)$$

The idea behind this definition is that when the validation error has increased not only once but during s consecutive strips, we assume that such increases indicate the beginning of final overfitting, independent of how large the increases actually are. The UP criteria have the advantage of measuring change locally so that they can be used in the context of pruning algorithms, where errors must be allowed to remain much higher than previous minima over long training periods.

None of these criteria alone can guarantee termination. We thus complement them by the rule that training is stopped when the progress drops below 0.1 or after at most 3000 epochs.

All stopping criteria are used in the same way: They decide to stop at some time t during training and the result of the training is then the set of weights that exhibited the lowest validation error $E_{opt}(t)$. Note that in order to implement this scheme, only one duplicate weight set is needed.

2.2.2 The trick: Criterion selection rules

These three classes of stopping criteria GL, UP, and PQ were evaluated on a variety of learning problems as described in Section 2.3 below. The results indicate that "slower" criteria, which stop later than others, on the average lead to improved generalization compared to "faster" ones. However, the training time that has to be expended for such improvements is rather large on average and also varies dramatically when slow criteria are used. The systematic differences between the criteria *classes* are only small.

For training setups similar to the one used in this work, the following rules can be used for selecting a stopping criterion:

1. Use fast stopping criteria unless small improvements of network performance (e.g. 4%) are worth large increases of training time (e.g. factor 4).
2. To maximize the probability of finding a "good" solution (as opposed to maximizing the average quality of solutions), use a GL criterion.
3. To maximize the average quality of solutions, use a PQ criterion if the network overfits only very little or an UP criterion otherwise.

2.3 Where and how well does this trick work?

As no mathematical analysis of the properties of stopping criteria is possible today (see Section 2.4 for the state of the art), we resort to an experimental evaluation.

We want to find out which criteria will achieve how much generalization using how much training time on which kinds of problems. To achieve broad coverage, we use 12 different network topologies, 12 different learning tasks, and 14 different stopping criteria. To keep the experiment feasible, only one training algorithm is used.

2.3.1 Concrete questions

To derive and evaluate the stopping criteria selection rules presented above we need to answer the following questions:

1. *Training time:* How long will training take with each criterion, i.e., how *fast* or *slow* are they?
2. *Efficiency:* How much of this training time will be redundant, i.e., will occur after the to-be-chosen validation error minimum has been seen?
3. *Effectiveness:* How good will the resulting network performance be?
4. *Robustness:* How sensitive are the above qualities of a criterion to changes of the learning problem, network topology, or initial conditions?
5. *Tradeoffs:* Which criteria provide the best time-performance tradeoff?
6. *Quantification:* How can the tradeoff be quantified?

The answers will directly lead to the rules already presented above in Section 2.2.2. To find the answers to the questions we record for a large number of runs when each criterion would stop and what the associated network performance would be.

2.3.2 Experimental setup

Approach: To measure network performance, we partition each dataset into two disjoint parts: *Training data* and *test data*. The training data is further subdivided into a *training set* of examples used to adjust the network weights and a *validation set* of examples used to estimate network performance during training as required by the stopping criteria. The validation set is never used for weight adjustment. This decision was made in order to obtain pure stopping criteria results. In contrast, in a real application after a reasonable stopping time has been computed, one would include the validation set examples in the training set and retrain from scratch.

Stopping criteria: The stopping criteria examined were GL_1, GL_2, GL_3, GL_5, $PQ_{0.5}$, $PQ_{0.75}$, PQ_1, PQ_2, PQ_3, UP_2, UP_3, UP_4, UP_6, and UP_8. All criteria where evaluated simultaneously, i.e., each single training run returned one result for each of the criteria. This approach reduces the variance of the estimation.

Learning tasks: Twelve different problems were used, all from the PROBEN1 NN benchmark set [19]. All problems are real datasets from realistic application domains; they form a sample of a broad class of domains, but none of them exhibits extreme nonlinearity. The problems have between 8 and 120 inputs, between 1 and 19 outputs, and between 214 and 7200 examples. All inputs and outputs are normalized to range 0...1. Nine of the problems are classification tasks using 1-of-n output encoding (*cancer, card, diabetes, gene, glass, heart, horse, soybean*, and *thyroid*), three are approximation tasks (*building, flare*, and *hearta*).

Datasets and network architectures: The examples of each problem were partitioned into training (50%), validation (25%), and test set (25% of examples) in three different random ways, resulting in 36 datasets. Each of these datasets was trained with 12 different feedforward network topologies: one hidden layer networks with 2, 4, 8, 16, 24, or 32 hidden nodes and two hidden layer networks with 2+2, 4+2, 4+4, 8+4, 8+8, or 16+8 hidden nodes in the first+second hidden layer, respectively; all these networks were fully connected including all possible shortcut connections. For each of the network topologies and each dataset, two runs were made with linear output units and one with sigmoidal output units using the activation function $f(x) = x/(1 + |x|)$.

Training algorithm: All runs were done using the RPROP training algorithm [21] using the squared error function and the parameters $\eta^+ = 1.1$, $\eta^- = 0.5$, $\Delta_0 \in 0.05...0.2$ randomly per weight, $\Delta_{max} = 50$, $\Delta_{min} = 0$, initial weights $-0.5...0.5$ randomly. RPROP is a fast backpropagation variant that is about as fast as quickprop [6] but more stable without adjustment of the parameters. RPROP requires epoch learning, i.e., the weights are updated only once per epoch. Therefore, the algorithm is fast without parameter tuning for small training sets but not recommendable for large training sets. Lack of parameter tuning helps to avoid the common methodological error of tuning parameters using the test error.

2.3.3 Experiment results

Altogether, 1296 training runs were made for the comparison, giving 18144 stopping criteria performance records for the 14 criteria. 270 of these records (or 1.5%) from 125 different runs reached the 3000 epoch limit instead of using the stopping criterion itself.

The results for each stopping criterion averaged over all 1296 runs are shown in Table 2.1. Figure 2.3 describes the variance embedded in the means given in the table. I will now explain and then interpret the entries in both, table and figure. Note that the discussion is biased by the particular collection of criteria chosen for the study.

Definitions: For each run, we define $E_{va}(C)$ as the minimum validation error found until criterion C indicates to stop; it is the error after epoch number $t_m(C)$ (read: "time of minimum"). $E_{te}(C)$ is the corresponding test error and characterizes network performance. Stopping occurs after epoch $t_s(C)$ (read: "time of stop"). A *best* criterion $\hat{C}$ of a particular run is one with minimum t_s of all those (among the examined) with minimum E_{va}, i.e., a criterion that found the best validation error fastest. There may be several best, because multiple criteria may stop at the same epoch. Note that there is no single criterion $\hat{C}$ because $\hat{C}$ changes from run to run. C is called *good* in a particular run if $E_{va}(C) = E_{va}(\hat{C})$, i.e., if it is among those that found the lowest validation set error, no matter how fast or slow.

2.3.4 Discussion: Answers to the questions

We now discuss the questions raised in Section 2.3.1.

C	training time		efficiency and effectiveness			
	$S_{\hat{c}}(C)$	$S_{GL_2}(C)$	$r(C)$	$B_{\hat{c}}(C)$	$B_{GL_2}(C)$	$P_g(C)$
UP_2	0.792	0.766	0.277	1.055	1.024	0.587
GL_1	0.956	0.823	0.308	1.044	1.010	*0.680
UP_3	1.010	1.264	0.419	*1.026	1.003	0.631
GL_2	1.237	1.000	0.514	1.034	1.000	*0.723
UP_4	1.243	1.566	0.599	*1.020	0.997	0.666
$PQ_{0.5}$	1.253	1.334	0.663	1.027	1.002	0.658
$PQ_{0.75}$	1.466	1.614	0.863	1.021	0.998	0.682
GL_3	1.550	1.450	*0.712	1.025	0.994	*0.748
PQ_1	1.635	1.796	1.038	1.018	0.994	0.704
UP_6	1.786	2.381	1.125	*1.012	0.990	0.737
GL_5	2.014	2.013	1.162	1.021	0.991	*0.772
PQ_2	2.184	2.510	1.636	1.012	0.990	0.768
UP_8	2.485	3.259	1.823	*1.010	0.988	0.759
PQ_3	2.614	3.095	2.140	1.009	0.988	0.800

Table 2.1. Behavior of stopping criteria. S_{GL_2} is normalized training time, B_{GL_2} is normalized test error (both relative to GL_2). r is the training time redundancy, P_g is the probability of finding a good solution. For further description please refer to the text.

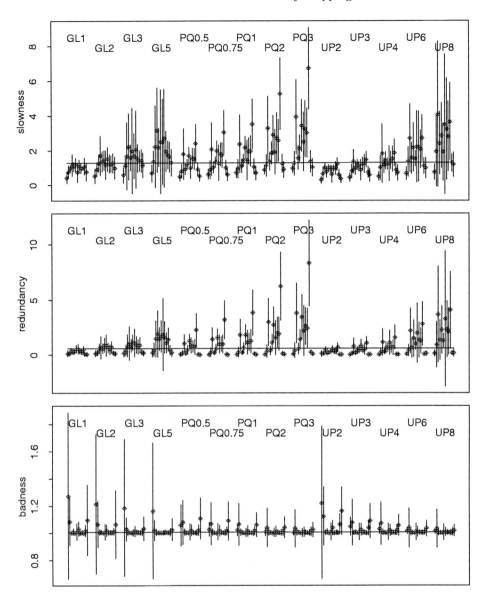

Fig. 2.3. Variance of slowness $S_{\hat{C}}(C)$ (top), redundancy $r(C)$ (middle), and badness $B_{\hat{C}}(C)$ (bottom) for each pair of learning problem and stopping criterion. In each of the 168 columns, the dot represents the mean computed from 108 runs: learning problem and stopping criterion are fixed, while three other parameters are varied (12 topologies × 3 runs × 3 dataset variants). The length of the line is twice the standard deviation within these 108 values. Within each block of dot-line plots, the plots represent (in order) the problems building, cancer, card, diabetes, flare, gene, glass, heart, hearta, horse, soybean, thyroid. The horizontal line marks the median of the means. Note: When comparing the criteria groups, remember that overall the PQ criteria chosen are slower than the others. It is unfair to compare, for example, $PQ_{0.5}$ to GL_1 and UP_2.

1. *Training time:* The *slowness* of a criterion C in a run, relative to another criterion x is $S_x(C) := t_s(C)/t_s(x)$, i.e., the relative total training time. As we see, the times relative to a fixed criterion as shown in column $S_{GL_2}(C)$ vary by more than factor 4. Therefore, the decision for a particular stopping criterion influences training times dramatically, even if one considers only the range of criteria used here. In contrast, even the slowest criteria train only about 2.5 times as long as the fastest criterion of each run that finds the same result, as indicated in column $S_{\hat{C}}(C)$. This shows that the training times are not completely unreasonable even for the slower criteria, but do indeed pay off to some degree.

2. *Efficiency:* The *redundancy* of a criterion can be defined as $r(C) := (t_s(C)/t_m(C)) - 1$. It characterizes how long the training continues after the final solution has been seen. $r(C) = 0$ would be perfect, $r(C) = 1$ means that the criterion trains twice as long as necessary. Low values indicate efficient criteria. As we see, the slower a criterion is, the less efficient it tends to get. Even the fastest criteria "waste" about one fifth of their overall training time. The slower criteria train twice as long as necessary to find the same solution.

3. *Effectiveness:* We define the *badness* of a criterion C in a run relative to another criterion x as $B_x(C) := E_{te}(C)/E_{te}(x)$, i.e., its relative error on the test set. $P_g(C)$ is the fraction of the 1296 runs in which C was a good criterion. This is an estimate of the probability that C is good in a run. As we see from the P_g column, even the fastest criteria are fairly effective. They reach a result as good as the best (of the same run) in about 60% of the cases. On the other hand, even the slowest criteria are not at all infallible; they achieve about 80%. However, P_g says nothing about how *far* from the optimum the non-good runs are. Columns $B_{\hat{C}}(C)$ and $B_{GL_2}(C)$ indicate that these differences are usually rather small: column $B_{GL_2}(C)$ shows that even the criteria with the lowest error achieve only about 1% lower error on the average than the relatively fast criterion GL_2. In column $B_{\hat{C}}(C)$ we see that several only modestly slow criteria have just about 2% higher error on the average than the best criteria of the same run. For obtaining the lowest possible generalization error, independent of training time, it appears that one has to use an extreme criterion such as GL_{50} or even use a conjunction of all three criteria classes with high parameter values.

4. *Robustness:* We call a criterion *robust* to the degree that its performance is independent of the learning problem and the learning environment (network topology, initial conditions etc.). Optimal robustness would mean that in Figure 2.3 all dots within a block are at the same height (problem independence) and all lines have length zero (environment independence). Note that slowness and badness are measured relative to the best criterion of the same program run. We observe the following:

- With respect to slowness and redundancy, slower criteria are much less robust than faster ones. In particular the PQ criteria are quite sensitive to the learning problem, with the card and horse problems being worst in this experimental setting.
- With respect to badness, the picture is completely different: slower criteria tend to be slightly *more* robust than slower ones. PQ criteria are a little

more robust than the others while GL criteria are significantly less robust. All criteria are more or less instable for the building, cancer, and thyroid problems. In particular, all GL criteria have huge problems with the building problem, whose dataset 1 is the only one that is partitioned non-randomly; it uses chronological order of examples, see [19]. The slower variants of the other criteria types are nicely robust in this case.
– Similar statements apply when one analyzes the influence of only large or only small network topologies separately (not shown in any figure or table). One notable exception was the fact that for networks with very few hidden nodes the PQ criteria are more cost-effective than both the GL and the UP criteria for minimizing $B_{\hat{C}}(C)$. The explanation may be that such small networks do not overfit severely; in this case it is advantageous to take training progress into account as an additional factor to determine when to stop training.

Overall, fast criteria improve the predictability of the training time, while slow ones improve the predictability of the solution quality.

5. Best tradeoffs: Despite the common overall trend, some criteria may be more cost-effective than others, i.e., provide better tradeoffs between training time and resulting network performance. Column $B_{\hat{c}}$ of the table suggests that the best tradeoffs between test error and training time are (in order of increasing willingness to spend lots of training time) UP_3, UP_4, and UP_6, if one wants to minimize the expected network performance from a single run. These criteria are also robust. If on the other hand one wants to make several runs and pick the network that seems to be best (based on its validation error), P_g is the relevant metric and the GL criteria are preferable. The best tradeoffs are marked with a star in the table. Figure 2.4 illustrates these results. The upper curve corresponds

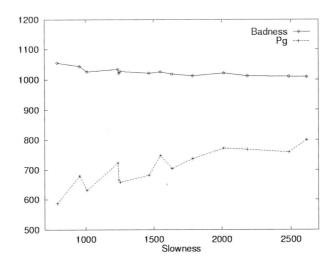

Fig. 2.4. Badness $B_{\hat{C}}(C)$ and P_g against slowness $S_{\hat{C}}(C)$ of criteria

to column $B_{\hat{C}}$ of the table (plotted against column $S_{\hat{C}}$); local minima indicate criteria with the best tradeoffs. The lower curve corresponds to column P_g; local maxima indicate the criteria with the best tradeoffs. All measurements are scaled by 1000.

   6. *Quantification:* From columns $S_{GL_2}(C)$ and $B_{GL_2}(C)$ we can quantify the tradeoff involved in the selection of a stopping criterion as follows: In the range of criteria examined we can roughly trade a 4% decrease in test error (from 1.024 to 0.988) for an about fourfold increase in training time (from 0.766 to 3.095). Within this range, some criteria are somewhat better than others, but there is no panacea.

2.3.5 Generalization of these results

It is difficult to say whether or how these results apply to different contexts than those of the above evaluation. Speculating though, I would expect that the behavior of the stopping criteria

- is similar for other learning rules, unless they frequently make rather extreme steps in parameter space,
- is similar for other error functions, unless they are discontinuous,
- is similar for other learning tasks, as long as they are in the same ballpark with respect to their nonlinearity, number of inputs and outputs, and amount of available training data.

Note however, that at least with respect to the learning task deviations do occur (see Figure 2.3). More research is needed in order to describe which properties of the learning tasks lead to which differences in stopping criteria behavior — or more generally: in order to understand how which features of tasks influence learning methods.

2.4 Why this works

Detailed theoretical analyses of the error curves cannot yet be done for the most interesting cases such as sigmoidal multi-layer perceptrons trained on a modest number of examples; today they are possible for restricted scenarios only [1–3, 24] and do usually not aim at finding the optimal stopping criterion in a way comparable to the present work. However, a simplification of the analysis performed by Wang et al. [24] or the alternative view induced by the bias/variance decomposition of the error as described by Geman et al. [10] can give some insights why early stopping behaves as it does.

   At the beginning of training (phase I), the error is dominated by what Wang et al. call the *approximation error* — the network has hardly learned anything and is still very biased. During training this part of the error is further and further reduced. At the same time, however, another component of the error increases: the *complexity error* that is induced by the increasing variance of the network model as the possible magnitude and diversity of the weights grows.

If we train long enough, the error will be dominated by the complexity error (phase III). Therefore, there is a phase during training, when the approximation and complexity (or: bias and variance) components of the error compete but none of them dominates (phase II). See Amari et al. [1, 2] for yet another view of the training process, using a geometrical interpretation. The task of early stopping as described in the present work is to detect when phase II ends and the dominance of the variance part begins.

Published theoretical results on early stopping appear to provide some nice techniques for practical application: Wang et al. [24] offer a method for computing the stopping point based on complexity considerations — without using a separate validation set at all. This could save precious training examples. Amari et al. [1, 2] compute the optimal split proportion of training data into training and validation set.

On the other hand, unfortunately, the practical applicability of these theoretical analyses is severely restricted. Wang et al.'s analysis applies to networks where only output weights are being trained; no hidden layer training is captured. It is unclear to what degree the results apply to the multi-layer networks considered here. Amari et al.'s analysis applies to the asymptotic case of very many training examples. The analysis does not give advice on stopping criteria; it shows that early stopping is not useful when very many examples are available but does not cover the much more frequent case when training examples are scarce.

There are several other theoretical works on early stopping, but none of them answers our practical questions. Thus, given these theoretic results, one is still left with making a good stopping decision for practical cases of multilayer networks with only few training examples and faced with a complicated evolution of the validation set error as shown in Figure 2.2. This is why the present empirical investigation was necessary.

The jagged form of the validation error curve during phase II arises because neither bias nor variance change monotonically, let alone smoothly. The bias error component may change abruptly because training algorithms never perform gradient descent, but take finite steps in parameter space that sometimes have severe results. The observed variance error component may change abruptly because, first, the validation set error is only an estimate of the actual generalization error and, second, the effect of a parameter change may be very different in different parts of parameter space.

Quantitatively, the different error minima that occur during phase II are quite close together in terms of size, but may be rather far apart in terms of training epoch. The exact validation error behavior seems rather unpredictable when only a short left section of the error curve is given. The behavior is also very different for different training situations.

For these reasons no class of stopping criteria has any big advantage over another (on average, for the mix of situations considered here), but scaling the same criterion to be slower always tends to gain a little generalization.

References

1. S. Amari, N. Murata, K.-R. Müller, M. Finke, and H. Yang. Statistical theory of overtraining - is cross-validation effective? In *[23]*, pages 176–182, 1996.
2. S. Amari, N. Murata, K.-R. Müller, M. Finke, and H. Yang. Asymptotic statistical theory of overtraining and cross-validation. *IEEE Trans. on Neural Networks*, 8(5):985–996, September 1997.
3. P. Baldi and Y. Chauvin. Temporal evolution of generalization during learning in linear networks. *Neural Computation*, 3:589–603, 1991.
4. J. D. Cowan, G. Tesauro, and J. Alspector, editors. *Advances in Neural Information Processing Systems 6*, San Mateo, CA, 1994. Morgan Kaufman Publishers Inc.
5. Y. Le Cun, J. S. Denker, and S. A. Solla. Optimal brain damage. In *[22]*, pages 598–605, 1990.
6. S. E. Fahlman. An empirical study of learning speed in back-propagation networks. Technical Report CMU-CS-88-162, School of Computer Science, Carnegie Mellon University, Pittsburgh, PA, September 1988.
7. S. E. Fahlman and C. Lebiere. The Cascade-Correlation learning architecture. In *[22]*, pages 524–532, 1990.
8. E. Fiesler (efiesler@idiap.ch). Comparative bibliography of ontogenic neural networks. (submitted for publication), 1994.
9. W. Finnoff, F. Hergert, and H. G. Zimmermann. Improving model selection by nonconvergent methods. *Neural Networks*, 6:771–783, 1993.
10. S. Geman, E. Bienenstock, and R. Doursat. Neural networks and the bias/variance dilemma. *Neural Computation*, 4:1–58, 1992.
11. S. J. Hanson, J. D. Cowan, and C. L. Giles, editors. *Advances in Neural Information Processing Systems 5*, San Mateo, CA, 1993. Morgan Kaufman Publishers Inc.
12. B. Hassibi and D. G. Stork. Second order derivatives for network pruning: Optimal brain surgeon. In *[11]*, pages 164–171, 1993.
13. A. Krogh and J. A. Hertz. A simple weight decay can improve generalization. In *[16]*, pages 950–957, 1992.
14. A. U. Levin, T. K. Leen, and J. E. Moody. Fast pruning using principal components. In *[4]*, 1994.
15. R. P. Lippmann, J. E. Moody, and D. S. Touretzky, editors. *Advances in Neural Information Processing Systems 3*, San Mateo, CA, 1991. Morgan Kaufman Publishers Inc.
16. J. E. Moody, S. J. Hanson, and R. P. Lippmann, editors. *Advances in Neural Information Processing Systems 4*, San Mateo, CA, 1992. Morgan Kaufman Publishers Inc.
17. N. Morgan and H. Bourlard. Generalization and parameter estimation in feedforward nets: Some experiments. In *[22]*, pages 630–637, 1990.
18. S. J. Nowlan and G. E. Hinton. Simplifying neural networks by soft weight-sharing. *Neural Computation*, 4(4):473–493, 1992.
19. L. Prechelt. PROBEN1 — A set of benchmarks and benchmarking rules for neural network training algorithms. Technical Report 21/94, Fakultät für Informatik, Universität Karlsruhe, Germany, September 1994. Anonymous FTP: /pub/papers/techreports/1994/1994-21.ps.gz on ftp.ira.uka.de.
20. R. Reed. Pruning algorithms — a survey. *IEEE Transactions on Neural Networks*, 4(5):740–746, 1993.

21. M. Riedmiller and H. Braun. A direct adaptive method for faster backpropagation learning: The RPROP algorithm. In *Proc. of the IEEE Intl. Conf. on Neural Networks*, pages 586–591, San Francisco, CA, April 1993.
22. D. S. Touretzky, editor. *Advances in Neural Information Processing Systems 2*, San Mateo, CA, 1990. Morgan Kaufman Publishers Inc.
23. D. S. Touretzky, M. C. Mozer, and M. E. Hasselmo, editors. *Advances in Neural Information Processing Systems 8*, Cambridge, MA, 1996. MIT Press.
24. C. Wang, S. S. Venkatesh, and J. S. Judd. Optimal stopping and effective machine complexity in learning. In *[4]*, 1994.
25. A. S. Weigend, D. E. Rumelhart, and B. A. Huberman. Generalization by weight-elimination with application to forecasting. In *[15]*, pages 875–882, 1991.

3
A Simple Trick for Estimating the Weight Decay Parameter

Thorsteinn S. Rögnvaldsson

Centre for Computer Architecture (CCA), Halmstad University, P.O. Box 823,
S-301 18 Halmstad, Sweden.
denni@cca.hh.se   http://www.hh.se/staff/denni/

Abstract. We present a simple trick to get an approximate estimate of the weight decay parameter λ. The method combines early stopping and weight decay, into the estimate

$$\hat{\lambda} = \|\nabla E(\boldsymbol{W}_{es})\|/\|2\boldsymbol{W}_{es}\|,$$

where $\boldsymbol{W}_{es}$ is the set of weights at the early stopping point, and $E(\boldsymbol{W})$ is the training data fit error.

The estimate is demonstrated and compared to the standard cross-validation procedure for λ selection on one synthetic and four real life data sets. The result is that $\hat{\lambda}$ is as good an estimator for the optimal weight decay parameter value as the standard search estimate, but orders of magnitude quicker to compute.

The results also show that weight decay can produce solutions that are significantly superior to committees of networks trained with early stopping.

3.1 Introduction

A regression problem which does not put constraints on the model used is ill-posed [21], because there are infinitely many functions that can fit a finite set of training data perfectly. Furthermore, real life data sets tend to have noisy inputs and/or outputs, which is why models that fit the data perfectly tend to be poor in terms of out-of-sample performance. Since the modeler's task is to find a model for the underlying function while not overfitting to the noise, models have to be based on criteria which include other qualities besides their fit to the training data.

In the neural network community the two most common methods to avoid overfitting are *early stopping* and *weight decay* [17]. Early stopping has the advantage of being quick, since it shortens the training time, but the disadvantage of being poorly defined and not making full use of the available data. Weight decay, on the other hand, has the advantage of being well defined, but the disadvantage of being quite time consuming. This is because much time is spent with selecting a suitable value for the weight decay parameter (λ), by searching over several values of λ and estimating the out-of-sample performance using e.g. cross validation [25].

In this paper, we present a very simple method for estimating the weight decay parameter, for the standard weight decay case. This method combines early stopping with weight decay, thus merging the quickness of early stopping with the more well defined weight decay method, providing a weight decay parameter which is essentially as good as the standard search method estimate when tested empirically.

We also demonstrate in this paper that the arduous process of selecting λ can be rewarding compared to simpler methods, like e.g. combining networks into committees [16].

The paper is organized as follows: In section 2 we present the background of how and why weight decay or early stopping should be used. In section 3 we review the standard method for selecting λ and also introduce our new estimate. In section 4 we give empirical evidence on how well the method works, and in section 5 we summarize our conclusions.

3.2 Ill-Posed Problems, Regularization, and Such Things...

3.2.1 Ill-Posed Problems

In what follows, we denote the input data by $x(n)$, the target data by $y(n)$, and the model (neural network) output by $f(W, x(n))$, where W denotes the parameters (weights) for the model. We assume a target data generating process of the form

$$y(n) = \phi[x(n)] + \varepsilon(n) \qquad (3.1)$$

where ϕ is the *underlying function* and $\varepsilon(n)$ are sampled from a stationary uncorrelated (IID) zero mean noise process with variance σ^2. We select models f from a model family F, e.g. multilayer perceptrons, to learn an approximation to the underlying function ϕ, based on the training data. That is, we are searching for

$$f^* \equiv f(W^*) \in F \text{ such that } E(f^*, \phi) \leq E(f, \phi) \; \forall \; f \in F, \qquad (3.2)$$

where $E(f, \phi)$ is a measure of the "distance" between the model f and the true model ϕ. Since we only have access to the target values y, and not the underlying function ϕ, $E(f, \phi)$ is often taken to be the mean square error

$$E(f, \phi) \rightarrow E(f, y) = E(W) = \frac{1}{2N} \sum_{n=1}^{N} [y(n) - f(W, x(n))]^2. \qquad (3.3)$$

Unfortunately, minimizing (3.3) is, more often than not, an ill-posed problem. That is, it does not meet the following three requirements [21]:

- The model (e.g. neural network) can learn the function training data, i.e. there *exists* a solution $f^* \in F$.
- The solution is *unique*.
- The solution is *stable* under small variations in the training data set. For instance, training with two slightly different training data sets sampled from the same process must result in similar solutions (similar when evaluated on e.g. test data).

The first and second of these requirements are often not considered serious problems. It is always possible to find a multilayer perceptron that learns the training data perfectly by using many internal units, since any continuous function can be constructed with a single hidden layer network with sigmoid units (see e.g. [6]), and we may be happy with any solution and ignore questions on uniqueness. However, a network that has learned the training data perfectly will be very sensitive to changes in the training data. Fulfilling the first requirement is thus usually in conflict with fulfilling the third requirement, which is a really important requirement. A solution which changes significantly with slightly different training sets will have very poor generalization properties.

3.2.2 Regularization

It is common to introduce so-called regularizers[1] in order to make the learning task well posed (or at least less ill-posed). That is, instead of only minimizing an error of fit measure like (3.3) we augment it with a regularization term $\lambda R(\boldsymbol{W})$ which expresses e.g. our prior beliefs about the solution.

The error functional then takes the form

$$E(\boldsymbol{W}) = \frac{1}{2N} \sum_{n=1}^{N} [y(n) - f(\boldsymbol{W}, \boldsymbol{x}(n))]^2 + \lambda R(\boldsymbol{W}) = E_0(\boldsymbol{W}) + \lambda R(\boldsymbol{W}), \quad (3.4)$$

where λ is the *regularization parameter* which weighs the importance of $R(\boldsymbol{W})$ relative to the error of fit $E_0(\boldsymbol{W})$.

The effect of the regularization term is to shrink the model family F, or make some models more likely than others. As a consequence, solutions become more stable to small perturbations in the training data.

The term "regularization" encompasses all techniques which make use of penalty terms added to the error measure to avoid overfitting. This includes e.g. weight decay [17], weight elimination [26], soft weight sharing [15], Laplacian weight decay [12] [27], and smoothness regularization [2] [9] [14]. Certain forms of "hints" [1] can also be called regularization.

3.2.3 Bias and Variance

The benefit of regularization is often described in the context of *model bias* and *model variance*. This originates from the separation of the expected generaliza-

---
[1] Called "stabilizers" by Tikhonov [21].

tion error $\langle E_{gen} \rangle$ into three terms [8]

$$\langle E_{gen} \rangle = \langle \int [y(\boldsymbol{x}) - f(\boldsymbol{x})]^2 p(\boldsymbol{x}) d\boldsymbol{x} \rangle$$
$$= \int [\phi(\boldsymbol{x}) - \langle f(\boldsymbol{x}) \rangle]^2 p(\boldsymbol{x}) d\boldsymbol{x} + \langle \int [f(\boldsymbol{x}) - \langle f(\boldsymbol{x}) \rangle]^2 p(\boldsymbol{x}) d\boldsymbol{x} \rangle +$$
$$\langle \int [y(\boldsymbol{x}) - \phi(\boldsymbol{x})]^2 p(\boldsymbol{x}) d\boldsymbol{x} \rangle$$
$$= \text{Bias}^2 + \text{Variance} + \sigma^2, \qquad (3.5)$$

where $\langle \rangle$ denotes taking the expectation over an ensemble of training sets. Here $p(\boldsymbol{x})$ denotes the input data probability density.

A high sensitivity to training data noise corresponds to a large model variance. A large bias term means either that $\phi \notin F$, or that ϕ is downweighted in favour of other models in F. We thus have a trade-off between model bias and model variance, which corresponds to the trade-off between the first and third requirements on well-posed problems.

Model bias is weighed versus model variance by selecting both a parametric form for $R(\boldsymbol{W})$ and an optimal[2] value for the regularization parameter λ.

Many neural network practitioners ignore the first part and choose weight decay by default, which corresponds to a Gaussian parametric form for the prior on $\boldsymbol{W}$. Weight decay is, however, not always the best choice (in fact, it is most certainly not the best choice for all problems). Weight decay does not for instance consider the function the network is producing, it only puts a constraint on the parameters. Another, perhaps more correct, choice would be to constrain the higher order derivatives of the network function (which is commonplace in statistics) like in e.g. [14].

3.2.4 Bayesian Framework

From a Bayesian and maximum likelihood perspective, *prior* information about the model (f) is weighed against the *likelihood* of the training data (D) through Bayes theorem (see [4] for a discussion on this). Denote the probability for observing data set D by $p(D)$, the prior distribution of models f by $p(f)$, and the likelihood for observing the data D, if f is the correct model, by $p(D|f)$. We then have for the posterior probability $p(f|D)$ for the model f given the observed data D

$$p(f|D) = \frac{p(D|f)p(f)}{p(D)} \Rightarrow$$
$$-\ln p(f|D) = -\log p(D|f) - \ln p(f) + \ln p(D) \Rightarrow$$
$$-\ln p(f|D) = \sum_{n=1}^{N} [y(n) - f(\boldsymbol{W}, \boldsymbol{x}(n))]^2 - \ln p(f) + \text{constant}, \qquad (3.6)$$

---
[2] Optimality is usually measured via cross-validation or some similar method.

where Gaussian noise ε is assumed in the last step. If we identify $2N\lambda R(\boldsymbol{W})$ with the negative logarithm of the model prior, $-\ln p(f)$, then maximizing $p(f|D)$ is equivalent to minimizing expression (3.4).

From this perspective, choosing $R(\boldsymbol{W})$ is equivalent to choosing a parameterized form for the model prior $p(f)$, and selecting a value for λ corresponds to estimating the parameters for the prior.

3.2.5 Weight Decay

Weight decay [17] is the neural network equivalent to the Ridge Regression [11] method. In this case $R(\boldsymbol{W}) = \|\boldsymbol{W}\|^2 = \sum_k w_k^2$ and the error functional is

$$E(\boldsymbol{W}) = E_0(\boldsymbol{W}) + \lambda R(\boldsymbol{W}) = \frac{1}{2N}\sum_{n=1}^{N}[y(n) - f(\boldsymbol{W},\boldsymbol{x}(n))]^2 + \lambda\|\boldsymbol{W}\|^2, \quad (3.7)$$

and λ is usually referred to as the *weight decay parameter*. In the Bayesian framework, weight decay means implicitly imposing the model prior

$$p[f(\boldsymbol{W})] = \sqrt{\frac{\lambda}{2\pi\sigma^2}}\exp\left(\frac{-\lambda\|\boldsymbol{W}\|^2}{2\sigma^2}\right) \quad (3.8)$$

where σ^2 is the variance of the noise in the data.

Weight decay often improves the generalization properties of neural network models, for reasons outlined above.

3.2.6 Early Stopping

Undoubtedly, the simplest and most widely used method to avoid overfitting is to stop training before the training set has been learned perfectly. This is done by setting aside a fraction of the training data for estimating the out-of-sample performance. This data set is called the validation data set. Training is then stopped when the error on the validation set starts to increase. Early stopping often shortens the training time significantly, but suffers from being ill-defined since there really is no well defined stopping point, and wasteful with data, since a part of the data is set aside.

There is a connection between early stopping and weight decay, if learning starts from small weights, since weight decay applies a potential which forces all weights towards zero. For instance, Sjöberg and Ljung [20] show that, if a constant learning rate η is used, the number of iterations n at which training is stopped is related to the weight decay parameter λ roughly as

$$\lambda \sim \frac{1}{2\eta n}. \quad (3.9)$$

This does not, however, mean that using early stopping is equivalent to using weight decay in practice. Expression (3.9) is based on a constant learning rate, a local expansion around the optimal stopping point, ignoring local minima, and assumes small input noise levels, which may not reflect the situation when overfitting is a serious problem. The choice of learning algorithm can also affect the early stopping point, and one cannot expect (3.9) to hold exactly in the practical case.

Inspired by this connection between early stopping and weight decay, we use early stopping in the following section to estimate the weight decay parameter λ.

3.3 Estimating λ

From a pure Bayesian point of view, the prior is something we know/assume in advance and do not use the training data to select (see e.g. [5]). There is consequently no such thing as "λ selection" in the pure Bayesian model selection scheme. This is of course perfectly fine if the prior is correct. However, if we suspect that our choice of prior is less than perfect, then we are better off if we take an "empirical Bayes" approach and use the data to tune the prior, through λ.

Several options for selecting λ have been proposed. Weigend et al. [26] present, for a slightly different weight cost term, a set of heuristic rules for changing λ during the training. Although Weigend et al. demonstrate the use of these heuristics on a couple of time series problems, we cannot get these rules to work consistently to our satisfaction. A more principled approach is to try several values of λ and estimate the out-of-sample error, either by correcting the training error, with some factor or term, or by using cross-validation. The former is done in e.g. [10], [23], and [24] (see also references therein). The latter is done by e.g. [25].

The method of using validation data for estimating the out-of-sample error is robust but slow since it requires training several models. We use cross-validation here because of its reliability.

3.3.1 Search Estimates

Finding the optimal λ requires the use of a search algorithm, which must be robust because the validation error can be very noisy. A simple and straightforward way is to start at some large λ where the validation error is large, due to the large model bias, and step towards lower values until the out-of-sample error becomes large again, due to the large model variance. In our experience, it often makes sense to do the search in $\log \lambda$ (i.e. with equally spaced increments in $\log \lambda$).

The result of such a search is a set of K values $\{\lambda_k\}$ with corresponding average n-fold cross validation errors $\{\log E_{nCV,k}\}$ and standard deviations $\{\sigma_{nCV,k}\}$ for the validation errors. These are defined as

$$\log E_{nCV,k} = \frac{1}{n} \sum_{j=1}^{n} \log E_{j,k} \qquad (3.10)$$

3. A Simple Trick for Estimating the Weight Decay Parameter

$$\sigma_{nCV,k}^2 = \frac{1}{n-1} \sum_{j=1}^{n} (\log E_{j,k} - \log E_{nCV,k})^2 \qquad (3.11)$$

when $\lambda = \lambda_k$. The number of validation data sets is n and $E_{j,k}$ denotes the validation error when $\lambda = \lambda_k$ and we use validation set j. Taking logarithms is motivated by our observation that the validation error distribution looks approximately log-normal and we use this in our selection of the optimal λ value below.

Once the search is finished, the optimal λ is selected. This is not necessarily trivial since a large range of values may look equally good, or one value may have a small average cross-validation error with a large variation in this error, and another value may have a slightly higher average cross-validation error with a small variation in this error. The simplest approach is to look at a plot of the validation errors versus λ and make a judgement on where the optimal λ is, but this adds an undesired subjectiveness to the choice. Another is to take a weighted average over the different λ values, which is what we use here (see Ripley [19] for a discussion on variants of λ selection methods).

Our estimate for the optimal λ is the value

$$\hat{\lambda}_{opt} = \frac{\sum_{k=1}^{K} n_k \lambda_k}{\sum_{k=1}^{K} n_k} \qquad (3.12)$$

where n_k is the number of times λ_k corresponds to the minimum validation error when we sample validation errors from K log-normal distributions with means $\log E_{nCV,k}$ and standard deviations $\sigma_{nCV,k}$, assuming that the validation errors are independent. This is illustrated on a hypothetical example in Figure 3.1. The choice (3.12) was done after confirming that it often agrees well with our subjective choice for λ. We refer to this below as a "Monte Carlo estimate" of λ.

3.3.2 Two Early Stopping Estimates

If $\boldsymbol{W}^*$ is the set of weights when $E(\boldsymbol{W})$ in eq. (3.4) is minimized, then

$$\nabla E(\boldsymbol{W}^*) = \nabla E_0(\boldsymbol{W}^*) + \lambda \nabla R(\boldsymbol{W}^*) = 0, \qquad (3.13)$$

which implies

$$\lambda = \frac{\|\nabla E_0(\boldsymbol{W}^*)\|}{\|\nabla R(\boldsymbol{W}^*)\|} \qquad (3.14)$$

for the regularization parameter λ. Thus, if we have a reasonable estimate of $\boldsymbol{W}^*$, or of $\|\nabla E_0(\boldsymbol{W}^*)\|$ and $\|\nabla R(\boldsymbol{W}^*)\|$, then we can use this to estimate λ. An appealingly simple way of estimating $\boldsymbol{W}^*$ is to use early stopping, because of its connection with weight decay.

Denoting the set of weights at the early stopping point by $\boldsymbol{W}_{es}$, we have

$$\hat{\lambda}_1 = \frac{\|\nabla E_0(\boldsymbol{W}_{es})\|}{\|\nabla R(\boldsymbol{W}_{es})\|}, \qquad (3.15)$$

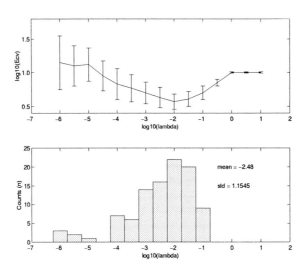

Fig. 3.1. Illustration of the procedure for estimating $\hat{\lambda}_{opt}$ on a hypothetical example. From the search we have a set of K lognormal distributions with means $\log E_{nCV,k}$ and variances $\sigma^2_{nCV,k}$, which is illustrated in the top plate. From these K distributions, we sample K error values and select the λ corresponding to the minimum error value as "winner". This is repeated several times (100 in the figure but 10,000 times in the experiments in the text) collecting statistics on how often different λ values are winners, and the mean $\log \lambda$ is computed. This is illustrated in the bottom plate, which shows the histogram resulting from sampling 100 times. From this we get $\log \hat{\lambda}_{opt} = -2.48 \pm 1.15$, which gives us $\lambda = 10^{-2.48} = 0.003$ for training the "best" network.

as a simple estimate for λ. A second possibility is to consider the whole set of linear equations defined by (3.13) and minimize the squared error

$$\|\nabla E_0(\boldsymbol{W}_{es}) + \lambda \nabla R(\boldsymbol{W}_{es})\|^2 =$$
$$\|\nabla E_0(\boldsymbol{W}_{es})\|^2 + 2\lambda \nabla E_0(\boldsymbol{W}_{es}) \cdot \nabla R(\boldsymbol{W}_{es}) + \lambda^2 \|\nabla R(\boldsymbol{W}_{es})\|^2 \quad (3.16)$$

with respect to λ. That is, solving the equation

$$\frac{\partial}{\partial \lambda} \left\{ \|\nabla E_0(\boldsymbol{W}_{es}) + \lambda \nabla R(\boldsymbol{W}_{es})\|^2 \right\} = 0 \quad (3.17)$$

which gives

$$\hat{\lambda}_2 = \max\left[0, \frac{-\nabla E_0(\boldsymbol{W}_{es}) \cdot \nabla R(\boldsymbol{W}_{es})}{\|\nabla R(\boldsymbol{W}_{es})\|^2}\right]. \quad (3.18)$$

The estimate is bound from below since λ must be positive.

The second estimate, $\hat{\lambda}_2$, corresponds to a linear regression without intercept term on the set of points $\{\partial_i E_0(\boldsymbol{W}_{es}), \partial_i R(\boldsymbol{W}_{es})\}$, whereas the first estimate, $\hat{\lambda}_1$, is closer to the ratio $\max[|\partial_i E_0(\boldsymbol{W}_{es})|]/\max[|\partial_i R(\boldsymbol{W}_{es})|]$. It follows from the Cauchy-Schwartz inequality that

$$\hat{\lambda}_1 \geq \hat{\lambda}_2. \quad (3.19)$$

For the specific case of weight decay, where $R(\boldsymbol{W}) = \|\boldsymbol{W}\|^2$, expressions (3.15) and (3.18) become

$$\hat{\lambda}_1 = \frac{\|\nabla E_0(\boldsymbol{W}_{es})\|}{2\|\boldsymbol{W}_{es}\|}, \quad (3.20)$$

$$\hat{\lambda}_2 = \max\left[0, \frac{-\nabla E_0(\boldsymbol{W}_{es}) \cdot \boldsymbol{W}_{es}}{2\|\boldsymbol{W}_{es}\|^2}\right]. \quad (3.21)$$

These estimates are sensitive to the particularities of the training and validation data sets used, and possibly also to the training algorithm. One must therefore average them over different validation and training sets. It is, however, still quicker to do this than to do a search since early stopping training often is several orders of magnitude faster to do than a full minimization of (3.7).

One way to view the estimates (3.15) and (3.18) is as the weight decay parameters that correspond to the early stopping point. However, our aim here is not to imitate early stopping with weight decay, but to use early stopping to estimate the weight decay parameter λ. We hope that using weight decay with this λ value will actually result in better out-of-sample performance than what we get from doing early stopping (the whole exercise becomes rather meaningless if this is not the case).

As a sidenote, we imagine that (3.15) and (3.18) could be used also to estimate weight decay parameters in cases when different weight decays are used for weights in different layers. This would then be done by considering these estimates for different groups of weights.

3.4 Experiments

3.4.1 Data Sets

We here demonstrate the performance of our algorithm on a set of five regression problems. For each problem, we vary either the number of inputs, the number of hidden units, or the amount of training data to study the effects of the numbers of parameters relative to the number of training data points. The five problems are:

Synthetic bilinear problem. The task is to model a bilinear function of the form
$$\phi(x_1, x_2) = x_1 x_2. \qquad (3.22)$$

We use three different sizes of training data sets, $M \in \{20, 40, 100\}$, but a constant validation set size of 10 patterns. The validation patterns are in addition to the M training patterns. The test error, or generalization error, is computed by numerical integration over 201×201 data points on a two-dimensional lattice $(x_1, x_2) \in [-1, 1]^2$. The target values (but not the inputs) are contaminated with three different levels of Gaussian noise with standard deviation $\sigma \in \{0.1, 0.2, 0.5\}$. This gives a total of $3 \times 3 = 9$ different experiments on this particular problem, which we refer to as setup A1, A2, ..., and A9 below.

This allows controlled studies w.r.t. noise levels and training set sizes, while keeping the network architecture constant (2 inputs, 8 tanh hidden, and one linear output).

Predicting Puget Sound Power and Light Co. power load between 7 and 8 a.m. the following day. This data set is taken from the Puget Sound Power and Light Co's power prediction competition [3]. The winner of this competition used a set of linear models, one for each hour of the day. We have selected the subproblem of predicting the load between 7 and 8 a.m. 24 hrs. in advance. This hour shows the largest variation in power load. The training set consists of 844 weekdays between January 1985 and September 1990. Of these, 150 days are randomly selected and used for validation. We use 115 winter weekdays, from between November 1990 and March 1992, for out-of-sample testing. The inputs are things like current load, average load during the last 24 hours, average load during the last week, time of the year, etc., giving a total of 15 inputs. Three different numbers of internal units are tried on this task: 15, 10, and 5, and we refer to these experiments as B1, B2, and B3 below.

Predicting daily riverflow in two Icelandic rivers. This problem is tabulated in [22], and the task is to model tomorrow's average flow of water in one of two Icelandic rivers, knowing today's and previous days' waterflow, temperature, and precipitation. The training set consists of 731 data points, corresponding to the years 1972 and 1973, out of which we randomly sample 150 datapoints for validation. The test set has 365 data points (the year 1974). We use two different lengths of lags, 8 or 4 days back, which correspond to 24 or 12 inputs, while the number of internal units is kept constant at 12. These experiments are referred to as C1, C2, C3, and C4 below.

Predicting the Wolf sunspots time series. This time series has been used several times in the context of demonstrating new regularization techniques, for instance by [15] and [26]. We try three different network architectures on this problem, always keeping 12 input units but using 4, 8, or 12 internal units in the network. These experiments are referred to as setup D1, D2, and D3 below. The training set size is kept constant at $M = 221$ (years 1700-1920), out of which we randomly pick 22 patterns for validation. We test our models under four different conditions: Single step prediction on "test set 1" with 35 data points (years 1921-1955), 4-step iterated prediction on "test set 1", 8-step iterated prediction on all 74 available test years (1921-1994), and 11-step iterated prediction on all available test years. These test conditions are coded as s1, m4, m8, and m11.

Estimating the peak pressure position in a combustion engine. This is a data set with 4 input variables (ignition time, engine load, engine speed, and air/fuel ratio) and only 49 training data points, out of which we randomly pick 9 patterns for validation. The test set consists of 35 data points, which have been measured under slightly different conditions than the training data. We try four different numbers of internal units on this task: 2, 4, 8, or 12, and refer to these experiments as E1, E2, E3, and E4.

3.4.2 Experimental Procedure

The experimental procedure is the same for all problems: We begin by estimating λ in the "traditional" way by searching over the region $\log \lambda \in [-6.5, 1.0]$ in steps of $\Delta \log \lambda = 0.5$. For each λ value, we train 10 networks using the Rprop training algorithm [3] [18]. Each network is trained until the total error (3.7) is minimized, measured by

$$\log \left[\frac{1}{100} \sum_{i=1}^{100} \frac{|\Delta E_i|}{\|\Delta \mathbf{W}_i\|} \right] < -5, \qquad (3.23)$$

where the sum runs over the most recent 100 epochs, or until 10^5 epochs have passed, whichever occurs first. The convergence criterion (3.23) is usually fulfilled within 10^5 epochs. New validation and training sets are sampled for each of the 10 networks, but the different validation sets are allowed to overlap. Means and standard deviations, $\log E_{nCV,k}$ and $\sigma_{nCV,k}$, for the errors are estimated

---

[3] Initial tests showed that the Rprop algorithm was considerably more efficient and robust than e.g. backprop or conjugate gradients in minimizing the error. We did not, however, try true second order algorithms like Levenberg-Marquardt or Quasi-Newton.

from these 10 network runs, assuming a lognormal distribution for the validation errors. Figure 3.2 shows an example of such a search for the Wolf sunspot problem, using a neural network with 12 inputs, 8 internal units, and 1 linear output.

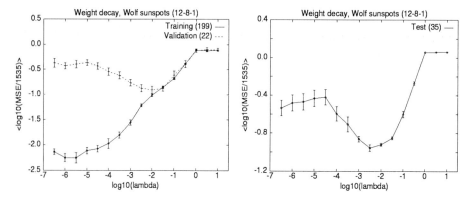

Fig. 3.2. Left panel: Training and validation errors on the Wolf sunspot time series, setup D2, plotted versus the weight decay parameter λ. Each point corresponds to an average over 10 runs with different validation and training sets. The error bars mark 95% confidence limits for the average validation and training errors, under the assumption that the errors are lognormally distributed. The objective Monte Carlo method gives $\log \hat{\lambda}_{opt} = -2.00 \pm 0.31$. Right panel: The corresponding plot for the test error on the sunpots "test set 1". The network architecture is 12 inputs, 8 tanh internal units, and 1 linear output.

Using the objective Monte Carlo method described above, we estimate an optimal $\hat{\lambda}_{opt}$ value from this search. This value is then used to train 10 new networks with all the training data (no validation set). The test errors for these networks are then computed using the held out test set.

A total of $16 \times 10 = 160$ network runs are thus done to select the $\hat{\lambda}_{opt}$ for each experiment. This corresponds to a few days' or a week's work, depending on available hardware and the size of the problem. Although this is in excess of what is really needed in practice (one could get away with about half as many runs in a real application) the time spent doing this is aggravating. The times needed for doing the searches described in this paper ranged from 10 up to 400 cpu-hours, depending on the problem and the computer[4]. For comparison, the early stopping experiments described below took between 10 cpu-minutes and 14 cpu-hours. There was typically a ratio of 40 between the time needed for a search and the time needed for an early stopping estimate.

---

[4] A variety of computers were used for the simulations, including NeXT, Sun Sparc, DEC Alpha, and Pentium computers running Solaris.

3. A Simple Trick for Estimating the Weight Decay Parameter

We then estimate $\hat{\lambda}_1$ and $\hat{\lambda}_2$, by training 100 networks with early stopping. One problem here is that the stopping point is ill-defined, i.e. the first observed minimum in the validation error is not necessarily the minimum where one should stop. The validation error quite often decreases again beyond this point. To avoid such problems, we keep a record of the weights corresponding to the latest minimum validation error and continue training beyond that point. The training is stopped when as many epochs have passed as it took to find the validation error minimum without encountering a new minimum. The weights corresponding to the last validation error minimum are then used as the early stopping weights. For example, if the validation error has a minimum at say 250 epochs, we then wait until a total of 500 epochs have passed before deciding on that particular stopping point. From the 100 networks, we get 100 estimates for $\hat{\lambda}_1$ and $\hat{\lambda}_2$. We take the logarithm of these and compute means $\langle \log \hat{\lambda}_1 \rangle$ and $\langle \log \hat{\lambda}_2 \rangle$, and corresponding standard deviations. The resulting arithmetic mean values are taken as the estimates for λ and the standard deviations are used as measures of the estimation error. The arithmetic means are then used to train 10 networks which use all the training data. Figure 3.3 shows the histograms corresponding to the problem presented in Figure 3.2.

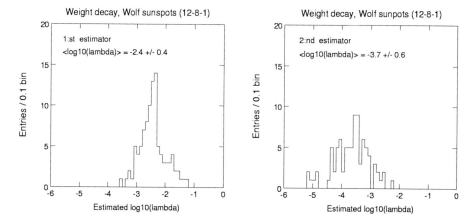

Fig. 3.3. Left panel: Histogram showing the estimated values $\hat{\lambda}_1$ for 100 different training runs, using different training and validation sets each time. Right panel: Similar histogram for $\hat{\lambda}_2$. The problem (D2) is the same as that depicted in Figure 3.2.

When comparing test errors achieved with different methods, we use the Wilcoxon rank test [13], also called the Mann-Whitney test, and report differences at 95% confidence level.

3.4.3 Quality of the λ Estimates

As a first test of the quality of the estimates $\hat{\lambda}_1$ and $\hat{\lambda}_2$, we check how well they agree with the $\hat{\lambda}_{opt}$ estimate, which can be considered a "truth". The estimates for all the problem setups are tabulated in table 3.1 and plotted in Figure 3.4.

Problem	$\log \hat{\lambda}_{opt}$	$\log \hat{\lambda}_1$	$\log \hat{\lambda}_2$
A1 ($M = 20, \sigma = 0.1$)	-2.82 ± 0.04	-2.71 ± 0.66	-3.44 ± 1.14
A2 ($M = 20, \sigma = 0.2$)	-2.67 ± 0.42	-2.32 ± 0.58	-3.20 ± 0.96
A3 ($M = 20, \sigma = 0.5$)	-0.49 ± 1.01	-1.93 ± 0.78	-3.14 ± 1.15
A4 ($M = 40, \sigma = 0.1$)	-2.93 ± 0.49	-2.85 ± 0.73	-3.56 ± 0.87
A5 ($M = 40, \sigma = 0.2$)	-2.53 ± 0.34	-2.41 ± 0.64	-2.91 ± 0.68
A6 ($M = 40, \sigma = 0.5$)	-2.43 ± 0.44	-2.13 ± 0.74	-2.85 ± 0.77
A7 ($M = 100, \sigma = 0.1$)	-3.45 ± 0.78	-3.01 ± 0.86	-3.74 ± 0.93
A8 ($M = 100, \sigma = 0.2$)	-3.34 ± 0.71	-2.70 ± 0.73	-3.33 ± 0.92
A9 ($M = 100, \sigma = 0.5$)	-3.31 ± 0.82	-2.34 ± 0.63	-3.13 ± 1.06
B1 (Power, 15 hidden)	-3.05 ± 0.21	-3.82 ± 0.42	-5.20 ± 0.70
B2 (Power, 10 hidden)	-3.57 ± 0.35	-3.75 ± 0.45	-4.93 ± 0.50
B3 (Power, 5 hidden)	-4.35 ± 0.66	-3.78 ± 0.52	-5.03 ± 0.74
C1 (Jökulsá Eystra, 8 lags)	-2.50 ± 0.10	-3.10 ± 0.33	-4.57 ± 0.59
C2 (Jökulsá Eystra, 4 lags)	-2.53 ± 0.12	-3.15 ± 0.40	-4.20 ± 0.59
C3 (Vatnsdalsá, 8 lags)	-2.48 ± 0.11	-2.65 ± 0.40	-3.92 ± 0.56
C4 (Vatnsdalsá, 4 lags)	-2.39 ± 0.55	-2.67 ± 0.45	-3.70 ± 0.62
D1 (Sunspots, 12 hidden)	-2.48 ± 0.12	-2.48 ± 0.50	-3.70 ± 0.42
D2 (Sunspots, 8 hidden)	-2.00 ± 0.31	-2.43 ± 0.45	-3.66 ± 0.60
D3 (Sunspots, 4 hidden)	-2.51 ± 0.44	-2.39 ± 0.48	-3.54 ± 0.65
E1 (Pressure, 12 hidden)	-3.13 ± 0.43	-3.03 ± 0.70	-4.69 ± 0.91
E2 (Pressure, 8 hidden)	-3.01 ± 0.52	-3.02 ± 0.64	-4.72 ± 0.82
E3 (Pressure, 4 hidden)	-3.83 ± 0.80	-3.07 ± 0.71	-4.50 ± 1.24
E4 (Pressure, 2 hidden)	-4.65 ± 0.78	-3.46 ± 1.34	-4.21 ± 1.40

Table 3.1. Estimates of λ for the 23 different problem setups. Code A corresponds to the synthetic problem, code B to the Power prediction, code C to the riverflow prediction, code D to the Sunspots series, and code E to the maximum pressure position problem. For the log $\hat{\lambda}_{opt}$ column, errors are the standard deviations of the Monte Carlo estimate. For the early stopping estimates, errors are the standard deviations of the estimates.

The linear correlation between $\log \hat{\lambda}_1$ and $\log \hat{\lambda}_{opt}$ is 0.71, which is more than three standard deviations larger than the expected correlation between 23 random points. Furthermore, a linear regression with intercept gives the result

$$\hat{\lambda}_{opt} = 0.30 + 1.13 \hat{\lambda}_1. \tag{3.24}$$

Thus, $\hat{\lambda}_1$ is a fairly good estimator of $\hat{\lambda}_{opt}$.

The linear correlation between $\hat{\lambda}_2$ and $\hat{\lambda}_{opt}$ is 0.48, more than two standard deviations from the random correlation. A linear regression gives

$$\hat{\lambda}_{opt} = -0.66 + 0.57\hat{\lambda}_2, \quad (3.25)$$

and the second estimator $\hat{\lambda}_2$ is clearly a less good estimator of $\hat{\lambda}_{opt}$.

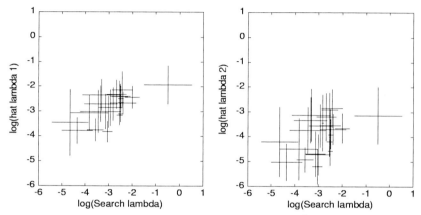

Fig. 3.4. Plot of the results in Table 3.1. Left plate: The $\hat{\lambda}_1$ estimate plotted versus $\hat{\lambda}_{opt}$. The linear correlation between $\log \hat{\lambda}_1$ and $\log \hat{\lambda}_{opt}$ is 0.71. Right plate: $\hat{\lambda}_2$ plotted versus $\hat{\lambda}_{opt}$. The linear correlation between $\log \hat{\lambda}_2$ and $\log \hat{\lambda}_{opt}$ is 0.48. The sizes of the crosses correspond to the error bars in Table 3.1.

We next compare the out-of-sample performances of these different λ estimates, which is what really matters to the practitioner. Table 3.2 lists the differences in out-of-sample performance when using the early stopping estimates or the search estimate. A "+" means that using the early stop estimate results in significantly (95% significance level) lower test error than if $\hat{\lambda}_{opt}$ is used. Similarly, a "–" means that the search estimate gives significantly lower test error than the early stopping estimates. A "0" means there is no significant difference. The conclusion from Table 3.2 is that $\hat{\lambda}_2$ is significantly worse than $\hat{\lambda}_{opt}$, but that there is no consistent difference between $\hat{\lambda}_1$ and $\hat{\lambda}_{opt}$. The two estimates are essentially equal, in terms of test error. In some cases, like the power prediction problem, it would have been beneficial to do a small search around the early stop estimate to check for a possibly better value.

The test errors for the combustion engine (setups E) are not included in Tables 3.2 (and 3.3) because the test set is too different from the training set to provide relevant results. In fact, no regularized network is significantly better than an unregularized network on this problem.

Table 3.2. Relative performance of single networks trained using the estimates $\hat{\lambda}_1$ and $\hat{\lambda}_2$, for the weight decay parameter, and the performance of single networks trained using the search estimate $\hat{\lambda}_{opt}$. The relative performances are reported as: "+" means that using $\hat{\lambda}_i$ results in a test error which is significantly lower than what the search estimate $\hat{\lambda}_{opt}$ gives, "0" means that the performances are equivalent, and "–" means that using $\hat{\lambda}_{opt}$ results in a lower test error than when using $\hat{\lambda}_i$. All results are reported for a 95% confidence level when using the Wilcoxon test. See the text on why the E results are left out.

Problem Setup	$\hat{\lambda}_1$ vs. $\hat{\lambda}_{opt}$	$\hat{\lambda}_2$ vs. $\hat{\lambda}_{opt}$
A1 ($M = 20, \sigma = 0.1$)	0	0
A2 ($M = 20, \sigma = 0.2$)	0	–
A3 ($M = 20, \sigma = 0.5$)	0	0
A4 ($M = 40, \sigma = 0.1$)	0	0
A5 ($M = 40, \sigma = 0.2$)	0	–
A6 ($M = 40, \sigma = 0.5$)	0	0
A7 ($M = 100, \sigma = 0.1$)	–	0
A8 ($M = 100, \sigma = 0.2$)	0	0
A9 ($M = 100, \sigma = 0.5$)	+	0
B1 (Power, 15 hidden)	–	–
B2 (Power, 10 hidden)	–	–
B3 (Power, 5 hidden)	+	–
C1 (Jökulsá Eystra, 8 lags)	–	–
C2 (Jökulsá Eystra, 4 lags)	0	–
C3 (Vatnsdalsá, 8 lags)	–	–
C4 (Vatnsdalsá, 4 lags)	0	–
D1.s1 (Sunspots, 12 hidden)	0	–
D2.s1 (Sunspots, 8 hidden)	+	–
D3.s1 (Sunspots, 4 hidden)	0	–
D1.m4 (Sunspots, 12 hidden)	0	–
D2.m4 (Sunspots, 8 hidden)	+	–
D3.m4 (Sunspots, 4 hidden)	+	–
D1.m8 (Sunspots, 12 hidden)	0	–
D2.m8 (Sunspots, 8 hidden)	–	–
D3.m8 (Sunspots, 4 hidden)	+	–
D1.m11 (Sunspots, 12 hidden)	0	0
D2.m11 (Sunspots, 8 hidden)	+	–
D3.m11 (Sunspots, 4 hidden)	0	–

3.4.4 Weight Decay versus Early Stopping Committees

Having trained all these early stopping networks, it is reasonable to ask if using them to estimate λ for a weight decay network is the optimal use of these networks? Another possible use is, for instance, to construct a committee [16] from them.

To test this, we compare the test errors for our regularized networks with those when using a committee of 10 networks trained with early stopping. The results are listed in Table 3.3.

Some observations from Table 3.3, bearing in mind that the set of problems is small, are: Early stopping committees seem like the better option when the problem is very noisy (setups A3, A6, and A9), and when the network does not have very many degrees of freedom (setups B3, C4, and D3). Weight decay networks, on the other hand, seem to work better than committees on problems with many degrees of freedom (setups B1 and C3), problems with low noise levels and much data (setup A7), and problems where the prediction is iterated through the network (m4, m8, and m11 setups). We emphasize, however, that these conclusions are drawn from a limited set of problems and that all problems tend to have their own set of weird characteristics.

We also check which model works best on each problem. On the power prediction, the best overall model is a large network (B1) which is trained with weight decay. On the river prediction problems, the best models are small (C2 and C4) and trained with either weight decay (Jökulsá Eystra) or early stopping and then combined into committees (Vatnsdalsá). On the sunspot problem, the best overall model is a large network (D1) trained with weight decay.

These networks are competitive with previous results on the same data sets. The performance of the power load B1 weight decay networks, using $\hat{\lambda}_{opt}$, are significantly better than what a human expert produces, and also significantly better than the results by the winner of the Puget Sound Power and Light Co. Power Load Competition [7], although the difference is small. The test results are summarized in Figure 3.5. The performance of the sunspot D1 weight decay network is comparable with the network by Weigend et al., listed in [26]. Figure 3.6 shows the performance of the D1 network trained with weight decay, $\lambda = \hat{\lambda}_1$, and compares it to the results by Weigend et al. [26]. The weight decay network produces these results using a considerably simpler λ selection method and regularization cost than the one presented in [26].

From these anecdotal results, one could be bold and say that weight decay shows a slight edge over early stopping committees. However, it is fair to say that it is a good idea to try both committees and weight decay when constructing predictor models.

It is emphasized that these results are from a small set of problems, but that these problems (except perhaps for the synthetic data) are all realistic in the sense that the datasets are small and noisy.

Table 3.3. Relative performance of single networks trained using weight decay and early stopping committees with 10 members. The relative performance of weight decay (WD) and 10 member early stopping committees are reported as: "+" means that weight decay is significantly better than committees, "0" means that weight decay and committees are equivalent, and "−" means that committees are better than weight decay. All results are reported for a 95% confidence level when using the Wilcoxon test. See the text on why the E results are left out.

Problem Setup	WD($\hat{\lambda}_{opt}$) vs. Comm.	WD($\hat{\lambda}_1$) vs. Comm.
A1 ($M = 20, \sigma = 0.1$)	0	+
A2 ($M = 20, \sigma = 0.2$)	0	0
A3 ($M = 20, \sigma = 0.5$)	−	−
A4 ($M = 40, \sigma = 0.1$)	0	0
A5 ($M = 40, \sigma = 0.2$)	+	+
A6 ($M = 40, \sigma = 0.5$)	−	−
A7 ($M = 100, \sigma = 0.1$)	+	+
A8 ($M = 100, \sigma = 0.2$)	0	0
A9 ($M = 100, \sigma = 0.5$)	−	0
B1 (Power, 15 hidden)	+	−
B2 (Power, 10 hidden)	0	−
B3 (Power, 5 hidden)	−	−
C1 (Jökulsá Eystra, 8 lags)	+	0
C2 (Jökulsá Eystra, 4 lags)	+	0
C3 (Vatnsdalsá, 8 lags)	+	+
C4 (Vatnsdalsá, 4 lags)	−	−
D1.s1 (Sunspots, 12 hidden)	0	0
D2.s1 (Sunspots, 8 hidden)	−	0
D3.s1 (Sunspots, 4 hidden)	−	−
D1.m4 (Sunspots, 12 hidden)	+	+
D2.m4 (Sunspots, 8 hidden)	+	+
D3.m4 (Sunspots, 4 hidden)	0	+
D1.m8 (Sunspots, 12 hidden)	+	+
D2.m8 (Sunspots, 8 hidden)	+	0
D3.m8 (Sunspots, 4 hidden)	0	0
D1.m11 (Sunspots, 12 hidden)	+	+
D2.m11 (Sunspots, 8 hidden)	0	+
D3.m11 (Sunspots, 4 hidden)	+	+

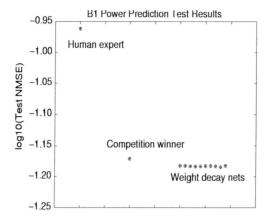

Fig. 3.5. The performance of the 10 neural networks with 15 inputs, 15 hidden units, and one output unit, trained with weight decay using $\lambda = \hat{\lambda}_{opt}$, on the power prediction problem. "Human expert" denotes the prediction result by the human expert at Puget Sound Power and Light Co., and "Competition winner" denotes the result by the model that won the Puget Sound Power and Light Co.'s Power Prediction Competition.

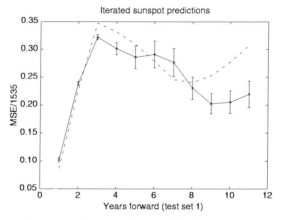

Fig. 3.6. The performance of a neural network with 12 inputs, 12 hidden units, and one output unit, trained with weight decay using $\lambda = \hat{\lambda}_1$, on iterated predictions for the sunspot problem. The error bars denote one standard deviation for the 10 trained networks. The dashed line shows the results when using the network listed in [26]. Note that these results are achieved with a simple weight decay cost and a very simple method for selecting λ, whereas [26] use weight elimination and a complicated heuristic scheme for setting λ.

3.5 Conclusions

The established connection between early stopping and weight decay regularization naturally leads to the idea of using early stopping to estimate the weight decay parameter. In this paper we have shown how this can be done and that the resulting λ results in as low test errors as achieved with the standard cross-validation method, although this varies between problems. In practical applications, this means replacing a search which may take days or weeks, with a computation that usually does not require more than a few minutes or hours. This value can also be used as a starting point for a more extensive cross-validation search.

We have also shown that using several early stopping networks to estimate λ can be smarter than combining the networks into committees. The conclusion from this is that although there is a correspondence between early stopping and weight decay under asymptotic conditions this does not mean that early stopping and weight decay give equivalent results in real life situations.

The method unfortunately only works for regularization terms that have a connection with early stopping, like quadratic weight decay or "weight decay like" regularizers where the weights are constrained towards the origin in weight space (but using e.g. a Laplacian prior instead of the usual Gaussian prior). The method does not carry over to regularizers which do not have any connection to early stopping (like e.g. Tikhonov smoothing regularizers).

Acknowledgements: David B. Rosen is thanked for a very inspiring dinner conversation during the 1996 "Machines that Learn" Workshop in Snowbird, Utah. Milan Casey Brace of Puget Sound Power and Light Co. is thanked for supplying the power load data. Financial support is gratefully acknowledged from NSF (grant CDA-9503968), Olle and Edla Ericsson's Foundation, the Swedish Institute, and the Swedish Research Council for Engineering Sciences (grant TFR-282-95-847).

References

1. Y. S. Abu-Mustafa. Hints. *Neural Computation*, 7:639–671, 1995.
2. C. M. Bishop. Curvature-driven smoothing: A learning algorithm for feedforward networks. *IEEE Transactions on Neural Networks*, 4(5):882–884, 1993.
3. M. C. Brace, J. Schmidt, and M. Hadlin. Comparison of the forecast accuracy of neural networks with other established techniques. In *Proceedings of the First International Form on Application of Neural Networks to Power System, Seattle WA.*, pages 31–35, 1991.
4. W. L. Buntine and A. S. Weigend. Bayesian back-propagation. *Complex Systems*, 5:603–643, 1991.
5. P. Cheeseman. On Bayesian model selection. In *The Mathematics of Generalization - The Proceedings of the SFI/CNLS Workshop on Formal Approaches to Supervised Learning*, pages 315–330, Reading, MA, 1995. Addison-Wesley.

6. G. Cybenko. Approximation by superpositions of a sigmoidal function. *Mathematics of Control, Signals and Systems*, 2:304–314, 1989.
7. R. Engle, F. Clive, W. J. Granger, R. Ramanathan, F. Vahid, and M. Werner. Construction of the puget sound forecasting model. EPRI Project # RP2919, Quantitative Economics Research Institute, San Diego, CA., 1991.
8. S. Geman, E. Bienenstock, and R. Doursat. Neural networks and the bias/variance dilemma. *Neural Computation*, 4(1):1–58, 1992.
9. F. Girosi, M. Jones, and T. Poggio. Regularization theory and neural networks architectures. *Neural Computation*, 7:219–269, 1995.
10. L. K. Hansen, C. E. Rasmussen, C. Svarer, and J. Larsen. Adaptive regularization. In J. Vlontzos, J.-N. Hwang, and E. Wilson, editors, *Proceedings of the IEEE Workshop on Neural Networks for Signal Processing IV*, pages 78–87, Piscataway, NJ, 1994. IEEE Press.
11. A. E. Hoerl and R. W. Kennard. Ridge regression: Biased estimation of nonorthogonal problems. *Technometrics*, 12:55–67, Feb. 1970.
12. M. Ishikawa. A structural learning algorithm with forgetting of link weights. Technical Report TR-90-7, Electrotechnical Laboratory, Information Science Division, 1-1-4 Umezono, Tsukuba, Ibaraki 305, Japan, 1990.
13. M. G. Kendall and A. Stuart. *The Advanced Theory of Statistics*. Hafner Publishing Co., New York, third edition, 1972.
14. J. E. Moody and T. S. Rögnvaldsson. Smoothing regularizers for projective basis function networks. In *Advances in Neural Information Processing Systems 9*, Cambridge, MA, 1997. MIT Press.
15. S. Nowlan and G. Hinton. Simplifying neural networks by soft weight-sharing. *Neural Computation*, 4:473–493, 1992.
16. M. P. Perrone and L. C. Cooper. When networks disagree: Ensemble methods for hybrid neural networks. In *Artificial Neural Networks for Speech and Vision*, pages 126–142, London, 1993. Chapman & Hall.
17. D. Plaut, S. Nowlan, and G. Hinton. Experiments on learning by backpropagation. Technical Report CMU-CS-86-126, Carnegie Mellon University, Pittsburg, PA, 1986.
18. M. Riedmiller and H. Braun. A direct adaptive method for faster backpropagation learning: The RPROP algorithm. In H. Ruspini, editor, *Proc. of the IEEE Intl. Conference on Neural Networks*, pages 586–591, San Fransisco, California, 1993.
19. B. D. Ripley. *Pattern Recognition and Neural Networks*. Cambridge University Press, Cambridge, 1996.
20. J. Sjöberg and L. Ljung. Overtraining, regularization, and searching for minimum with application to neural nets. *Int. J. Control*, 62(6):1391–1407, 1995.
21. A. N. Tikhonov and V. Y. Arsenin. *Solutions of Ill-Posed problems*. V. H. Winston & Sons, Washington D.C., 1977.
22. H. Tong. *Non-linear Time Series: A Dynamical System Approach*. Clarendon Press, Oxford, 1990.
23. J. Utans and J. E. Moody. Selecting neural network architectures via the prediction risk: Application to corporate bond rating prediction. In *Proceedings of the First International Conference on Artificial Intelligence Applications on Wall Street*. IEEE Computer Society Press, Los Alamitos, CA, 1991.
24. G. Wahba, C. Gu, Y. Wang, and R. Chappell. Soft classification, a.k.a. risk estimation, via penalized log likelihood and smoothing spline analysis of variance. In *The Mathematics of Generalization - The Proceedings of the SFI/CNLS Workshop on Formal Approaches to Supervised Learning*, pages 331–359, Reading, MA, 1995. Addison-Wesley.

25. G. Wahba and S. Wold. A completely automatic french curve. *Communications in Statistical Theory & Methods*, 4:1–17, 1975.
26. A. Weigend, D. Rumelhart, and B. Hubermann. Back-propagation, weight-elimination and time series prediction. In T. Sejnowski, G. Hinton, and D. Touretzky, editors, *Proc. of the Connectionist Models Summer School*, San Mateo, California, 1990. Morgan Kaufmann Publishers.
27. P. M. Williams. Bayesian regularization and pruning using a Laplace prior. *Neural Computation*, 7:117–143, 1995.

4
Controlling the Hyperparameter Search in MacKay's Bayesian Neural Network Framework

Tony Plate

School of Mathematical and Computing Sciences,
Victoria University, Wellington, New Zealand
tap@mcs.vuw.ac.nz   http://www.mcs.vuw.ac.nz/~tap/

Abstract. In order to achieve good generalization with neural networks overfitting must be controlled. Weight penalty factors are one common method of providing this control. However, using weight penalties creates the additional search problem of finding the optimal penalty factors. MacKay [5] proposed an approximate Bayesian framework for training neural networks, in which penalty factors are treated as hyperparameters and found in an iterative search. However, for classification networks trained with cross-entropy error, this search is slow and unstable, and it is not obvious how to improve it. This paper describes and compares several strategies for controlling this search. Some of these strategies greatly improve the speed and stability of the search. Test runs on a range of tasks are described.

4.1 Introduction

Neural networks can provide useful flexible statistical models for non-linear regression and classification. However, as with all such models, the flexibility must be controlled to avoid overfitting. One way of doing this in neural networks is to use weight penalty factors (regularization parameters). This creates the problem of finding the values of the penalty factors which will maximize performance on new data. As various researchers have pointed out, including MacKay [5], Neal [10] and Bishop [1], it is generally advantageous to use more than one penalty factor, in order to differentially penalize weights between different layers of the network. However, doing this makes it computationally infeasible to choose optimal penalty factors by k-fold cross validation.

MacKay [5] describes a Bayesian framework for training neural networks and choosing optimal penalty factors (which are hyperparameters in his framework). In this framework, we choose point estimates of hyperparameters to maximize the "evidence" of the network. Parameters (i.e., weights) can be assigned into different groups, and each controlled by a separate hyperparameter. This allows weights between different layers to be penalized differently. MacKay [6, 8] and Neal [10] have shown that it also provides a way of implementing "Automatic Relevance Detection" (ARD), in which connections emerging from different units in the input layer are assigned to different regularization groups. The idea is that hyperparameters controlling weights for irrelevant inputs should become large, driving those weights to zero, while hyperparameters for relevant inputs stabilize at small to moderate values. This can help generalization by causing the network to ignore irrelevant inputs and also makes it possible to see at a glance which inputs are important.

In this framework the search for an optimal network has two levels. The inner level is a standard search for weights which minimize error on the training data, with fixed hyperparameters. The outer level is a search for hyperparameters which maximize the evidence. For the Bayesian theory to apply, the inner level search should be allowed to converge to a local minima at each step of the outer level search. However, this can be expensive and slow. Problems with speed and stability of the search seem especially severe with classification networks trained with cross-entropy error.

This paper describes experiments with different control strategies for updating the hyperparameters in the outer level search. These experiments show that the simple "let it run to convergence and then update" strategy often does not work well, and that other strategies can generally work better. In previous work, the current author successfully employed one of these strategies in an application of neural networks to epidemiological data analysis [11]. The experiments reported here confirm the necessity for update strategies and also demonstrate that although the strategy used in this previous work is reasonably effective in some situations, there are simpler and better strategies which work in a wider range of situations. These experiments also furnish data on the relationship between the evidence and the generalization error. This data confirms theoretical expectations about when the evidence should and should not be a good indication of generalization error.

In the second section of this chapter, the update formulas for hyperparameters are given. Network propagation and weight update formulas are not given, as they are well known and available elsewhere, e.g., in Bishop [1]. Different control strategies for the outer level hyperparameter search are described in the third section. In the fourth section, the simulation experiments are described, and the results are reported in the fifth section. The experimental relationships between evidence and generalization error are reported in the sixth section.

4.2 Hyperparameter updates

The update formulas for the hyperparameters (weight penalty factors) in the outer level search are quite simple. Before describing them we need some terminology. For derivations and background theory see Bishop [1], MacKay [5, 7], or Thodberg [13].

- n is the total number of weight in the network.
- w_i the value of the ith weight.
- K is the number of hyperparameters.
- $\mathcal{I}_c$ is the set of indices of the weights in the cth hyperparameter group.

4. Hyperparameter Search in MacKay's Bayesian Framework

- α_c is the value of the hyperparameter controlling the cth hyperparameter group; it specifies the prior distribution on the weights in that group. $\alpha_{[i]}$ denotes the value of the hyperparameter controlling the group to which weight i belongs.
- n_c is the number of weights in the cth hyperparameter group.
- C is the weight cost (penalty term) for the network: $C = \frac{1}{2} \sum_{i=1}^{n} \alpha_{[i]} w_i^2$.
- m is the total number of training examples.
- y^j and t^j are the network outputs and target values, respectively, for the jth training example.
- E is the error term of the network. For the classification networks described here, the modified cross-entropy (Bishop [1], p.232) is used:

$$E = -\sum_{j=1}^{m} \{t^j \log \frac{y^j}{t^j} + (1 - t^j) \log \frac{1 - y^j}{1 - t^j}\}.$$

Note that all graphs and tables of test set performance use the "deviance", which is twice the error.

- $\mathbf{H}$ is the Hessian of the network (the second partial derivatives of the sum of the error and weight cost). h_{ij} denotes the ijth element of this matrix, and h_{ij}^{-1} denotes the ijth element of $\mathbf{H}^{-1}$:

$$h_{ij} = \frac{\partial^2 (E + C)}{\partial w_i \partial w_j}.$$

$\mathbf{H}^E$ is the matrix of second partial derivatives of just the error, and $\mathbf{H}^C$ is the matrix of second partial derivatives of just the weight cost.
- $\mathrm{Tr}(\mathbf{H}^{-1})$ is the trace of the inverse of $\mathbf{H}$: $\mathrm{Tr}(\mathbf{H}^{-1}) = \sum_{i=1}^{n} h_{ii}^{-1}$.
- $\mathrm{Tr}_c(\mathbf{H}^{-1})$ is the trace of the inverse Hessian for just those elements of the cth regularization group: $\mathrm{Tr}_c(\mathbf{H}^{-1}) = \sum_{i \in \mathcal{I}_c} h_{ii}^{-1}$.
- γ_c is a derived parameter which can be seen as an estimate of the number of well-determined parameters in the cth regularization group, i.e., the number of parameters determined by the data rather than by the prior.

The overall training procedure is shown in Figure 4.1.

---

set the α_c to initial values
set w_i to initial random values
repeat
    repeat
        make an optimization step for weights to minimize $E + C$
    until finished weight optimization
    re-estimate the α_c
until finished max number of passes through training data

Fig. 4.1. The training procedure

The updates for the hyperparameters α_c depend on the estimate γ_c (the number of well-determined parameters in group c) which is calculated as follows (Eqn 27 in [7]; derivable from Eqn 10.140 in [1]):

$$\gamma_c = n_c - \alpha_c \mathrm{Tr}_c(\mathbf{H}^{-1}). \tag{4.1}$$

If a Gaussian distribution is a reasonable approximation to the posterior weight distribution, γ_c should be between 0 and n_c. Furthermore, we expect each parameter in group c to contribute between 0 and 1 to γ_c. Hence, we expect h_{ii}^{-1} to always be in the range $[0, 1/\alpha_{[i]}]$.

The updates for the α_c is as follows (Eqn 22 in [7]; Eqn 10.74 in [1]):

$$\alpha'_c = \frac{\gamma_c}{\sum_{i \in \mathcal{I}_c} w_i^2} \tag{4.2}$$

MacKay [7] remarks that this formula can be seen as matching the prior to the data: $1/\alpha_c$ is an estimate of the variance for the weights in group c, taking into account the effective number of well determined parameters (effective degrees of freedom) in that group.

4.2.1 Difficulties with using the update formulas

The difficulties with using these update formulas arise when the assumption that the error plus cost surface is a quadratic bowl is false. This assumption can fail in two ways: the error plus cost surface may not be quadratic, or it may not be a bowl (i.e., the Hessian is not positive definite). In either of these situations, it is possible for γ_c to be out of the range $[0, n_c]$. To illustrate, consider a single diagonal element of the Hessian in the situation where off-diagonal elements are zero:

$$\mathbf{H} = \begin{bmatrix} \ddots & & 0 \\ & h_{ii} & \\ 0 & & \ddots \end{bmatrix} = \begin{bmatrix} \ddots & & 0 \\ & h_{ii}^E + \alpha_{[i]} & \\ 0 & & \ddots \end{bmatrix}$$

Since the off-diagonal elements are zero, the inverse Hessian is simple to write down:

$$\mathbf{H}^{-1} = \begin{bmatrix} \ddots & & 0 \\ & \frac{1}{h_{ii}^E + \alpha_{[i]}} & \\ 0 & & \ddots \end{bmatrix}$$

Suppose the i parameter is in the cth regularization group, by itself. Then the number of well-determined parameters in this group is given by:

$$\gamma_{[i]} = 1 - \alpha_{[i]} h_{ii}^{-1} = 1 - \frac{\alpha_{[i]}}{h_{ii}^E + \alpha_{[i]}} = \frac{h_{ii}^E}{h_{ii}^E + \alpha_{[i]}} \tag{4.3}$$

If h_{ii}^E is positive, $\gamma_{[i]}$ will be between 0 and 1. $\gamma_{[i]}$ will be large if h_{ii}^E is large relative to $\alpha_{[i]}$, which means that w_i is well determined by the data, i.e., small moves of w_i will make a large difference to E. $\gamma_{[i]}$ will be small if h_{ii}^E is small relative to $\alpha_{[i]}$, which means that w_i is poorly determined by the data.

The expectation that h_{ii}^{-1} is in the range $[0, 1/\alpha_c]$ (and hence contributes between 0 and 1 well determined parameter to $\gamma_{[i]}$) can fail even if the model is at a local minima of $E + C$. Being at a local minimum of $E + C$ does not guarantee that h_{ii}^E will be positive: it is possible for the hyperparameter to "pin" the weight value to a convex portion of a non-quadratic E surface. Consider the case where the Hessian is diagonal and positive definite, but h_{ii}^E is negative. From Eqn 4.3, we can see that h_{ii}^{-1} can make a negative contribution[1] to γ_c, which makes little sense in terms of "numbers of well-determined parameters". This situation[2] is illustrated in Figure 4.2: at the minimum of the $E + C$ function the E function is convex ($\frac{d^2 E}{dw^2}$ is negative). Here, negative degrees of freedom would be calculated under the (incorrect) assumption that error plus cost is quadratic. This is important for neural networks, because even if sum-squared error is used, non-linearities in the sigmoids can cause the the error plus cost function to be not a quadratic function of weights.

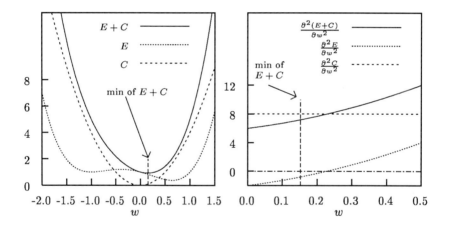

Fig. 4.2. In minimizing $E + C$, a weight cost function C can pin a weight value to a convex portion of the error surface E. The plot on the left shows the surfaces, the plot on the right shows the derivatives in the region of the minimum.

If the model is not at a local minimum of $E + C$ all bets are off. **H** may not even be positive definite (i.e., the Hessian of a quadratic bowl), and if this is the case it is almost certain that some h_{ii}^{-1} will be out of the range $[0, 1/\alpha_{[i]}]$. Even

---

[1] With general matrices it is possible that $h_{ii}^{-1} < -\alpha_{[i]}$, in which case the contribution will be an unbounded positive number.
[2] In Figure 4.2, $E = w(w-1)(w+1)^2 + 1$, $C = 4w^2$, $\frac{d(E+C)}{dw}\big|_{0.152645} \approx 0$, and $\frac{d^2 E}{dw^2}\big|_{0.152645} \approx -0.8036$.

if **H** is positive definite, and $\mathbf{H}^E$ is also positive definite, it can still be the case that some h_{ii}^{-1} are out-of-bounds, and thus make contributions of less than zero or more than one "well-determined parameter" each.

These difficulties leave us with two problems:

1. When should the hyperparameters be updated?
2. What should be done when h_{ii}^{-1} is not in $[0, 1/\alpha_{[i]}]$ (i.e., is "out-of-bounds").

Bishop [1] suggests updating hyperparameters after every few weight updates. Thodberg [13] suggests updating the α_c after every weight update, but only recalculating the γ_c occasionally (at five evenly-spaced intervals throughout the whole process). While updating hyperparameters after every weight update is not feasible when using conjugate gradient or other second-order training methods, common practice seems to be more or less in line with Thodberg's recommendations: train the network more or less to convergence before each hyperparameter update. However, this strategy can result in an extremely slow overall search. Furthermore, training to convergence does not eliminate the problem of out-of-bounds h_{ii}^{-1} values. In the remainder of this chapter, various strategies for choosing when to update hyperparameters, and for dealing with out-of-bounds h_{ii}^{-1} values are described and compared in experiments.

4.3 Control strategies

The strategies tested here fall into three groups: strategies for when to update hyperparameters, strategies for dealing with out-of-bounds h_{ii}^{-1} values, and special strategies for dealing with exceptional cases. For each task, searches were run a fixed number of minimization steps, with different combinations of control strategies. Each step of the inner minimization loop was a step of a conjugate gradient method, and could involve a number of passes through the training data, though the average was just over two.

4.3.1 Choosing when to update hyperparameters

Four different strategies for choosing when to break out of the inner loop and update hyperparameters were employed:

rare: Update at 10 evenly spaced intervals.
medium: Update at 30 evenly spaced intervals.
often: Update at 100 evenly spaced intervals.
patience: Update when the improvement in the last n steps was less than 1% of the improvement since the start of the inner loop, or when the improvement in the last n steps was less than 0.01% of the null error (the minimum error that can be achieved with a constant output value). At least n steps of the inner loop are taken.

4. Hyperparameter Search in MacKay's Bayesian Framework

Convergence was difficult to test when using a number of different datasets and networks. One standard way of detecting convergence is to test whether the ratio $|\bar{g}|/|\bar{w}|$ is less than some threshold ($|\bar{g}|$ is the Euclidean length of the vector of weight derivatives, and $|\bar{w}|$ is the Euclidean length of the vector of weights.) However, the appropriate threshold varied greatly among different tasks. In any case, with all strategies, if a convergence test was met ($|\bar{g}|/|\bar{w}| < 10^{-6}$), the inner loop terminated, and the hyperparameters were updated. This did not occur often in the experiments described here.

The "patience" strategy is intended to be a surrogate for convergence: the inner loop runs out of patience when the improvement achieved in the last n steps is minuscule. With "patience", the inner loop is guaranteed to terminate if the error is bounded below. In practice, the inner loop runs out of patience reasonable quickly: the update rate is somewhere between "rare" and "medium" depending on the difficulty of the optimization problem.

4.3.2 Dealing with out-of-bounds estimates of numbers of well-determined parameters

In each of the experiments, one of the following strategies was used to deal with out-of-bounds h_{ii}^{-1} values. In describing these strategies, $h_{ii}^{-1'}$ or γ_c' are used to denote values which are used instead of the originally calculated ones.

none: This strategy allows γ_c to take on unreasonably large values, but not negative ones (h_{ii}^{-1} values are not checked):

$$\gamma_c' = \begin{cases} 0 & \text{if } \gamma_c < 0 \\ \gamma_c & \text{otherwise} \end{cases}$$

group: This strategy forces the total number of well-determined parameters in a regularization group to be reasonable:

$$\gamma_c' = \begin{cases} 0 & \text{if } \gamma_c < 0 \\ n_c & \text{if } \gamma_c > n_c \\ \gamma_c & \text{otherwise} \end{cases}$$

trim: This strategy forces the contribution of each h_{ii}^{-1} to be reasonable. If h_{ii}^{-1} is out-of-bounds, it is assumed to represent zero well-determined parameters:

$$h_{ii}^{-1'} = \begin{cases} 1/\alpha_{[i]} & \text{if } h_{ii}^{-1} \text{ is not in } [0, 1/\alpha_{[i]}] \\ h_{ii}^{-1} & \text{otherwise} \end{cases}$$

snip: This strategy forces the contribution of each h_{ii}^{-1} to be reasonable. If h_{ii}^{-1} is out-of-bounds, it is assumed to represent one well-determined parameter:

$$h_{ii}^{-1'} = \begin{cases} 0 & \text{if } h_{ii}^{-1} \text{ is not in } [0, 1/\alpha_{[i]}] \\ h_{ii}^{-1} & \text{otherwise} \end{cases}$$

useold: This strategy forces the contribution of each h_{ii}^{-1} to be reasonable. If h_{ii}^{-1} is out-of-bounds the last good estimate of the well-determinedness of parameter i is used:

$$\gamma'_c = \sum_{i=1}^{n_c} \begin{cases} 1 - \alpha_c^* h_{ii}^{-1*} & \text{if } h_{ii}^{-1} \text{ is not in } [0, 1/\alpha_{[i]}] \\ 1 - \alpha_c h_{ii}^{-1} & \text{otherwise} \end{cases},$$

where α_c^* and h_{ii}^{-1*} are most recent values such that h_{ii}^{-1} is in $[0, 1/\alpha_{[i]}]$, or if there are no such values, $h_{ii}^{-1*} = 0$.

cond: This strategy only updates α_c, using γ_c or a snipped version of γ_c, under the following conditions:
  (a) all eigenvalues of **H** are positive, **and**
  (b) for all $i \in \mathcal{I}_c$, h_{ii}^{-1} is in $[0, 1/\alpha_{[i]}]$, **or** a snipped version of γ_c will result in a change in α_c in the same direction as the last change when all the eigenvalues of **H** were positive and all the h_{ii}^{-1} in group c were in bounds, and there have not been more than five such changes since all of the h_{ii}^{-1} for group c have been in range.

cheap: This is Mackay's [7] "cheap and cheerful" method, in which all parameters are assumed to be well determined, i.e.,

$$\gamma'_c = n_c.$$

The advantage of this method is that Hessian need not be calculated. Mackay remarks that this method can be expected to perform poorly when there are a large number of poorly determined parameters.

4.3.3 Further generally applicable strategies

Several further strategies which could be combined with any of the ones already mentioned were also employed:

nza: (no zero alphas) Do not accept an updated alpha value of zero (retain the old value).

limit: Limit the the change in an alpha value to have a magnitude of no more than 10, i.e., round α'_c to be in the interval $[0.1\alpha_c, 10\alpha_c]$.

omit: If there are any h_{ii}^{-1} not in $[0, 1/\alpha_{[i]}]$, omit the corresponding rows and columns from **H** to give the smaller matrix **H'**, and use the diagonal elements of $\mathbf{H'}^{-1}$ for h_{ii}^{-1}. The idea is to omit troublesome components of the model so that they do not interfere with estimates of well-determinedness for well-behaved parameters. Strategies from the previous section are used to assign well-determinedness values for parameters with out-of-bound h_{ii}^{-1} values in $\mathbf{H}^{-1}$ or $\mathbf{H'}^{-1}$.

4.4 Experimental setup

The experiments reported here used seven different test functions, based on the five 2-dimensional functions used by Hwang et al. [4] and Roosen and Hastie [12]:

linear function:
$$f_0(x_1, x_2) = (2x_1 + x_2)/0.6585$$
simple interaction:
$$f_1(x_1, x_2) = 10.391((x_1 - 0.4)(x_2 - 0.6) + 0.36)$$
radial function:
$$f_2(x_1, x_2) = 24.234((x_1 - 0.5)^2 + (x_2 - 0.5)^2)(0.75 - ((x_1 - 0.5)^2 \\ + (x_2 - 0.5)^2))$$
harmonic function:
$$f_3(x_1, x_2) = 42.659(0.1 + (x_1 - 0.5)(0.05 - 10(x_1 - 0.5)^2(x_2 - 0.5)^2 \\ + (x_1 - 0.5)^4 + 5(x_2 - 0.5)^4))$$
additive function:
$$f_4(x_1, x_2) = 1.3356(1.5(1 - x_1) + e^{(2x_1 - 1)}\sin(3\pi(x_1 - 0.6)^2) \\ + e^{3(x_2 - 0.5)}\sin(4\pi(x_2 - 0.9)^2))$$
complicated interaction:
$$f_5(x_1, x_2) = 1.9(1.35 + e^{x_1}\sin(13(x_1 - 0.6)^2)e^{-x_2}\sin(7x_2))$$
interaction plus linear:
$$f_6(x_1, x_2) = 0.83045(f_0(x_1, x_2) + f_1(x_1, x_2))$$

Training data for the binary classification tasks was generated by choosing 250 x_1^j and x_2^j points from a uniform distribution over $[0, 1]$. Another 250 s^j points were chosen from the uniform $[0, 1]$ distribution to determine whether the target should be 0 or 1. The $\{0, 1\}$ target t_i^j for case j for function i depended on the probability p_i^j of a 1 calculated as the sigmoid of the function value (with the mean of the function subtracted):

$$p_i^j = \frac{1}{1 + e^{-(f_i(x_1^j, x_2^j) - \mu_i)}}$$

$$t_i^j = \begin{cases} 0 & \text{if } p_i^j < s^j \\ 1 & \text{otherwise} \end{cases}$$

where μ_i is the mean of each function over the unit square ($\mu_0 = 2.28, \mu_1 = 3.6, \mu_2 = 2.3, \mu_3 = 4.4, \mu_4 = 2.15, \mu_5 = 2.7, \mu_6 = 4.92$).

A further 5 random distractor points chosen from a uniform $[0, 1]$ distribution were concatenated with each (x_1^j, x_2^j) pair, to give a 7 dimensional function-approximation task. These extra points were added so that in order for the neural network to generalize well it would be necessary that the automatic relevance determination set the weights from irrelevant inputs to zero (by driving the α's for those weights to high values.)

For testing, the probabilities were used as targets, rather than stochastic binary values. This was done to reduce the noise in measuring the test error. The test set inputs consisted of 400 (x_1, x_2) points from the uniform grid over $[0, 1]$, and 400 vectors of distractors chosen from a uniform distribution over $[0, 1]$. In order that test and training errors be comparable the test error was calculated as the expected error over the test points, assuming actual targets had been chosen randomly with the given target probabilities. The expected error for one test case was calculated as follows:

$$E_i^j = - \left(p_i^j \log y_i^j + (1 - p_i^j) \log(1 - y_i^j) \right)$$

where y_i^j was the prediction of the network for test case j.

The seven learning tasks are called I0 through I6 (the "I" is for "impure" in Roosen and Hastie's terminology [12].) It should be noted that these learning tasks are much harder than the tasks used by Hwang et al. [4], as the binomial outputs make the training data much noisier, and irrelevant inputs are present.

4.4.1 Targets for "good" performance

Various simple modeling strategies were applied to the data to give an indication of what level of test error performance was achievable, and to demonstrate how important it was to ignore the distractor inputs. Three different models were tried: linear and quadratic logistic models, and generalized additive models (GAMs) [3], each with and without the distractor inputs. No attempt was made to prevent overfitting, as the intention of these models was to show how much overfitting can occur if distractor are not ignored.

null: No inputs are used. The predicted output is the average of the targets (1 parameter).
true: The actual function value. This is included to indicate the level of noise in the data.
lin: A logistic (linear) model fit using only x_1 and x_2 as inputs (3 parameters).
lin.D: A logistic (linear) model fit using all inputs, including the distractors (8 parameters).
quad: A logistic model (with quadratic terms) fit using only x_1 and x_2 as inputs (6 parameters).
quad.D: A logistic model (with quadratic terms) fit using all inputs, including the distractors (36 parameters).
gam: A generalized additive model fit using only x_1 and x_2 as inputs, with three degrees of freedom for each dimension (approx 7 parameters).
gam.D: A generalized additive model fit using all inputs, with three degrees of freedom for each dimension (approx 22 parameters).
T: The target for "good" network performance.

4. Hyperparameter Search in MacKay's Bayesian Framework   103

The test set deviances for the various models and tasks are shown in Table 4.1, ordered by error for each task. Some tasks are easy, while others are very difficult (finding good solutions for task I3 appears to extremely difficult, as Roosen and Hastie [12] also discovered.) Targets for "good" neural network performance were derived from the errors achieved by any model. The targets were chosen to be achievable by neural network models and yet lower than the test set deviance achieved by any of the above simple models using all the inputs (except for I3, which neural networks had great difficulty with.)

I0	I1	I2	I3	I4	I5	I6
true 478	true 488	true 475	true 495	true 480	true 491	true 478
lin 487	quad 495	quad 494	gam 546	gam 498	gam 539	quad 485
T (165) 493	T (258) 520	gam 507	T (7) 555	T (114) 530	T (58) 545	T (233) 500
lin.D 498	lin 545	T (44) 530	lin 556	quad 533	quad 547	lin 517
quad 501	gam 548	gam.D 551	lin.D 557	gam.D 537	lin 547	lin.D 523
gam 502	null 556	null 555	null 558	lin 554	lin.D 555	gam 529
gam.D 519	lin.D 558	lin 561	quad 566	null 556	null 555	quad.D 540
null 555	quad.D 595	quad.D 575	gam.D 574	lin.D 558	gam.D 572	gam.D 550
quad.D 556	gam.D 608	lin.D 577	quad.D 669	quad.D 597	quad.D 620	null 557

Table 4.1. Test set deviances (twice the error) for various models. Names ending in ".D" are those of models which used both the distractor and relevant inputs. The number in parentheses beside the target for good network performance (T) is the number of networks which achieved this target at the end of training (out of 540.)

4.4.2 Network architecture and training

Standard feed-forward networks were used, with details as follows. All networks had 3 to 15 hidden units, which computed the tanh function (a symmetric sigmoid). Inputs were in the range 0 to 1. The output unit computed the logistic function of its total input. Weights were initialized to random values drawn from a Gaussian distribution with variance 0.5.

Networks had nine hyperparameters (weight penalties): one for the weights for each input, one for hidden unit biases, and one for hidden to output weights. There was no penalty on the output bias. All hyperparameters were initialized to 0.5. Hyperparameters were allowed to increase to maximum value of 10,000.

Networks were trained using a conjugate algorithm for the number of steps specified in Table 4.2. These numbers of steps were chosen give ample time for reasonably good methods to converge on some solution (the harder problems required more steps). Each step of the conjugate gradient algorithm involved one or more passes through the training set (the average was just over two). Training was terminated if the total number of passes through the training set exceeded 2.3 times the maximum allowed number of conjugate gradient steps.

The Hessian was calculated using the exact analytical method described in Buntine and Weigend [2]. This requires $h + 1$ passes through the training data, where h is the number of hidden units. This is usually far faster than a finite-differences method, which requires $n+1$ for the forward differences method and $2n+1$ passes for the more accurate central differences method. The total number of floating point operations involved in the exact calculation of the Hessian is dominated by the update of the Hessian matrix (Eqn 15c in [2]). For a network with one output unit it is approximately $3.5(h+1)Nn^2$ (there are 7 operations in each Hessian element update, but only half the elements need be computed as the matrix is symmetric). Eigendecomposition and inversion of the Hessian takes approximately $4/3n^3$ operations, but this is usually small compared to the calculation of the Hessian matrix. As long as hyperparameters are not updated too frequently, the time taken by Hessian evaluation and inversion is generally not an excessive amount on top of the time taken by the standard weight-training part of the procedure. For example, in the easier tasks (I0, I1, and I6) with the most frequent Hessian calculations (the "often" updates: 100 during), approximately one-third of the computation time was spent in Hessian calculations. In the more difficult tasks, relatively less time was spent in Hessian calculations because the training times were longer.

Thirty six different combinations of the hyperparameter update strategies discussed in Section 4.3 were tested. Training on each problem was repeated five times with different initial weights. The same five sets of random initial weights were used for each strategy. This means that there were a total of 160 attempts to train each sized network for each problem.

Task	Number of hidden units			
	3	5	10	15
I0, I1, I6	1800	3000	6000	–
I2, I3, I4	3000	6000	12000	–
I5	–	6000	12000	30000

Table 4.2. Number of conjugate gradient steps allowed for different sized networks on the different tasks.

4.5 Effectiveness of control strategies

Effective of various combinations of control strategies was judged by whether or not the test error at the end of training was acceptable (using the deviance targets in Table 4.1). Good performance on any of the tasks was not possible without setting the hyperparameters in the appropriate ranges: low for the weights coming from the x_1 and x_2 inputs, and high for the weights coming from the distractor inputs. Figure 4.3 shows example plots of hyperparameter values versus test deviance for networks trained on task I6 (with jitter added to make dense clouds of points visible). Task I6 was a reasonably easy task: nearly all networks ended with appropriate low values for the relevant-input hyperparameters. Finding appropriate high values for distractor hyperparameters was more difficult, and those networks which did not did not perform well. All such plots had the same tendencies as those in Figure 4.3: low test set deviance was achieved only be networks with low values for relevant-input hyperparameters and high values for distractor hyperparameters. Some of the other tasks were more difficult, e.g., I5, and poor search strategies would set relevant-input hyperparameters to high values, resulting in poor performance from ignoring the relevant inputs.

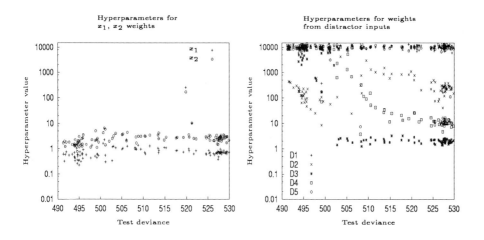

Fig. 4.3. Final hyperparameter values versus test set deviances for networks with 5 hidden units trained on task I6.

The number of successes for each of the 36 (combinations of) strategies on each task is shown in Table 4.3. Each asterisk represents one success, i.e., a network which ended up with test set performance lower than the target for good performance (Table 4.1.) The maximum number of successes for any cell is 15, as 5 random starts were used for each of three different sized networks. The row totals are out of 105, and the column totals are out of 540. The grand total (987 successes) is out of 3780.

The clear overall best strategy is "snip+often". The special strategies "nza" and "limit" seem to help a little. These conclusions are strengthened by examining plots of the evolution of test deviance during training. Figures 4.4 and 4.5 show plots of test deviance during training for networks with 5 hidden units on task I6, and for networks with 10 hidden units on task I4. There are five lines in each plot because five random starts were used for each network. The ideal network has a test deviance which descends rapidly and then stays below the target performance (the dotted line) – this shows that the strategy quickly found good values for hyperparameters (the tasks were set up so that it was not

	total	I0	I1	I2	I3	I4	I5	I6
cheap+often	4				****			
group+medium	10	***	***			*		***
group+patience	10	*	***		*			*****
none+patience	10		****			*		*****
none+rare	11		***			**		******
group+rare	12		***			***		******
snip+rare	14	**	**			****		******
trim+often	16	*****	****	****				***
trim+patience	17	***	******	*		*	*	*****
none+medium	18	****	******	*	*	**		****
omit+snip+patience	18		*******			******		*****
omit+snip+patience+limit	18		*******			******		*****
trim+medium	20	*****	******	***		***	***	***
trim+rare	20	**	******			***	****	*****
useold+patience	20	****	******	*	*	**	*	*****
omit+useold+patience	21	****	*******			***		*****
omit+useold+patience+limit	22	****	******			****	***	*****
cheap+medium	22	*****	*****	***			*	*********
cheap+patience	23	*****	*****	*		**	**	**********
none+often	26	********	**********	*		***	**	***
snip+medium	26	******	**********			*****	**	*****
snip+patience	26	******	*****	*		*****	****	*****
omit+trim+patience	27	***********	******	**		**	**	*****
snip+patience+nza+limit	27	*****	*******	*		*****	****	*****
omit+useold+often	28	********	********	***		**		******
omit+useold+often+limit	29	*******	*******	**		***		*******
snip+patience+nza	29	*******	*******	*		*****	****	*****
group+often	31	********	*******			*	**	*********
cond+often+limit+nza	32		*************			**********		**********
omit+snip+often+limit	32		*************			**********		************
useold+often	32	************	********	****		**		*******
cheap+rare	34	*********	*********	*****		********	**	*********
omit+snip+often	36	**	****************			********	*****	****************
snip+often	50	*********	****************	**	**	****	*****	****************
snip+often+nza	54	**********	***************	**	**	*****	******	***************
snip+often+nza+limit	54	**********	****************	*		*******	********	***************
	987	165	258	44	7	114	58	233

Table 4.3. Number of successes for each strategy on each task.

possible to achieve low test deviance without having appropriate values for the hyperparameters). Lines which flatten out at around 557 are for networks whose hyperparameters have all been set to high values, so that the network ignores all inputs and thus has the same performance as the null model (see Table 4.1). The search was terminated early for many of these networks, because no hyperparameters or weights were changing. Note that few of the networks with 10 hidden units reached the target for good performance on task I4 though networks with 5 hidden units did much better. This indicates that the redundancy in these networks could not be effectively controlled.

Update strategies other than "snip" tend to be very unstable. Frequent updates seem essential for achieving good final performance within a reasonable amount of time. The more complex strategies for getting or retaining good estimates of γ's, i.e., omit and useold, seemed to be of little benefit.

Finding good hyperparameter values is difficult. If updates are too frequent, α's can rise uncontrollably or become unstable. If updates are too infrequent, the search is too slow. Uncontrollable rises in α's can occur when the network has not started using the weights it needs, and those weights are small. In these circumstances α is overestimated, which forces weights to become smaller in a runaway-feedback process. Instability results because of the feedback between γ and α: a change in γ causes a same direction change in α, and a change in *alpha* causes an opposite direction change in γ. Thus, if γ is overestimated, this leads to an overestimation in α, which lead to a lower reestimate of γ, which in turn can lead to a lower reestimate of α. While this process sometimes results in stable self-correcting behavior, other times it results in uncontrolled oscillations (which is what is happening with the "none+often" strategy: second row, third column in Figures 4.4 and 4.5). In general, while it is slow to raise an α from a value that is too low, it is more difficult to lower an α from a value which is too high (because the weights are forced to zero immediately). The reasons for the differing behavior of the various strategies appear to be as follows:

none, group: Frequently give out-of-bounds or very poor estimates for γ, and cause instability in the search.
cond: Results in fairly stable behavior, but often does not find the optimal values for all α's because it stops updating them.
cheap: Overestimates γ and kills weights too quickly.
trim: Sometimes underestimates γ because it assumes zero degrees of freedom when the γ estimates from the Hessian are out-of-bounds, and thus reduces α values which should be large, resulting in instability.
snip: Sometimes overestimates γ because it assumes one degree of freedom for a parameter whose γ estimate from the Hessian is out-of-bounds. However, this keeps α values high without raising them too much: overestimation of γ appears to be much less harmful than underestimation.

Update frequency was also important. With good α calculation strategies, updating frequently (i.e., "often") gave fastest convergence to good α values and best overall performance. Waiting for some degree of convergence in the conjugate gradient search before updating α values (i.e., "patience") was of no benefit at all.

4.6 Relationship of test set error to evidence

If we have trained a number of networks, we often want to know which will have the lowest generalization error. The "evidence" value calculated for each network can be used choose networks which will perform well. The evidence is the log likelihood of the data given the values of the hyperparameters, integrated over weight values based on the assumption of a Gaussian distribution for the posterior of the weights. The evidence for a network evaluated with cross-entropy error is as follows (Eqn 10.67 in [1]; Eqn 30 in [7]):

$$\ln p(D|\alpha) = -\frac{1}{2}\sum_{i=1}^{n} \alpha_{[i]} w_i^2 - E - \frac{1}{2}|H| + \sum_c \frac{n_c}{2}\ln \alpha_c - \frac{N}{2}\ln(2\Pi) \quad (4.4)$$

Whether or not the evidence is a good guide to which networks will perform well is questionable, as various assumptions on which evidence calculations are based are often violated in particular networks. The simulations performed offer a good opportunity to examine how accurately high evidence indicates good test set performance.

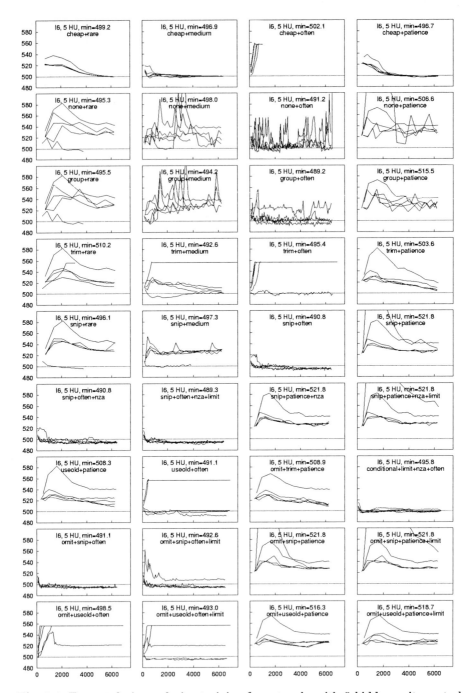

Fig. 4.4. Test set deviance during training for networks with 5 hidden units on task I6. The horizontal axis is the number of passes through the training set. The dotted line is the target for good performance. The minimum value quoted is the minimum deviance at any point in training.

4. Hyperparameter Search in MacKay's Bayesian Framework 109

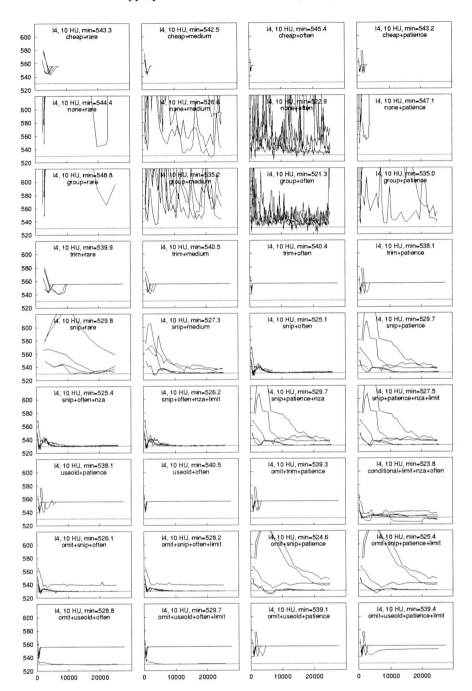

Fig. 4.5. Test set deviance during training for networks with 10 hidden units on task I4. The horizontal axis is the number of passes through the training set. The dotted line is the target for good performance. The minimum value quoted is the minimum deviance at any point in training.

The evidence is particularly sensitive to low eigenvalues in the Hessian of the network. The validity of the evidence value is doubtful in cases where the Hessian has low or negative eigenvalues. Bishop [1] recommends omitting eigenvalues which are lower than some threshold from the calculation of the evidence.

Figure 4.6 shows plots of test deviances that would be achieved by selecting twenty networks based on their evidence values. Two different methods of calculating the evidence were used: (a) ignore negative eigenvalues of the Hessian, and (b) if an eigenvalue is lower than the smallest non-zero α_c, replace it with that α_c ("clipped evidence"). Because the evidence is sensitive to low eigenvalues, four different filters were applied to networks: (a) use all networks; (b) throw out networks with negative eigenvalues; (c) throw out networks with eigenvalues lower than the the smallest α_c; and (d) throw out networks with out-of-bounds h_{ii}^{-1} values.

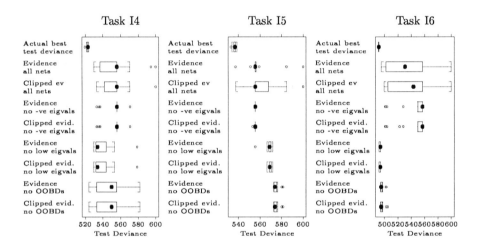

Fig. 4.6. Distribution of test errors for the 20 "best" nets selected according to evidence and network diagnostics.

The first row in each plot of Figure 4.6 shows the distribution of the actual best 20 test deviances (this would normally be impossible to calculate, but we can calculate it here because we have artificial tasks.) If evidence were a perfect predictor of test set performance, the 20 networks with highest evidence would be the same 20 networks, but it is not, as the rest of the figure shows. The remaining eight rows show the distribution of test errors for nets with the highest evidence, with evidence calculated in both of the ways described above. Lower pairs of rows rules more nets out of consideration, based on diagnostics of the Hessian and inverse Hessian.

The two different ways of calculating the evidence did not make any discernible difference. The box-and-whisker plots show the median as a solid circle. The box shows the upper and lower quartiles (i.e., 50% of points are in the box).

The whiskers show the furthest points within 1.5 times the interquartile range (the length of the box) of the quartiles. Points outside the whiskers are plotted individually with hollow circles.

Plots of the evidence versus test deviance show a strong correlation, which seems strongest and most reliable for networks which have no low eigenvalues. However, the plots in Figure 4.6 show that using the evidence to select networks is not always reliable. Networks with high evidence but with poorly conditioned Hessians have a wide range of test error, but some networks with the lowest test error have poorly conditioned Hessians. This means that allowing networks with poorly conditioned Hessians resulted in choosing too many networks with high test error, whereas rejecting networks with poorly conditioned Hessians rejected too many of the networks with low test error. This seemed to be more of a problem for the more difficult tasks. On the easier tasks (e.g., I6), networks with high evidence and no low eigenvalues had very low test deviance.

4.7 Conclusion

For classification networks, the search for optimal hyperparameters can be slow and unstable. However, the search can be improved by using the strategies summarized below. Similar experiments with regression networks (with linear output units and sum-square errors, on tasks with continuous targets and Gaussian noise) revealed that the hyperparameter search in such tasks is generally faster and more stable than in classification tasks. Only when the tasks had very high noise in the training cases (with variance of 4 or greater) did the hyperparameter search become at all difficult. Under these conditions, none of the strategies tried was clearly superior.

One of the main causes of instability in the hyperparameter search is low eigenvalues in the Hessian. In turn, one of the main causes of this is redundancies in the network. It is easily verified that the Hessian for a network with redundancies can have zero or close to zero eigenvalues – the only thing that prevents the eigenvalues being zero is non-zero hyperparameters. In neural networks both additive (parallel) and multiplicative (serial) redundancies can occur. Larger networks are more likely to have redundancies. Hence we could expect the search to be more stable for smaller networks, and this is what was observed (though of course some small networks did not perform well as they did not have sufficient capacity to model the task).

The initial hyperparameter and weight values are important. If the initial hyperparameters are too high, they can force all the weights to zero before the network has a chance to learn anything. If the initial hyperparameters are too low the network can get trapped in an overfitting mode. Thodberg [13] makes the reasonable suggestion that the initial hyperparameters should be set so that the weight cost is 10% of the error at the beginning of training.

The results described in this chapter lead to the following recommendations for hyperparameter updates:

- update hyperparameters frequently

- if any h_{ii}^{-1} values are out-of-bounds, replace them with zero in the calculation of $\gamma_{[i]}$ (so that each out-of-bounds h_{ii}^{-1} contributes one well-determined parameter)
- ignore updates for α_c which suggest a zero value
- limit changes in α_c to have a maximum magnitude of 10.
- ignore negative eigenvalues in calculations of the evidence
- regard evidence values as unreliable for networks with eigenvalues lower than the lowest α_c

Those interested in Bayesian approaches to neural network modeling should also consider Neal's [9] Markov-chain Monte Carlo methods. Although these Monte Carlo methods sometimes require longer computation times than methods based on Mackay's approximate Bayesian framework, they have some theoretical advantages and also require less tweaking.

References

1. C. Bishop. *Neural Networks for Pattern Recognition*. Oxford University Press, 1995.
2. W. L. Buntine and A. S. Weigend. Computing second derivatives in feed-forward networks: A review. *IEEE Transactions on Neural Networks*, 5(3):480–488, 1994.
3. T. J. Hastie and R. J. Tibshirani. *Generalized additive models*. Chapman and Hall, London, 1990.
4. J. N. Hwang, S. -R. Lay, M. Maechler, R. D. Martin, and J. Schimert. Regression modeling in back-propagation and projection pursuit learning. *IEEE Transactions on Neural Networks*, 5(3):342–353, 1994.
5. D. J. C. MacKay. A practical Bayesian framework for backpropagation networks. *Neural Computation*, 4(3):448–472, 1992.
6. D. J. C. MacKay. Bayesian methods for backpropagation networks. In E. Domany, J. L. van Hemmen, and K. Schulten, editors, *Models of Neural Networks III*, chapter 6. Springer-Verlag, New York, 1994.
7. D. J. C. MacKay. Probable networks and plausible predictions - a review of practical Bayesian methods for supervised neural networks. *Network: Computation in Neural Systems*, 6:469–505, 1995.
8. D. J. C. MacKay. Bayesian non-linear modelling for the 1993 energy prediction competition. In G. Heidbreder, editor, *Maximum Entropy and Bayesian Methods, Santa Barbara 1993*, pages 221–234, Dordrecht, 1996. Kluwer.
9. R. M. Neal. Monte carlo implementation of gaussian process models for bayesian regression and classification. Technical Report TR9702, Dept. of Statistics, University of Toronto, 1997. Software available at http://www.cs.utoronto.ca/~radford/.
10. R. M. Neal. *Bayesian Learning for Neural Networks*. Springer-Verlag, New York, 1996.
11. T. Plate, J. Bert, J. Grace, and P. Band. A comparison between neural networks and other statistical techniques for modeling the relationship between tobacco and alcohol and cancer. In M.C. Mozer, M.I. Jordan, and T. Petsche, editors, *Advances in Neural Information Processing 9 (NIPS*96)*. MIT Press, 1997.
12. C. Roosen and T. Hastie. Logistic response projection pursuit. Technical report, AT&T Bell Laboratories, 1993.
13. H. H. Thodberg. A review of bayesian neural networks with an application to near infrared spectroscopy. *IEEE Transactions on Neural Networks*, 7(1):56–72, 1996.

5
Adaptive Regularization in Neural Network Modeling

Jan Larsen[1], Claus Svarer[2], Lars Nonboe Andersen[1], and Lars Kai Hansen[1]

[1] Department of Mathematical Modeling, Building 321
Technical University of Denmark
DK-2800 Lyngby, Denmark
[2] Neurobiology Research Unit
Department of Neurology, Building 9201
Copenhagen University Hospital
Blegdamsvej 9
DK-2100 Copenhagen, Denmark
{jl, lna, lkhansen}@imm.dtu.dk, csvarer@pet.rh.dk
http://eivind.imm.dtu.dk, http://neuro.pet.rh.dk

Abstract. In this paper we address the important problem of optimizing regularization parameters in neural network modeling. The suggested optimization scheme is an extended version of the recently presented algorithm [25]. The idea is to minimize an empirical estimate – like the cross-validation estimate – of the generalization error with respect to regularization parameters. This is done by employing a simple iterative gradient descent scheme using virtually no additional programming overhead compared to standard training. Experiments with feed-forward neural network models for time series prediction and classification tasks showed the viability and robustness of the algorithm. Moreover, we provided some simple theoretical examples in order to illustrate the potential and limitations of the proposed regularization framework.

5.1 Introduction

Neural networks are flexible tools for time series processing and pattern recognition. By increasing the number of hidden neurons in a 2-layer architecture any relevant target function can be approximated arbitrarily close [19]. The associated risk of overfitting on noisy data is of major concern in neural network design, which find expression in the ubiquitous bias-variance dilemma, see e.g., [9].

The need for regularization is two-fold: First, it remedies numerical problems in the training process by smoothing the cost function and by introducing curvature in low (possibly zero) curvature regions of cost function. Secondly, regularization is a tool for reducing variance by introducing extra bias. The overall objective of architecture optimization is to minimize the generalization error.

The architecture can be optimized *directly* by stepwise selection procedures (including pruning techniques) or *indirectly* using regularization. In general, one would prefer a hybrid scheme; however, a very flexible regularization may substitute the need for selection procedures. The numerical experiments we consider mainly hybrid pruning/adaptive regularization schemes.

The trick presented in this communication addresses the problem of adapting regularization parameters.

The trick consists in formulating a simple iterative gradient descent scheme for adapting the regularization parameters aiming at minimizing the generalization error.

We suggest to use an empirical estimate[1] of the generalization error, viz. the K-fold cross-validation [8], [39]. In [25] and [3] the proposed scheme was studied using the hold-out validation estimator.

In addition to empirical estimators for the generalization error a number of *algebraic* estimators like FPE [1], FPER [23], GEN [21], GPE [31] and NIC [33] have been developed in recent years. These estimates, however, depend on a number of statistical assumptions which can be quite hard to justify. In particular, they are $o(1/N_t)$ estimators where N_t is the number of training examples. However, for many practical modeling set-ups it is hard to meet the large training set assumption.

In [14] properties of adaptive regularization is studied in the simple case of estimating the mean of a random variable using an algebraic estimate of the average[2] generalization error and [15] proposed an adaptive regularization scheme for neural networks based on an algebraic estimate. However, experiments indicate that this scheme has a drawback regarding robustness. In addition, the requirement of a large training set may not be met.

The Bayesian approach to adapt regularization parameters is to minimize the so-called evidence [5, Ch. 10], [30]. The evidence, however, does not in a simple way relate to the generalization error which is our primary object of interest.

Furthermore [2] and [38] consider the use of a validation set to tune the amount of regularization, in particular when using the early-stop technique.

Section 5.2 considers training and empirical generalization assessment. In Section 5.3 the framework for optimization of regularization parameters is presented. The experimental section 5.4 deals with examples of feed-forward neural networks models for classification and time series prediction. Further, in order to study the theoretical potential/limitations of the proposed framework, we include some simulations on a simple toy problem.

5.2 Training and Generalization

Suppose that the neural network is described by the vector function $f(x;w)$ where x is the input vector and w is the vector of network weights and thresholds

---

[1] For further discussion on empirical generalization assessment, see e.g., [24].
[2] Average w.r.t. to different training sets.

with dimensionality m. The objective is to use the network model for approximating the true conditional input-output distribution $p(y|x)$, or some moments hereof. For regression and signal processing problems we normally model the conditional expectation $E\{y|x\}$.

Assume that we have available a data set $\mathcal{D} = \{x(k); y(k)\}_{k=1}^{N}$ of N input-output examples. In order to both train and empirically estimate the generalization performance we follow the idea of K-fold cross-validation [8], [39] and split the data set into K randomly chosen disjoint sets of approximately equal size, i.e., $\mathcal{D} = \cup_{j=1}^{K} \mathcal{V}_j$ and $\forall i \neq j : \mathcal{V}_i \cap \mathcal{V}_j = \emptyset$. Training and validation is replicated K times, and in the j'th run training is done on the set $\mathcal{T}_j = \mathcal{D} \setminus \mathcal{V}_j$ and validation is performed on $\mathcal{V}_j$.

The cost function, $C_{\mathcal{T}_j}$, for network training on $\mathcal{T}_j$, is supposed to be the sum of a loss function (or training error), $S_{\mathcal{T}_j}(w)$, and a regularization term $R(w, \kappa)$ parameterized by a set of regularization parameters κ, i.e.,

$$C_{\mathcal{T}_j}(w) = S_{\mathcal{T}_j}(w) + R(w, \kappa) = \frac{1}{N_{tj}} \sum_{k=1}^{N_{tj}} \ell(y(k), \widehat{y}(k); w) + R(w, \kappa) \quad (5.1)$$

where $\ell(\cdot)$ measures the distance between the output $y(k)$ and the network prediction $\widehat{y}(k) = f(x(k); w)$. In section 5.4 we will consider log-likelihood and the square error loss function $\ell = |y - \widehat{y}|^2$. $N_{tj} \equiv |\mathcal{T}_j|$ defines the number of training examples in $\mathcal{T}_j$ and k is used to index the k'th example $[x(k), y(k)]$. Training provides the estimated weight vector $\widehat{w}_j = \arg\min_w C_{\mathcal{T}_j}(w)$.

The j'th validation set $\mathcal{V}_j$ consist of $N_{vj} = N - N_{tj}$ examples and the validation error[3] of the trained network reads

$$S_{\mathcal{V}_j}(\widehat{w}_j) = \frac{1}{N_{vj}} \sum_{k=1}^{N_{vj}} \ell(y(k), \widehat{y}(k); \widehat{w}_j) \quad (5.2)$$

where the sum runs over the N_{vj} validation examples. $S_{\mathcal{V}_j}(\widehat{w}_j)$ is thus an estimate of the generalization error, defined as the expected loss,

$$G(\widehat{w}_j) = E_{x,y}\{\ell(y, \widehat{y}; \widehat{w}_j)\} = \int \ell(y, \widehat{y}; \widehat{w}_j) \cdot p(x, y) \, dx dy \quad (5.3)$$

where $p(x, y)$ is the unknown joint input-output probability density. Generally, $S_{\mathcal{V}_j}(\widehat{w}_j) = G(\widehat{w}_j) + O(1/\sqrt{N_{vj}})$ where $O(\cdot)$ is the Landau order function[4]. Thus we need large N_{vj} to achieve an accurate estimate of the generalization error. On the other hand, this leaves only few data for training thus the true generalization $G(\widehat{w}_j)$ increases. Consequently there exist a trade-off among the two conflicting aims which calls for finding an optimal split ratio. The optimal split ratio[5] is an interesting open and difficult problem since it depends on the total algorithm in

---
[3] That is, the loss function on the validation set.
[4] If $h(x) = O(g(x))$ then $|h(x)|/|g(x)| < \infty$ for $x \to 0$.
[5] For more elaborations on the split of data, see e.g., [2], [20], [24] and [26].

which the validation error enters. Moreover, it depends on the learning curve[6] [17].

The final K-fold cross-validation estimate is given by the average validation error estimates,

$$\widehat{\Gamma} = \frac{1}{K} \sum_{j=1}^{K} S_{\mathcal{V}_j}(\widehat{\boldsymbol{w}}_j). \qquad (5.4)$$

$\widehat{\Gamma}$ is an estimate of the *average* generalization error over all possible training sets of size N_{tj},

$$\Gamma = E_{\mathcal{T}}\{G(\widehat{\boldsymbol{w}}_j)\}. \qquad (5.5)$$

$\widehat{\Gamma}$ is an unbiased estimate of Γ if the data of $\mathcal{D}$ are independently distributed[7], see e.g., [16].

The idea is now to optimize the amount of regularization by minimizing $\widehat{\Gamma}$ w.r.t. the regularization parameters $\boldsymbol{\kappa}$. An algorithm for this purpose is described in Section 5.3. Furthermore, we might consider optimizing regularization using the hold-out validation estimate corresponding to $K = 1$. In this case one has to choose a split ratio. Without further ado, we will recommend a 50/50 splitting.

Suppose that we found the optimal $\boldsymbol{\kappa}$ using the cross-validation estimate. Replications of training result in K different weight estimates $\widehat{\boldsymbol{w}}_j$ which might be viewed as an ensemble of networks. In [16] we showed under certain mild conditions that when considering a $o(1/N)$ approximation, the average generalization error of the ensemble network $f_{\text{ens}}(\boldsymbol{x}) = \sum_{j=1}^{K} \beta_j \cdot f(\boldsymbol{x}, \widehat{\boldsymbol{w}}_j)$ equals that of the network trained on all examples in $\mathcal{D}$ where β_j weights the contribution from the j'th network and $\sum_j \beta_j = 1$. If K is a divisor in N then $\forall j$, $\beta_j = 1/K$, otherwise $\beta_j = (N - N_{vj})/N(K-1)$. Consequently, one might use the ensemble network to compensate for the increase in generalization error due to only training on $N_{tj} = N - N_{vj}$ data. Alternatively, one might retrain on the full data set $\mathcal{D}$ using the optimal $\boldsymbol{\kappa}$. We use the latter approach in the experimental section.

A minimal necessary requirement for a procedure which estimates the network parameters on the training set and optimizes the amount of regularization from a cross-validation set is: the generalization error of the regularized network should be smaller than that of the unregularized network trained on the full data set $\mathcal{D}$. However, this is not always the case, and is the quintessence of various "no free lunch" theorems [12], [44], [46]:

- If the regularizer is parameterized using many parameters, $\boldsymbol{\kappa}$, there is a potential risk of over-fitting on the cross-validation data. A natural way to avoid this situation is to limit the number of regularization parameters. Another recipe is to impose constraints on $\boldsymbol{\kappa}$ (hyper regularization).
- The specific choice of the regularizers functional form impose prior constraints on the functions to be implemented by the network[8]. If the prior

---

[6] Defined as the average generalization error as a function of the number of training examples.
[7] That is, $[\boldsymbol{x}(k_1), \boldsymbol{y}(k_1)]$ is independent of $[\boldsymbol{x}(k_2), \boldsymbol{y}(k_2)]$ for all $k_1 \neq k_2$.
[8] The functional constraints are through the penalty imposed on the weights.

information is mismatched to the actual problem it might be better not to use regularization.
- The de-biasing procedure described above which compensate for training only on $N_{tj} < N$ examples might fail to yield better performance since the weights now are optimized using all data, including those which where left out exclusively for optimizing regularization parameters.
- The split among training/validation data, and consequently the number of folds, K, may not be chosen appropriately.

These problems are further addressed in Section 5.4.1.

5.3 Adapting Regularization Parameters

The choice of regularizer may be motivated by

- the fact that the minimization of the cost function is normally an ill- posed task. Regularization smoothens the cost function and thereby facilitates the training. The weight decay regularizer[9], originally suggested by Hinton in the neural networks literature, is a simple way to accomplish this task, see e.g., [35].
- a priori knowledge of the weights, e.g., in terms of a prior distribution (when using a Bayesian approach). In this case the regularization term normally plays the role of a log-prior distribution. Weight decay regularization may be viewed as a Gaussian prior, see e.g., [2]. Other types of priors, e.g., the Laplacian [13], [43] and soft weight sharing [34] has been considered. Moreover, priors have been developed for the purpose of restricting the number of weights (pruning), e.g., the so-called weight elimination [42].
- a desired characteristics of the functional mapping performed by the network. Typically, a smooth mapping is preferred. Regularizers which penalizes curvature of the mapping has been suggested in [4], [7], [32], [45], [10].

In the experimental section we consider weight decay regularization and some generalizations hereof. Without further ado, weight decay regularization has proven to be useful in many neural network applications.

The standard approach for estimation of regularization parameters is more and less systematic search and evaluation of the cross-validation error. However, this is not viable for multiple regularization parameters. On the other hand, as will be demonstrated, it is possible to derive an optimization algorithm based on gradient descent.

Consider a regularization term $R(\boldsymbol{w}, \boldsymbol{\kappa})$ which depends on q regularization parameters contained in the vector $\boldsymbol{\kappa}$. Since the estimated weights $\widehat{\boldsymbol{w}}_j = \arg\min_{\boldsymbol{w}} C_{\mathcal{T}_j}(\boldsymbol{w})$ are controlled by the regularization term, we may in fact consider the cross-validation error Eq. (5.4) as an *implicit function* of the

---
[9] Also known as ridge regression.

regularization parameters, i.e.,

$$\widehat{\Gamma}(\kappa) = \frac{1}{K} \sum_{j=1}^{K} S_{\mathcal{V}_j}(\widehat{w}_j(\kappa)) \tag{5.6}$$

where $\widehat{w}_j(\kappa)$ is the κ-dependent vector of weights estimated from training set $\mathcal{T}_j$. The optimal regularization can be found by using gradient descent[10],

$$\kappa_{(n+1)} = \kappa_{(n)} - \eta \frac{\partial \widehat{\Gamma}}{\partial \kappa}(\widehat{w}(\kappa_{(n)})) \tag{5.7}$$

where $\eta > 0$ is a step-size (learning rate) and $\kappa_{(n)}$ is the estimate of the regularization parameters in iteration n.

Suppose the regularization term is linear in the regularization parameters,

$$R(w, \kappa) = \kappa^\top r(w) = \sum_{i=1}^{q} \kappa_i r_i(w) \tag{5.8}$$

where κ_i are the regularization parameters and $r_i(w)$ the associated regularization functions. Many suggested regularizers are linear in the regularization parameters, this includes the popular weight decay regularization as well as regularizers imposing smooth functions such as the Tikhonov regularizer [4], [2] and the smoothing regularizer for neural networks [32], [45]. However, there exist exceptions such as weight-elimination [42] and soft weight sharing [34]. In this case the presented method needs few modifications.

Using the results of the Appendix, the gradient of the cross-validation error equals

$$\frac{\partial \widehat{\Gamma}}{\partial \kappa}(\kappa) = \frac{1}{K} \sum_{j=1}^{K} \frac{\partial S_{\mathcal{V}_j}}{\partial \kappa}(\widehat{w}_j), \tag{5.9}$$

$$\frac{\partial S_{\mathcal{V}_j}}{\partial \kappa}(\widehat{w}_j) = -\frac{\partial r}{\partial w^\top}(\widehat{w}_j) \cdot J_j^{-1}(\widehat{w}_j) \cdot \frac{\partial S_{\mathcal{V}_j}}{\partial w}(\widehat{w}_j). \tag{5.10}$$

where $J_j = \partial^2 C_{\mathcal{T}_j}/\partial w \partial w^\top$ is the Hessian of the cost function. As an example, consider the case of weight decay regularization with separate weight decays for two group of weights, e.g., the input-to-hidden and hidden-to output weights, i.e.,

$$R(w, \kappa) = \kappa^I \cdot |w^I|^2 + \kappa^H \cdot |w^H|^2 \tag{5.11}$$

where $\kappa = [\kappa^I, \kappa^H]$, $w = [w^I, w^H]$ with w^I, w^H denoting the input-to-hidden and hidden-to output weights, respectively. The gradient of the validation error then yields,

$$\frac{\partial S_{\mathcal{V}_j}}{\partial \kappa^I}(\widehat{w}_j) = -2(\widehat{w}_j^I)^\top \cdot g_j^I, \quad \frac{\partial S_{\mathcal{V}_j}}{\partial \kappa^H}(\widehat{w}_j) = -2(\widehat{w}_j^H)^\top \cdot g_j^H \tag{5.12}$$

---

[10] We have recently extended this algorithm incorporating second order information via the Conjugate Gradient technique [11].

where g_j is the vector

$$g_j = [g_j^I, g_j^H] = J_j^{-1}(\widehat{w}_j) \cdot \frac{\partial S_{\mathcal{V}_j}}{\partial w}(\widehat{w}_j). \tag{5.13}$$

In summary, the algorithm for adapting regularization parameters consists of the following 8 steps:

1. Choose the split ratio; hence, the number of folds, K.
2. Initialize κ and the weights of the network[11].
3. Train the K networks with fixed κ on $\mathcal{T}_j$ to achieve $\widehat{w}_j(\kappa)$, $j = 1, 2, \cdots, K$. Calculate the validation errors $S_{\mathcal{V}_j}$ and the cross-validation estimate $\widehat{\Gamma}$.
4. Calculate the gradients $\partial S_{\mathcal{V}_j}/\partial \kappa$ and $\partial \widehat{\Gamma}/\partial \kappa$ cf. Eq. (5.9) and (5.10). Initialize the step-size η.
5. Update κ using Eq. (5.7).
6. Retrain the K networks from the previous weight estimates and recalculate the cross-validation error $\widehat{\Gamma}$.
7. If no decrease in cross-validation error then perform a bisection of η and go to step 5; otherwise, continue.
8. Repeat steps 4–7 until the relative change in cross-validation error is below a small percentage or, e.g., the 2-norm of the gradient $\partial \widehat{\Gamma}/\partial \kappa$ is below a small number.

Compared to standard neural network training the above algorithm does generally not lead to severe computational overhead. First of all, the standard approach of tuning regularization parameters by, more or less systematic search, requires a lot of training sessions. The additional terms to be computed in the adaptive algorithm are: 1) the derivative of the regularization functions w.r.t. the weights, $\partial r/\partial w$, 2) the gradient of the validation errors, $\partial S_{\mathcal{V}_j}/\partial w$, and 3) the inverse Hessians, J_j^{-1}. The first term is often a simple function of the weights[12] and computationally inexpensive. In the case of feed-forward neural networks, the second term is computed by one pass of the validation examples through a standard back-propagation algorithm. The third term is computationally more expensive. However, if the network is trained using a second order scheme, which requires computation of the inverse Hessian[13], there is no computational overhead.

The adaptive algorithm requires of the order of $K \cdot itr_\kappa \cdot itr_\eta$ weight retrainings. Here itr_κ is the number of iterations in the gradient descent scheme for κ and itr_η is the average number of bisections of η in step 7 of the algorithm. In the experiments carried out the number of retrainings is approx. 100–300 times K. Recall, since we keep on retraining from the current weight estimate, the number of training epochs is generally small.

---

[11] In Sec. 5.4.1 a practical initialization procedure for κ is described.
[12] For weight decay, it is $2w$.
[13] Often the computations are reduced by using a Hessians approximation, e.g., the Gauss-Newton approximation. Many studies have reported significant training speed-up by using second order methods, see e.g., [22], [35].

The number of weight retrainings is somewhat higher than that involved when optimizing the network by using a pruning technique like validation set based Optimal Brain Damage (vOBD) [25], [27]. vOBD based on K-fold cross-validation requires of the order of $K \cdot m$ retrainings, where $m = \dim(\boldsymbol{w})$. The adaptive regularization algorithm is easily integrated with the pruning algorithm as demonstrated in the experimental section.

5.4 Numerical Experiments

5.4.1 Potentials and Limitations in the Approach

The purpose of the section is to demonstrate the potential and limitations of the suggested adaptive regularization framework. We consider the simple linear data generating *system*, viz. estimating the mean of a Gaussian variable,

$$y(k) = w^\circ + \varepsilon(k) \tag{5.14}$$

where w° is the true mean and the noise $\varepsilon(k) \sim \mathcal{N}(0, \sigma_\varepsilon^2)$.

We employ 2-fold cross-validation, i.e., $\mathcal{D} = \mathcal{T}_1 \cup \mathcal{T}_2$, where $\mathcal{T}_j$, $j = 1, 2$ denote the two training sets in the validation procedure containing approximately half the examples[14]. The linear *model* $y(k) = w + e(k)$ is trained using the mean square cost function augmented by simple weight decay, as shown by

$$C_{\mathcal{T}_j}(w) = \frac{1}{N_{tj}} \sum_{k=1}^{N_{tj}} (y(k) - w)^2 + \kappa \cdot w^2 \tag{5.15}$$

where k runs over examples of the data set in question. The estimated weights are $\widehat{w}_j = \bar{y}_j/(1 + \kappa)$ where $\bar{y}_j = N_{tj}^{-1} \sum_{k=1}^{N_{tj}} y(k)$ are the estimated mean. For this simple case, the minimization of the cross-validation error given by,

$$\widehat{\Gamma}(\kappa) = \frac{1}{2} \sum_{j=1}^{2} S_{\mathcal{V}_j}(\widehat{w}_j(\kappa)), \quad S_{\mathcal{V}_j}(\widehat{w}_j(\kappa)) = \frac{1}{N_{vj}} \sum_{k=1}^{N_{vj}} (y(k) - \widehat{w}_j)^2, \tag{5.16}$$

can be done exactly. The optimal κ is given by

$$\kappa_{\text{opt}} = \frac{\bar{y}_1^2 + \bar{y}_2^2}{2\bar{y}_1\bar{y}_2} - 1. \tag{5.17}$$

Assuming N to be even, the ensemble average of the estimated weights[15], $\widehat{w}_j(\kappa_{\text{opt}})$, leads to the final estimate

$$\widehat{w}_{\text{reg}} = \frac{1}{2}(\widehat{w}_1(\kappa_{\text{opt}}) + \widehat{w}_2(\kappa_{\text{opt}})) = \frac{\bar{y}_1\bar{y}_2(\bar{y}_1 + \bar{y}_2)}{\bar{y}_1^2 + \bar{y}_2^2}. \tag{5.18}$$

---

[14] That is, $N_{t1} = \lfloor N/2 \rfloor$ and $N_{t2} = N - N_{t1}$. Note that these training sets are also the two validation sets, $\mathcal{V}_1 = \mathcal{T}_2$, and vice versa.

[15] The ensemble average corresponds to retraining on all data using κ_{opt}. The weighting of the two estimates is only valid for N even (see Sec. 5.2 for the general case).

Notice two properties: First, the estimate is self-consistent as $\lim_{N\to\infty} \widehat{w}_{\text{reg}} = \lim_{N\to\infty} \widehat{w}_D = w°$ where $\widehat{w}_D = N^{-1}\sum_{k=1}^{N} y(k) = (\bar{y}_1 + \bar{y}_2)/2$ is the unregularized estimate trained on all data. Secondly, it is easy to verify that $\bar{y}_j \sim \mathcal{N}(w°, 2\sigma_\varepsilon^2/N)$. That is, if the *normalized true weight* $\theta \equiv w°/A$ where $A = \sqrt{2/N} \cdot \sigma_\varepsilon$ is large then $\bar{y}_j \approx w°$ which means, $\widehat{w}_{\text{reg}} \approx \widehat{w}_D$.

The objective is now to test whether using $\widehat{w}_{\text{reg}}$ results in lower generalization error than employing the unregularized estimate $\widehat{w}_D$. The generalization error associated with using the weight w is given by

$$G(w) = \sigma_\varepsilon^2 + (w - w°)^2. \qquad (5.19)$$

Further define the generalization error improvement,

$$Z = G(\widehat{w}_D) - G(\widehat{w}_{\text{reg}}) = (\widehat{w}_D - w°)^2 - (\widehat{w}_{\text{reg}} - w°)^2. \qquad (5.20)$$

Note that Z merely is a function of the random variables $\bar{y}_1, \bar{y}_2$ and the true weight $w°$, i.e., it suffices to get samples of $\bar{y}_1, \bar{y}_2$ when evaluating properties of Z. Define the normalized variables

$$\widetilde{y}_j = \frac{\bar{y}_j}{A} \sim \mathcal{N}\left(\frac{w°}{\sigma_\varepsilon} \cdot \sqrt{\frac{N}{2}}, 1\right) = \mathcal{N}(\theta, 1). \qquad (5.21)$$

It is easily shown that the normalized generalization error improvement Z/A^2 is a function of $\widetilde{y}_1, \widetilde{y}_2$ and θ; hence, the distribution of Z/A^2 is parameterized solely by θ.

As a quality measure we consider the *probability of improvement* in generalization error given by $\text{Prob}\{Z > 0\}$. Note that $\text{Prob}\{Z > 0\} = 1/2$ corresponds to equal preference of the two estimates. The probability of improvement depends only on the normalized weight θ since $\text{Prob}\{Z > 0\} = \text{Prob}\{Z/A^2 > 0\}$.

Moreover, we consider the *relative generalization error improvement*, defined as

$$\text{RGI} = 100\% \cdot \frac{Z}{G(\widehat{w}_D)}. \qquad (5.22)$$

In particular, we focus on the probability that the relative improvement in generalization is bigger than[16] x, i.e., $\text{Prob}(\text{RGI} > x)$. Optimally $\text{Prob}(\text{RGI} > x)$ should be close to 1 for $x \leq 0\%$ and slowly decaying towards zero for $0\% < x \leq 100\%$. Using the notation $\widetilde{w}_{\text{reg}} = \widehat{w}_{\text{reg}}/A$, $\widetilde{w}_D = \widehat{w}_D/A$, RGI can be written as

$$\text{RGI} = 100\% \cdot \frac{(\widetilde{w}_D - \theta)^2 - (\widetilde{w}_{\text{reg}} - \theta)^2}{N/2 + (\widetilde{w}_D - \theta)^2}. \qquad (5.23)$$

Thus, the distribution of RGI is parameterized by θ and N.

The quality measures are computed by generating Q independent realizations of $\widetilde{y}_1, \widetilde{y}_2$, i.e., $\{\widetilde{y}_1^{(i)}, \widetilde{y}_2^{(i)}\}_{i=1}^{Q}$. E.g., the probability of improvement is estimated by $P_{\text{imp}} = Q^{-1}\sum_{i=1}^{Q} \mu(Z^{(i)})$ where $\mu(x) = 1$ for $x > 0$, and zero otherwise.

The numerical results of comparing $\widehat{w}_{\text{reg}}$ to the unregularized estimate $\widehat{w}_D$ is summarized in Fig. 5.1.

---
[16] Note that, $\text{Prob}(\text{RGI} > 0) = \text{Prob}(Z > 0)$.

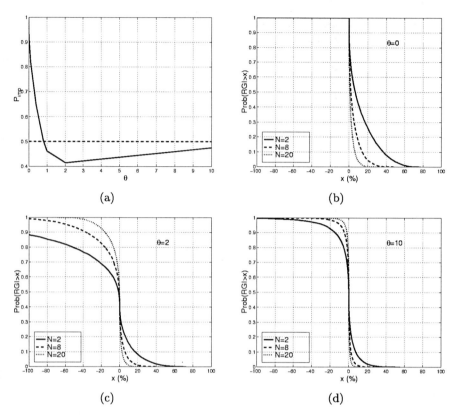

Fig. 5.1. Result of comparing the optimally regularized estimate $\widehat{w}_{\text{reg}}$ of the mean of a Gaussian variable to the unregularized estimate $\widehat{w}_D$. The results are based on $Q = 10^5$ independent realizations. The probability of improvement P_{imp}, shown in panel (a), is one for when the normalized true weight $\theta = \sqrt{N/2} \cdot w^\circ / \sigma_\epsilon = 0$, and above 0.5 for $\theta \lesssim 0.8$. That is, when the prior information of the weight decay regularizer is correct (true weight close to zero), when N is small or when σ_ϵ is large. As θ becomes large P_{imp} tends to 0.5 due to the fact that $\widetilde{w} \approx \widehat{w}_D$. Panel (b)–(d) display $\text{Prob}(\text{RGI} > x)$ for $\theta \in \{0, 2, 10\}$. The ideal probability curve is 1 for $x < 0$ and a slow decay towards zero for $x > 0$. The largest improvement is attained for small θ and small N. Panel (c) and (d) indicate that small N gives the largest probability for $x > 0$; however, also the smallest probability for negative x. That is, a higher chance of getting a good improvement also increases the change of deterioration. Notice, even though $P_{\text{imp}} < 0.5$ for $\theta = 2, 10$ there is still a reasonable probability of getting a significant improvement.

5.4.2 Classification

We test the performance of the adaptive regularization algorithm on a vowel classification problem. The data are based on the Peterson and Barney database [36]. The classes are vowel sounds characterized by the first four formant frequencies. 76 persons (33 male, 28 female and 15 children) have pronounced $c = 10$

different vowels (IY IH EH AE AH AA AO UH UW ER) two times. This results in a data base of totally 1520 examples. The database is the verified database described in [41] where all data[17] are used, including examples where utterance failed of unanimous identification in the listening test (26 listeners). All examples were included to make the task more difficult.

The regularization was adapted using a hold-out validation error estimator, thus the examples were split into a data set, $\mathcal{D}$, consisting of $N = 760$ examples (16 male, 14 female and 8 children) and an independent test set of the remaining 760 examples. The regularization was adapted by splitting the data set $\mathcal{D}$ equally into a validation set of $N_v = 380$ examples and a training set of $N_t = 380$ examples (8 male, 7 female and 4 children in each set).

We used a feed-forward 2-layer neural network with hyperbolic tangent neurons in the hidden layer and modified SoftMax normalized outputs, $\hat{y}_i$, see e.g., [2], [18], [3]. Thus, the outputs estimates the posterior class probabilities $p(\mathcal{C}_i|\boldsymbol{x})$, where $\mathcal{C}_i$ denotes the i'th class, $i = 1, 2, \cdots, c$. Bayes rule (see e.g., [2]) is used to assign $\mathcal{C}_i$ to input $\boldsymbol{x}$ if $i = \arg\max_j p(\mathcal{C}_j|\boldsymbol{x})$. Suppose that the network weights are given by $\boldsymbol{w} = [\boldsymbol{w}^I, \boldsymbol{w}^I_{\text{bias}}, \boldsymbol{w}^H, \boldsymbol{w}^H_{\text{bias}}]$ where $\boldsymbol{w}^I$, $\boldsymbol{w}^H$ are input-to-hidden and hidden-to-output weights, respectively, and the bias weights are assembled in $\boldsymbol{w}^I_{\text{bias}}$ and $\boldsymbol{w}^H_{\text{bias}}$. Suppose that the targets $y_i(k) = 1$ if $\boldsymbol{x}(k) \in \mathcal{C}_i$, and zero otherwise. The network is optimized using a log-likelihood loss function augmented by a weight decay regularizer using 4 regularization parameters,

$$C(\boldsymbol{w}) = \frac{1}{N_t} \sum_{k=1}^{N_t} \sum_{i=1}^{c} y_i(k) \log(\hat{y}_i(k, \boldsymbol{w}))$$
$$+ \kappa^I \cdot |\boldsymbol{w}^I|^2 + \kappa^I_{\text{bias}} \cdot |\boldsymbol{w}^I_{\text{bias}}|^2 + \kappa^H \cdot |\boldsymbol{w}^H|^2 + \kappa^H_{\text{bias}} \cdot |\boldsymbol{w}^H_{\text{bias}}|^2. \quad (5.24)$$

We further define unnormalized weight decays as $\alpha \equiv \kappa \cdot N_t$. This regularizer is motivated by the fact that the bias, input and hidden layer weights play a different role, e.g., the input, hidden and bias signals normally have different scale (see also [2, Ch. 9.2]).

The simulation set-up is:

- Network: 4 inputs, 5 hidden neurons, 9 outputs[18].
- Weights are initialized uniformly over $[-0.5, 0.5]$, regularization parameters are initialized at zero. One step in a gradient descent training algorithm (see e.g., [29]) is performed and the weight decays are re-initialized at $\lambda_{\max}/10^2$, where $\lambda_{\max}$ is the max. eigenvalue of the Hessian matrix of the cost function. This initialization scheme is motivated by the following observations:
    • Weight decays should be so small that they do not reduce the approximation capabilities of the network significantly.
    • They should be so large that the algorithm is prevented from being trapped in a local optimum and numerical instabilities are eliminated.

---

[17] The database can be retrieved from ftp://eivind.imm.dtu.dk/dist/data/vowel/PetersonBarney.tar.Z

[18] We only need 9 outputs since the posterior class probability of the 10th class is given by $1 - \sum_{j=1}^{9} p(\mathcal{C}_j|\boldsymbol{x})$.

- Training is now done using a Gauss-Newton algorithm (see e.g., [29]). The Hessian is inverted using the Moore-Penrose pseudo inverse ensuring that the eigenvalue spread[19] is less than 10^8.
- The regularization step-size η is initialized at 1.
- When the adaptive regularization scheme has terminated 3% of the weights are pruned using a validation set based version of the Optimal Brain Damage (vOBD) recipe [25], [27].
- Alternation between pruning and adaptive regularization continues until the validation error has reached a minimum.
- Finally, remaining weights are retrained on all data using the optimized weight decay parameters.

Table 5.1. Probability of misclassification (pmc) and log-likelihood cost function (without reg. term, see Eq. (5.24)) for the classification example. The neural network averages and standard deviations are computed from 10 runs. In the case of small fixed regularization, weight decays were set at initial values equal to $\lambda_{max}/10^6$ where λ_{max} is the largest eigenvalue of the Hessian matrix of the cost function. Optimal regularization refers to the case of optimizing 4 weight decay parameters. Pruning refers to validation set based OBD. KNN refers to k-nearest-neighbor classification.

Probability of Misclassification (pmc)

	NN small fixed reg.	NN opt. reg.+prun.	KNN ($k=9$)
Training Set	0.075 ±0.026	0.107 ± 0.008	0.150
Validation Set	0.143 ± 0.014	0.115 ± 0.004	0.158
Test Set	0.146 ± 0.010	0.124 ± 0.006	0.199
Test Set (train. on all data)	0.126 ± 0.010	0.119 ± 0.004	0.153

Log-likelihood Cost Function

	NN small fixed reg.	NN opt. reg.+prun.
Training Set	0.2002 ± 0.0600	0.2881 ± 0.0134
Validation Set	0.7016 ± 0.2330	0.3810 ± 0.0131
Test Set	0.6687 ± 0.2030	0.3773 ± 0.0143
Test Set (train. on all data)	0.4426 ± 0.0328	0.3518 ± 0.0096

Table 5.1 reports the average and standard deviations of the probability of misclassification (pmc) and log-likelihood cost function over 10 runs for pruned networks using the optimal regularization parameters. Note that retraining on

---
[19] Eigenvalue spread should not be larger than the square root of the machine precision [6].

the full data set decreases the test *pmc* slightly on the average. In fact, improvement was noticed in 9 out of 10 runs. The table further shows the gain of the combined adaptive regularization/pruning algorithm relative to using a small fixed weight decay. However, recall, cf. Sec. 5.4.1, that the actual gain is *very* dependent on the noise level, data set size, etc. The objective is not to demonstrate high gain for a specific problem, rather to demonstrate that algorithm runs fairly robust in a classification set-up. For comparison we used a k-nearest-neighbor (KNN) classification (see e.g., [2]) and found that $k = 9$ neighbors was optimal by minimizing *pmc* on the validation set. The neural network performed significantly better. Contrasting the obtained results to other work is difficult. In [37] results on the Peterson-Barney vowel problem are reported, but their data are not exactly the same; only the first 2 formant frequencies were used. Furthermore, different test sets have been used for the different methods presented. The best result reported [28] is obtained by using KNN and reach $pmc = 0.186$ which is significantly higher than our results.

Fig. 5.2 shows the evolution of the adaptive regularization as well as the pruning algorithm.

5.4.3 Time Series Prediction

We tested the performance of the adaptive regularization schemes on the Mackey-Glass chaotic time series prediction problem, see e.g., [22], [40]. The goal is to predict the series 100 steps ahead based on previous observations. The feed-forward net configuration is an input lag-space $\boldsymbol{x}(k) = [x(k), x(k-6), x(k-12), x(k-18)]$ of 4 inputs, 25 hidden hyperbolic tangent neurons, and a single linear output unit $\widehat{y}(k)$ which predicts $y(k) = x(k+100)$. The cost function is the squared error, $N_t^{-1} \sum_{k=1}^{N_t} (y(k) - \widehat{y}(k, \boldsymbol{w}))^2$, augmented by a weight decay regularizer using 4 different weight decays as described in Section 5.4.2.

The simulation set-up is:

- The data set, $\mathcal{D}$, has $N = 500$ examples and an independent test has 8500 examples.
- The regularization parameters are optimized using a hold-out validation error with an even split[20] of the data set into training and validation sets each having 250 examples.
- Weight decays are initialized at zero and one Gauss-Newton iteration is performed, then weight decays were re-initialized at $\lambda_{\max}/10^6$, where $\lambda_{\max}$ is the max. eigenvalue of the Hessian matrix of the cost function.
- The network is trained using a Gauss-Newton training scheme. The Hessian is inverted using the Moore-Penrose pseudo inverse ensuring that the eigenvalue spread is less than 10^8.
- The regularization step-size η is initialized at 10^{-2}.

---

[20] The sensitivity to different splits are considered in [25].

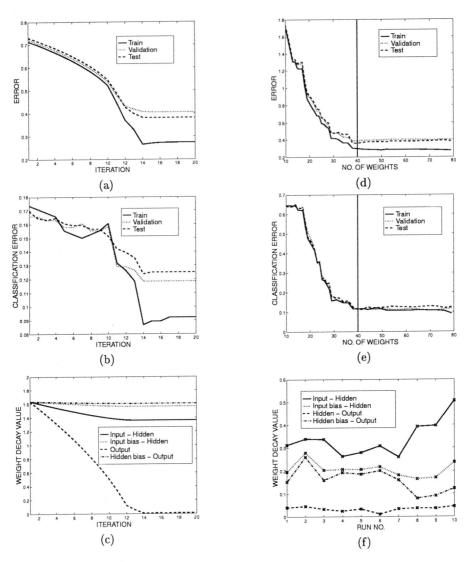

Fig. 5.2. Classification example. Panels (a), (b) and (c) show the evolution of the adaptive regularization algorithm in a typical run (fully connected network). The weight decays are optimized aiming at minimizing the validation error in panel (a). Note that also the test error decreases. This tendency is also evident in panel (b) displaying pmc even though a small increase noticed. In panel (c) the convergence unnormalized weight decays, $\alpha = \kappa \cdot N_t$, are depicted. Panels (d) and (e) show the evolution of errors and pmc during the pruning session. The optimal network is chosen as the one with minimal validation error, as indicated by the vertical line. There is only a marginal effect of pruning in this run. Finally, in panel (f), the variation of the optimal (end of pruning) α's in different runs is demonstrated. A clear similarity over runs is noticed.

- Alternation between adapting the 4 weight decays and validation set based pruning [25].
- The pruned network is retrained on all data using the optimized weight decay parameters.

Table 5.2 reports the average and standard deviations of the normalized squared error (i.e., the squared error normalized with the estimated variance of $x(k)$, denoted $\hat{\sigma}_x^2$) over 10 runs for optimal regularization parameters. Retraining on the full data set decreases the test error somewhat on the average. Improvement was noticed in 10 out of 10 runs. We tested 3 different cases: small fixed regularization, small fixed regularization assisted by pruning and combined adaptive regularization/pruning. It turns that pruning alone does not improve performance; however, supplementing by adaptive regularization gives a test error reduction.

We furthermore tried a flexible regularization scheme, viz. individual weight decay where $R(w, \kappa) = \sum_{i=1}^{m} \kappa_i w_i^2$ and $\kappa_i \geq 0$ are imposed. In the present case it turned out that the flexible regularizer was not able to outperform the joint adaptive regularization/pruning scheme; possibly due to training and validation set sizes.

Table 5.2. Normalized squared error performance for the time series prediction examples. All figures are in units of $10^{-3}\hat{\sigma}_x^2$ and averages and standard deviations are computed from 10 runs. In the case of small fixed regularization, weight decays were set at initial values equal to $\lambda_{\max}/10^6$ where $\lambda_{\max}$ is the largest eigenvalue of the Hessian matrix of the cost function. Optimal regularization refers to the case of optimizing 4 weight decay parameters. Pruning refers to validation set based OBD.

	NN small fixed reg.	NN small fixed reg.+prun.	NN opt. reg.+prun.
Training Set	0.17 ± 0.07	0.12 ± 0.04	0.10 ± 0.07
Validation Set	0.53 ± 0.26	0.36 ± 0.07	0.28 ± 0.14
Test Set	1.91 ± 0.68	1.58 ± 0.21	1.29 ± 0.46
Test Set (train. on all data)	1.33 ± 0.43	1.34 ± 0.26	1.17 ± 0.48

Fig. 5.3 demonstrates adaptive regularization and pruning in a typical case using 4 weight decays.

5.5 Conclusions

In this paper it was suggested to adapt regularization parameters by minimizing the cross-validation error or a simple hold-out validation error. We derived a simple gradient descent scheme for optimizing regularization parameters which has a small programming overhead and an acceptable computational overhead compared to standard training. Numerical examples with a toy linear model showed limitations and advantages of the adaptive regularization approach. Moreover,

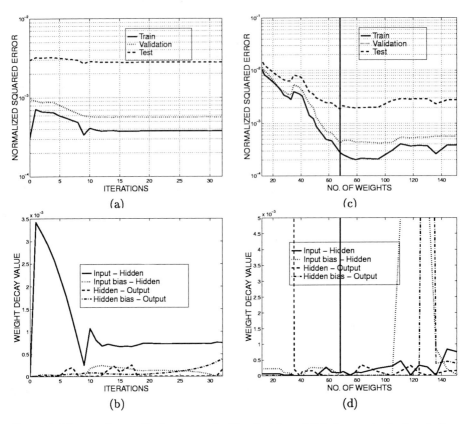

Fig. 5.3. Time series prediction example. Panels (a) and (b) show a typical evolution of errors and unnormalized weight decay values α when running the adaptive regularization algorithm using 4 weight decays. The normalized validation error drops approx. a factor of 2 when adapting weight decays. It turns out that some regularization of the input-to-hidden and output bias weights are needed whereas the other weights essentially requires no regularization[22]. In panel (c) and (d) it is demonstrated that pruning reduces the test error slightly. The optimal network is chosen as the one with minimal validation error, as indicated by the vertical line.

numerical experiments on classification and time series prediction problems successfully demonstrated the functionality of the algorithm. Adaptation of regularization parameters resulted in lower generalization error; however, it should be emphasized that the actual yield is very dependent on the problem and the choice of the regularizers functional form.

---

[22] Recall that if a weight decay κ is below $\lambda_{\max}/10^8$ it does not influence the Moore-Penrose pseudo inversion of the Hessian.

Acknowledgments

This research was supported by the Danish Natural Science and Technical Research Councils through the Computational Neural Network Center. JL furthermore acknowledge the Radio Parts Foundation for financial support. Mads Hintz-Madsen and Morten With Pedersen are acknowledged for stimulating discussions.

Appendix

Assume that the regularization term is linear in the regularization parameters, i.e.,

$$R(w, \kappa) = \kappa^\top r(w) = \sum_{i=1}^{q} \kappa_i r_i(w) \quad (5.25)$$

The gradient of the cross-validation error Eq. (5.4) is

$$\frac{\partial \widehat{\Gamma}}{\partial \kappa}(\kappa) = \frac{1}{K} \sum_{j=1}^{K} \frac{\partial S_{\mathcal{V}_j}}{\partial \kappa}(\widehat{w}_j(\kappa)) \quad (5.26)$$

Using the chain rule the gradient vector of the validation error, $S_{\mathcal{V}_j}$, can be written as

$$\frac{\partial S_{\mathcal{V}_j}}{\partial \kappa}(\widehat{w}_j(\kappa)) = \frac{\partial w^\top}{\partial \kappa}(\widehat{w}_j(\kappa)) \cdot \frac{\partial S_{\mathcal{V}_j}}{\partial w}(\widehat{w}_j(\kappa)) \quad (5.27)$$

where $\partial w^\top/\partial \kappa$ is the $q \times m$ derivative matrix of the estimated weights w.r.t. the regularization parameters and $m = \dim(w)$. In order to find this derivative matrix, consider the gradient of the cost function w.r.t. to the weights as a function of κ and use the following expansion around the current estimate $\kappa_{(n)}$,

$$\frac{\partial C_{\mathcal{T}_j}}{\partial w}(\kappa) = \frac{\partial C_{\mathcal{T}_j}}{\partial w}(\kappa_{(n)}) + \frac{\partial^2 C_{\mathcal{T}_j}}{\partial w \partial \kappa^\top}(\kappa_{(n)}) \cdot (\kappa - \kappa_{(n)}) + o(|\kappa - \kappa_{(n)}|). \quad (5.28)$$

Requiring $\widehat{w}(\kappa_{(n+1)})$ in the next iteration to be an optimal weight vector, i.e., $\partial C_{\mathcal{T}_j}/\partial w(\kappa_{(n+1)}) = \mathbf{0}$ implies

$$\frac{\partial^2 C_{\mathcal{T}_j}}{\partial w \partial \kappa^\top}(\widehat{w}(\kappa_{(n)})) = \mathbf{0}. \quad (5.29)$$

Recall that $\partial C_{\mathcal{T}_j}/\partial w(\kappa_{(n)}) = \mathbf{0}$ by assumption. Eq. (5.29) can be used for determining $\partial w^\top/\partial \kappa$. Recognizing that the cost function $C_{\mathcal{T}_j}(\widehat{w}(\kappa)) = S_{\mathcal{T}_j}(\widehat{w}(\kappa)) + R(\widehat{w}(\kappa), \kappa)$ depends *implicitly* (thorough $\widehat{w}(\kappa)$) and *explicitly* on κ it is possible, by using Eq. (5.25), to derive the following relation[23]:

$$\frac{\partial w^\top}{\partial \kappa}(\widehat{w}_j) = -\frac{\partial r}{\partial w^\top}(\widehat{w}_j) \cdot \mathbf{J}_j^{-1}(\widehat{w}_j) \quad (5.30)$$

---

[23] For convenience, here $\widehat{w}$'s explicit κ-dependence is omitted.

where $\boldsymbol{J}_j = \partial^2 C_{\mathcal{T}_j}/\partial \boldsymbol{w}\partial \boldsymbol{w}^\top$ is the Hessian of the cost function which e.g., might be evaluated using the Gauss-Newton approximation [29]. Finally, substituting Eq. (5.30) into (5.27) gives

$$\frac{\partial S_{\mathcal{V}_j}}{\partial \kappa}(\widehat{\boldsymbol{w}}_j) = -\frac{\partial \boldsymbol{r}}{\partial \boldsymbol{w}^\top}(\widehat{\boldsymbol{w}}_j) \cdot \boldsymbol{J}_j^{-1}(\widehat{\boldsymbol{w}}_j) \cdot \frac{\partial S_{\mathcal{V}_j}}{\partial \boldsymbol{w}}(\widehat{\boldsymbol{w}}_j) \tag{5.31}$$

$\partial S_{\mathcal{V}_j}/\partial \boldsymbol{w}$ is found by ordinary back-propagation on the validation set while $\partial \boldsymbol{r}/\partial \boldsymbol{w}^\top$ is calculated from the specific assumptions on the regularizer.

References

1. H. Akaike: Fitting Autoregressive Models for Prediction. Annals of the Institute of Statistical Mathematics **21** (1969) 243–247
2. S. Amari, N. Murata, K.R. Müller, M. Finke and H. Yang: Asymptotic Statistical Theory of Overtraining and Cross-Validation. Technical report METR 95-06 (1995) and IEEE Transactions on Neural Networks, 8, 5, 985-996 (1997)
3. L. Nonboe Andersen, J. Larsen, L.K. Hansen and M. Hintz-madsen: Adaptive Regularization of Neural Classifiers. In J. Principe et al. (eds.), Proceedings of the IEEE Workshop on Neural Networks for Signal Processing VII, Piscataway, New Jersey: IEEE, (1997) 24–33
4. C.M. Bishop: Curvature-Driven Smoothing: A Learning Algorithm for Feedforward Neural Networks. IEEE Transactions on Neural Networks 4(4) (1993) 882–884
5. C.M. Bishop: Neural Networks for Pattern Recognition. Oxford, UK: Oxford University Press (1995)
6. J.E. Dennis and R.B. Schnabel: Numerical Methods for Unconstrained Optimization and Non-linear Equations. Englewood Cliffs, NJ: Prentice-Hall, (1983)
7. H. Drucker and Y. Le Cun: Improving Generalization Performance in Character Recognition. In B.H. Juang et al. (eds.), Neural Networks for Signal Processing: Proceedings of the 1991 IEEE-SP Workshop, Piscataway, New Jersey: IEEE (1991) 198–207
8. S. Geisser: The Predictive Sample Reuse Method with Applications. Journal of the American Statistical Association **50** (1975) 320–328
9. S. Geman, E. Bienenstock and R. Doursat: Neural Networks and the Bias/Variance Dilemma. Neural Computation 4 (1992) 1–58
10. F. Girosi, M. Jones, T. Poggio, Regularization Theory and Neural Networks Architectures, Neural Computation 7, 2, (1995) 219–269
11. C. Goutte and J. Larsen: Adaptive Regularization of Neural Networks using Conjugate Gradient, in Proceedings of ICASSP'98, Seattle USA **2** (1998) 1201-1204
12. C. Goutte: Note on Free Lunches and Cross-Validation. Neural Computation **9**(6) (1997) 1211–1215
13. C. Goutte: Regularization with a Pruning Prior. To appear in Neural Networks (1997)
14. L.K. Hansen and C.E. Rasmussen: Pruning from Adaptive Regularization. Neural Computation **6** (1994) 1223–1232
15. L.K. Hansen, C.E. Rasmussen, C. Svarer and J. Larsen: Adaptive Regularization. In J. Vlontzos, J.-N. Hwang and E. Wilson (eds.), Proceedings of the IEEE Workshop on Neural Networks for Signal Processing IV, Piscataway, New Jersey: IEEE (1994) 78–87

16. L.K. Hansen and J. Larsen: Linear Unlearning for Cross-Validation. Advances in Computational Mathematics **5** (1996) 269–280
17. J. Hertz, A. Krogh and R.G. Palmer: Introduction to the Theory of Neural Computation. Redwood City, California: Addison-Wesley Publishing Company (1991)
18. M. Hintz-Madsen, M. With Pedersen, L.K. Hansen, and J. Larsen: Design and Evaluation of Neural Classifiers. In S. Usui, Y. Tohkura, S. Katagiri and E. Wilson (eds.), Proceedings of the IEEE Workshop on Neural Networks for Signal Processing VI, Piscataway, New Jersey: IEEE, (1996) 223–232
19. K. Hornik: Approximation Capabilities of Multilayer Feedforward Networks. Neural Networks **4** (1991) 251–257
20. M. Kearns: A Bound on the Error of Cross Validation Using the Approximation and Estimation Rates, with Consequences for the Training-Test Split. Neural Computation **9**(5) (1997) 1143–1161
21. J. Larsen: A Generalization Error Estimate for Nonlinear Systems. In S.Y. Kung et al. (eds.), Neural Networks for Signal Processing 2: Proceedings of the 1992 IEEE-SP Workshop, Piscataway, New Jersey: IEEE (1992) 29–38
22. J. Larsen: Design of Neural Network Filters, Ph.D. Thesis, Electronics Institute, Technical University of Denmark (1993). Available via ftp://eivind.imm.dtu.dk/dist/PhD_thesis/jlarsen.thesis.ps.Z
23. J. Larsen and L.K. Hansen: Generalization Performance of Regularized Neural Network Models. In J. Vlontzos et al. (eds.), Proceedings of the IEEE Workshop on Neural Networks for Signal Processing IV, Piscataway, New Jersey: IEEE (1994) 42–51
24. J. Larsen and L.K. Hansen: Empirical Generalization Assessment of Neural Network Models. In F. Girosi et al. (eds.), Proceedings of the IEEE Workshop on Neural Networks for Signal Processing V, Piscataway, New Jersey: IEEE (1995) 30–39
25. J. Larsen, L.K. Hansen, C. Svarer and M. Ohlsson: Design and Regularization of Neural Networks: The Optimal Use of a Validation Set. In S. Usui, Y. Tohkura, S. Katagiri and E. Wilson (eds.), Proceedings of the IEEE Workshop on Neural Networks for Signal Processing VI, Piscataway, New Jersey: IEEE, (1996) 62–71
26. J. Larsen et al. : Optimal Data Set Split Ratio for Empirical Generalization Error Estimates. In preparation.
27. Y. Le Cun, J.S. Denker and S.A. Solla: Optimal Brain Damage. In D.S. Touretzky (ed.), Advances in Neural Information Processing Systems 2, Proceedings of the 1989 Conference, San Mateo, California: Morgan Kaufmann Publishers (1990) 598–605
28. D. Lowe: Adaptive Radial Basis Function Nonlinearities and the Problem of Generalisation. Proc. IEE Conf. on Artificial Neural Networks, (1989) 171–175
29. L. Ljung: System Identification: Theory for the User. Englewood Cliffs, New Jersey: Prentice-Hall (1987)
30. D.J.C. MacKay: A Practical Bayesian Framework for Backprop Networks. Neural Computation **4**(3) (1992) 448–472
31. J. Moody: Prediction Risk and Architecture Selection for Neural Networks. In V. Cherkassky et al. (eds.), From Statistics to Neural Networks: Theory and Pattern Recognition Applications, Berlin, Germany: Springer-Verlag Series F **136** (1994)
32. J. Moody, T. Rögnvaldsson: Smoothing Regularizers for Projective Basis Function Networks. In Advances in Neural Information Processing Systems 9, Proceedings of the 1996 Conference, Cambridge, Massachusetts: MIT Press (1997)

33. N. Murata, S. Yoshizawa and S. Amari: Network Information Criterion — Determining the Number of Hidden Units for an Artificial Neural Network Model. IEEE Transactions on Neural Networks 5(6) (1994) 865–872
34. S. Nowlan and G. Hinton: Simplifying Neural Networks by Soft Weight Sharing. Neural Computation 4(4) (1992) 473–493
35. M. With Pedersen: Training Recurrent Networks. In Proceedings of the IEEE Workshop on Neural Networks for Signal Processing VII, Piscataway, New Jersey: IEEE, (1997)
36. G.E. Peterson and H.L. Barney: Control Methods Used in a Study of the Vowels. JASA **24** (1952) 175–184
37. R.S. Shadafan and M. Niranjan: A Dynamic Neural Network Architecture by Sequential Partitioning of the Input Space. Neural Computation **6**(6) (1994) 1202–1222
38. J. Sjöberg: Non-Linear System Identification with Neural Networks, Ph.D. Thesis no. 381, Department of Electrical Engineering, Linköping University, Sweden, (1995)
39. M. Stone: Cross-validatory Choice and Assessment of Statistical Predictors. Journal of the Royal Statistical Society B **36**(2) (1974) 111–147
40. C. Svarer, L.K. Hansen, J. Larsen and C. E. Rasmussen: Designer Networks for Time Series Processing. In C.A. Kamm *et al.* (eds.), Proceedings of the IEEE Workshop on Neural Networks for Signal Processing 3, Piscataway, New Jersey: IEEE (1993) 78–87
41. R.L. Watrous: Current Status of PetersonBarney Vowel Formant Data. JASA **89** (1991) 2459–2460
42. A.S. Weigend, B.A. Huberman and D.E. Rumelhart: Predicting the Future: A Connectionist Approach. International Journal of Neural Systems **1**(3) (1990) 193–209
43. P.M. Williams: Bayesian Regularization and Pruning using a Laplace Prior. Neural Computation **7**(1) (1995) 117–143
44. D.H. Wolpert and W.G. Macready: The Mathematics of Search. Technical Report SFI-TR-95-02-010, Santa Fe Instute (1995)
45. L. Wu and J. Moody: A Smoothing Regularizer for Feedforward and Recurrent Neural Networks. Neural Computation **8**(3) 1996
46. H. Zhu and R. Rohwer. No Free Lunch for Cross Validation. Neural Computation **8**(7) (1996) 1421–1426

6
Large Ensemble Averaging

David Horn[1], Ury Naftaly[1], and Nathan Intrator[2]

[1] School of Physics and Astronomy
[2] School of Mathematical Sciences
Raymond and Beverly Sackler Faculty of Exact Sciences
Tel Aviv University, Tel Aviv 69978, Israel
horn@neuron.tau.ac.il   http://neuron.tau.ac.il/~horn/

Abstract. Averaging over many predictors leads to a reduction of the variance portion of the error. We present a method for evaluating the mean squared error of an infinite ensemble of predictors from finite (small size) ensemble information. We demonstrate it on ensembles of networks with different initial choices of synaptic weights. We find that the optimal stopping criterion for large ensembles occurs later in training time than for single networks. We test our method on the suspots data set and obtain excellent results.

6.1 Introduction

Ensemble averaging has been proposed in the literature as a means to improve the generalization properties of a neural network predictor[3, 11, 7]. We follow this line of thought and consider averaging over a set of networks that differ from one another just by the initial values of their synaptic weights.

We introduce a method to extract the performance of large ensembles from that of finite size ones. This is explained in the next section, and is demonstrated on the sunspots data set. Ensemble averaging over the initial conditions of the neural networks leads to a lower prediction error, which is obtained for a later training time than that expected from single networks. Our method outperforms the best published results for the sunspots problem [6].

The theoretical setting of the method is provided by the bias/variance decomposition. Within this framework, we define a particular bias/variance decomposition for networks differing by their initial conditions only. While the bias of the ensemble of networks with different initial conditions remains unchanged, the variance error decreases considerably.

6.2 Extrapolation to Large-Ensemble Averages

The training procedure of neural networks starts out with some choice of initial values of the connection weights. We consider ensembles of networks that differ from one another just by their initial values and average over them. Since the space of initial conditions is very large we develop a technique which allows us to approximate averaging over the whole space.

Our technique consists of constructing groups of a fixed number of networks, Q. All networks differ from one another by the random choice of their initial weights. For each group we define our predictor to be the average of the output of all Q networks. Choosing several different groups of the same size Q, and averaging over their predictions for the test set, defines the finite size average that is displayed in Fig. 1. Then we perform a parametric estimate of the limit $Q \to \infty$. A simple regression in $1/Q$ suffices to obtain this limit in the suspots problem, as shown in Fig. 2. In general one may encounter a more complicated inverse power behavior, indicating correlations between networks with different initial weights.

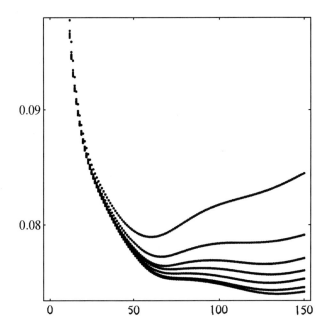

Fig. 6.1. *ARV of test set*

Prediction error (ARV) is plotted *vs.* training time in kilo epochs (KE). The curves correspond to different choices of group sizes: $Q = 1, 2, 4, 5, 10, 20$ from top to bottom. The lowest curve is the extrapolation to $Q \to \infty$.

6.2.1 Application to the Sunspots Problem

Yearly sunspot statistics have been gathered since 1700. These data have been extensively studied and have served as a benchmark in the statistical literature [9, 10, 4]. Following previous publications [10, 6, 8] we choose the training set to

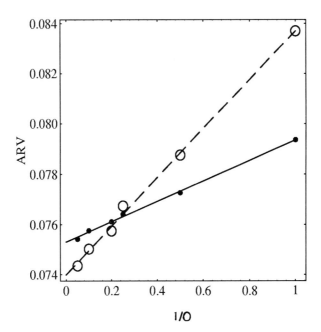

Fig. 6.2. *Extrapolation method used for extracting the $Q \to \infty$ prediction.*
The results for different ensemble size Q at two different training periods, $t = 70KE$ (dots) and 140KE (circles) lie on straight lines as a function of $1/Q$. For each curve, the first three points from the right represent ensemble sizes of 1, 2, and 4 respectively. While the three points of 140KE all lie above the corresponding ones of 70KE, an extrapolation to larger ensemble sizes suggests that the overall performance will be better for 140KE as is observed from the fit to the line.

contain the period between 1701 and 1920, and the test-set to contain the years 1921 to 1955. Following [10], we calculate the prediction error according to the average relative variance (ARV) of the data set S:

$$\text{ARV} = \frac{\sum_{k \in S} (y_k - f(\boldsymbol{x}_k))^2}{\sum_{k \in S} (y_k - E[y_k])^2} \tag{6.1}$$

$y_k(\boldsymbol{x}_k)$ are the data values and $f(\boldsymbol{x}_k)$ is the predictor. In our time series problem, for any given time point $t = k$, the input vector $\boldsymbol{x}_k$ has component values taken from the series at times $t - 1, t - 2, \cdots, t - 12$ (as in [10]). The denominator in Eq. 1 is $\sigma^2 = 1535$ for the training set. The same value is used for the test set. We use neural networks with 12 inputs, one sigmoidal hidden layer consisting of 4 units and a linear output. They are then enlarged to form recurrent networks (SRN) [1] in which the input layer is increased by adding to it the hidden layer of the previous point in the time series. The learning algorithm consists of back propagation applied to an error function which is the MSE of the training set. A validation set containing 35 randomly chosen points was left out during training to serve for performance validation.

Fig. 6.1 displays our results on the test set as a function of the number of training epochs. We observe a descending order of $Q = 1, 2, 4, 5, 10, 20$ followed by the extrapolation to $Q \to \infty$. All of these curves correspond to averages over groups of size Q, calculated by running 60 networks. To demonstrate how the extrapolation is carried out we display in Fig. 6.2 the points obtained for $t = 70$ and $t = 140$ KE as a function of $\frac{1}{Q}$. It is quite clear that a linear extrapolation is very satisfactory. Moreover, the results for $Q = 20$ are not far from the extrapolated $Q \to \infty$ results. Note that the minimum of the $Q \to \infty$ curve in Fig.1 occurs at a much higher training time than that of the $Q = 1$ single network curve. This is also evident from the crossing of the $t = 70$ and $t = 140$ KE lines on Fig. 2. An important conclusion is that the stopping criterion for ensemble training (to be applied, of course, to every network in the group) is very different from that of single network training.

6.2.2 Best Result

The curves shown in Fig. 1 were obtained with a learning rate of 0.003. Lower learning rates lead to lower errors. In that case the effect of ensemble averaging is not as dramatic. We obtained our best result by changing our input vector into the six dimensional choice of Pi & Peterson [8] that consists of $x_{t-1}, x_{t-2}, x_{t-3}, x_{t-4}, x_{t-9}$ and x_{t-10}. Using a learning rate of 0.0005 on the otherwise unchanged SRN described above, we obtain the minimum of the prediction error at 0.0674, which is better than any previously reported result.

6.3 Theoretical Analysis

The theoretical setting of the method is provided by the bias/variance decomposition. Within this framework, we define a particular bias/variance decomposition for networks differing by their initial conditions only. This is a particularly useful subset of the general set of all sources of variance.

The performance of an estimator is commonly evaluated by the Mean Square Error (MSE) defined as

$$\mathrm{MSE}(f) \equiv E\left[(y - f(x))^2\right] \tag{6.2}$$

where the average is over test sets for the predictor f, and y are the target values of the data in x. Assuming the expectation E is taken with respect to the true probability of x and y, the MSE can be decomposed into

$$E\left[(y - f(x))^2\right] = E\left[(y - E[y|x])^2\right] + E\left[(f(x) - E[y|x])^2\right]. \tag{6.3}$$

The first RHS term represents the variability or the noise in the data and is independent of the estimator f. It suffices therefore to concentrate on the second term. Any given predictor $f(x)$ is naturally limited by the set of data on which

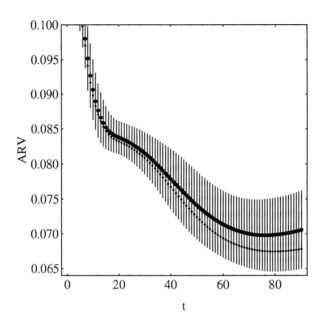

Fig. 6.3. Our best results for the test set of the sunspots problem. Plotted here are $Q = 1$ results for various choices of initial conditions, represented by their averages with error-bars extending over a standard deviation, and $Q = 20$ results (the thinner points), as a function of training time in K-epochs. The network is based on the Pi & Peterson variables, and the learning rate is 0.0005.

it is trained. Considering a typical error one may average over all data space [2] and decompose this error into Bias and Variance components:

$$E_\mathcal{D}\left[(f(\mathbf{x}) - E[y|\mathbf{x}])^2\right] = B_\mathcal{D} + V_\mathcal{D} \tag{6.4}$$

where

$$B_\mathcal{D}(f(\mathbf{x})) = (E_\mathcal{D}[f(\mathbf{x})] - E[y|\mathbf{x}])^2 \tag{6.5}$$

$$V_\mathcal{D}(f(\mathbf{x})) = E_\mathcal{D}\left[(f(\mathbf{x}) - E_\mathcal{D}[f(\mathbf{x})])^2\right]. \tag{6.6}$$

In our application we use as our predictor $E_\mathcal{I}[f(\mathbf{x})]$ where the subscript $\mathcal{I}$ denotes the space of initial weights of the neural network that serves as $f(\mathbf{x})$. To understand the effect of averaging over initial weights let us construct a space $\mathcal{R}$ by the direct product of $\mathcal{D}$ and $\mathcal{I}$. It can then be shown that

$$B_\mathcal{R}(f(\mathbf{x})) = B_\mathcal{D}(E_\mathcal{I}[f(\mathbf{x})]), \tag{6.7}$$

and

$$V_\mathcal{R}(f(\mathbf{x})) \geq V_\mathcal{D}(E_\mathcal{I}[f(\mathbf{x})]). \tag{6.8}$$

This means that using $E_\mathcal{I}[f(x)]$ as the predictor, the characteristic error has reduced variance but unchanged bias.

The bias term may also be represented as $B_\mathcal{R} = E_\mathcal{D}[B_\mathcal{I}]$ where

$$B_\mathcal{I}(f(x)) = (E_\mathcal{I}[f(x)] - E[y|x])^2. \tag{6.9}$$

$B_\mathcal{I}$ is unaffected when $f(x)$ is replaced by its average. The analogously defined variance term, $V_\mathcal{I}$, gets eliminated by such averaging. In other words, by averaging over all $\mathcal{I}$ we eliminated all variance due to the choice of initial weights. The difference between the $Q = 1$ and $Q \to \infty$ curves in Fig. 1 represents this reduction of variance.

To understand the Q dependence of Fig. 2 consider an average defined by $\bar{f}(x)$ over functions that represent independent identically distributed random variables. It can then be shown that

$$B(\bar{f}) = \overline{B(f)} \qquad V(\bar{f}) = \overline{V(f)}/Q. \tag{6.10}$$

Hence we can interpret the $1/Q$ behavior displayed in Fig. 2 as a demonstration that the choice of initial conditions in that analysis acts as effective random noise. In general this is not necessarily the case, since networks with different initial conditions may have non-trivial correlations. For a more thorough discussion of this and other points see [5].

In conclusion, we see that averaging over networks with different initial weights is helpful in reducing the prediction error by eliminating the variance induced by initial conditions. Performing this average over groups of finite size Q one finds out from the Q dependence if the errors induced by initial conditions are correlated or not. Moreover, one may estimate the Q needed to eliminate this source of variance.

References

1. J.L. Elman and D. Zipser. *Learning the Hidden Structure of Speech*. J. Acoust. Soc. Amer. 83, 1615–1626. 1988.
2. S. Geman, E. Bienenstock and R. Doursat. *Neural networks and the bias/variance dilemma*. Neural Comp., 4(1):1–58. 1992.
3. W.P. Lincoln and J. Skrzypek. *Synergy of clustering multiple back propagation networks*. In Touretzky, D. S., editors, *Advances in Neural Information Processing Systems 2*, pages 650–657, SanMateo, CA. Morgan Kaufmann. 1990.
4. J. Morris. *Forecasting the sunspot cycle*. J. Roy. Stat. Soc. Ser. A, 140, 437–447. 1977.
5. U. Naftaly, N. Intrator and D. Horn. *Optimal Ensemble Averaging of Neural Networks*. Network, Comp. Neural Sys., 8, 283–296. 1997.
6. S.J. Nowlan and G.E. Hinton. *Simplifying neural networks by soft weight-sharing*. Neural Computation. 4, 473–493. 1992.
7. P.M. Perrone. *Improving regression estimation: averaging methods for variance reduction with extensions to general convex measure optimization*. PhD thesis, Brown University, Institute for Brain and Neural Systems, 1993.

8. H. Pi and C. Peterson. *Finding the Embedding Dimension and Variable Dependencies in Time Series.* Neural Comp. 6, 509–520. 1994.
9. M.B. Priestley. *Spectral Analysis and Time Series.* Academic Press. 1981.
10. A.S. Weigend, B.A. Huberman and D. Rumelhart. *Predicting the future: A connectionist approach.* Int. J. Neural Syst. 1, 193–209. 1990.
11. D. H. Wolpert. *Stacked generalization.* Neural Networks 5:241-259. 1992

Improving Network Models and Algorithmic Tricks

Preface

This section contains 5 chapters presenting easy to implement tricks which modify either the architecture and/or the learning algorithm so as to enhance the network's modeling ability. Better modeling means better solutions in less time.

In chapter 7, Gary Flake presents a trick that gives an MLP the additional power of an RBF. Trivial to implement, one simply adds extra inputs whose values are the **square of the original inputs** (p. 146). While adding higher order terms as inputs is not a new idea, this chapter contributes new insight by providing (1) a good summary of previous work, (2) simple clear examples illustrating this trick, (3) a theoretical justification showing that one need only add the higher order squared terms, and (4) a thorough comparison with numerous other network models. The need for *only* the squared terms is significant because it means that we gain this extra power without having the number of inputs grow excessively large. We remark that this idea can be extended further by including relevant features other than squared inputs e.g. by using kernel PCA [2] to obtain the non-linear features.

Rich Caruana in chapter 8 presents **multi-task learning (MTL)** (p. 165) where extra outputs are added to a network to predict tasks separate but related to the primary task. To introduce the trick, the chapter begins with an example and detailed discussion of a simple boolean function of binary inputs. The author then presents several of what one might use as extra outputs in practice. These include, among others: (1) features that are available only **after predictions** must be made, but which are available offline at *training* time (p. 173), (2) the same task but with a **different metric** (p. 178), and (3) the same task but with **different output representations** (p. 178). Empirical results are presented for (1) mortality rankings for pneumonia where the extra outputs are test results not available when the patient first enters the hospital but which are available a posteriori to complement the training data, and (2) a vehicle steering task where other outputs include location of centerline and road edges, etc. The last part of the chapter is devoted to topics for implementing MTL effectively, such as size of hidden layers (p. 183), early stopping (p. 184), and learning rates (p. 187).

The next chapter by Patrick van der Smagt and Gerd Hirzinger presents a trick to reduce the problem of ill-conditioning in the Hessian. If a unit has a very small outgoing weight then the influence of the incoming weights to that unit will be severely diminished (see chapter 1 for other sources of Hessian ill-conditioning). This results in flat spots in the error surface, which translates into slow training (see also [1]). The trick is to add linear **shortcut connections** (p. 198) from the input to the output nodes to create what the authors refer to as a **linearly augmented feed-forward network**. These connections **share** the weights with the input to hidden connections so that no new weight parameters are added. This trick enhances the sensitivity of the network to those incoming weights thus removing or reducing the flat spots in the error surface. The improvement in the quality of the error surface is illustrated in a toy example. Simulations with data from a robot arm are also shown.

The trick discussed by Nicol Schraudolph in chapter 10 is to center the various factors comprising the neural network's gradient (p. 210): input and hidden unit activities (see chapter 1), error signals, and the slope of the hidden units' nonlinear activation functions (p. 210). To give an example: *activity centering* (p. 209) is done by simply transforming the values of the components x_i into

$$\check{x}_i = x_i - \langle x_i \rangle,$$

where $\langle \cdot \rangle$ denotes averaging over training samples. All different centering strategies can be implemented efficiently for a stochastic, batch or mini batch learning scenario (p. 211). He also uses **shortcut connections** (p. 210) but quite differently from the previous chapter: the shortcut connections contain new weights (not shared) which complement slope centering by carrying the linear component of the signal, making it possible for the rest of the network to concentrate on the nonlinear component of the problem. So, with respect to shortcut connections, the approaches in chapters 9 and 10 appear complementary. Centering gives a nice speed-up without much harm to the generalization error, as seen in the simulations on toy and vowel data (p. 213).

In chapter 11, Tony Plate presents a trick requiring only minimal memory overhead that reduces **numerical round-off** in backpropagation networks. Round-off error can occur in the standard method for computing the derivative of the logistic function since it requires calculating the product

$$y(1-y)$$

where y is the output of either a hidden or output unit. When the value of y is close to 1 then the limited precision of single or even double precision floating point numbers can result in the product being zero. This may not be a serious problem for on-line learning but can cause significant problems for networks using batch mode, particularly when second order methods are used. Such round-off can occur in other types of units as well. This chapter provides formulas for reducing such round-off errors in the computation of

- derivatives of the error for logistic units or tanh units (p. 228 and 231)
- derivatives in a one-of-k classification problem with cross-entropy error and softmax (p. 231)
- derivatives and errors in a two-class classification problem using a single logistic output unit with cross entropy error and 0/1 targets (p. 230).

Jenny & Klaus

References

1. S. Hochreiter and J. Schmidhuber. Flat minima. *Neural Computation*, 9(1):1–42, 1997.

2. B. Schölkopf, A. Smola, and K.-R. Müller. Nonlinear component analysis as a kernel eigenvalue problem. *Neural Computation*, 10:1299 – 1319, 1998.

7
Square Unit Augmented, Radially Extended, Multilayer Perceptrons

Gary William Flake

Siemens Corporate Research, Inc.
755 College Road East
Princeton, NJ 08540
flake@scr.siemens.com
http://mitpress.mit.edu/books/FLAOH/cbnhtml/author.html

Abstract. Consider a multilayer perceptron (MLP) with d inputs, a single hidden sigmoidal layer and a linear output. By adding an additional d inputs to the network with values set to the square of the first d inputs, properties reminiscent of higher-order neural networks and radial basis function networks (RBFN) are added to the architecture with little added expense in terms of weight requirements. Of particular interest, this architecture has the ability to form localized features in a d-dimensional space with a single hidden node but can also span large volumes of the input space; thus, the architecture has the localized properties of an RBFN but does not suffer as badly from the curse of dimensionality. I refer to a network of this type as a SQuare Unit Augmented, Radially Extended, MultiLayer Perceptron (SQUARE-MLP or SMLP).

7.1 Introduction and Motivation

When faced with a new and challenging problem, the most crucial decision that a neural network researcher must make is in the choice of the model class to pursue. Several different types of neural architectures are commonly found in most model building tool-boxes, with two of the more familiar, radial basis function networks (RBFNs) [15, 16, 13, 1, 17] and multilayer perceptrons (MLPs) [25, 20], exemplifying the differences found between global and local model types. Specifically, an MLP is an example of a global model that builds approximations with features that alter the entire input-output response, while an RBFN is a local model that uses features confined to a finite region in the input-space. This single difference between the two model types has major implications for many architectural and algorithmic issues. While MLPs are slow learners, have low memory retention, typically use homogeneous learning rules and are relatively less troubled by the curse of dimensionality, RBFNs are nearly opposite in every way: they are fast learners, have high memory retention, typically use heterogeneous learning rules and are greatly troubled by the curse of dimensionality.

Because of these differences, it is often tempting to use one or more heuristics to make the choice, e.g., RBFNs (MLPs) for low (high) dimensional problems, or RBFNs (MLPs) for continuous function approximation (pattern classification) problems. While these rules-of-thumb are often sufficient for simple problems there are many exceptions that defy the rules (e.g., see [12]). Moreover, more challenging problems from industrial settings often have high-dimensional input-spaces that are locally well-behaved and may not be clearly defined as either function approximation or pattern classification problems. This means that choosing the best architectures for a particular problem can be a nontrivial problem in itself.

Ironically, a good compromise to this dilemma has been known for quite some time but only recently has the elegance of the trick been appreciated. For years, researchers have commonly used augmented MLPs by adding the squares of the inputs as auxiliary inputs. The justification for this has always been fairly casual and has usually boiled down to the argument that using this trick couldn't possibly hurt. But as it turns out, an MLP augmented in this way with n hidden nodes can almost perfectly approximate an RBFN with n basis functions. The "almost" comes from the fact that the radial basis function in the augmented MLP is not Gaussian but quasi-Gaussian (which is an admittedly undefined term that I simply use to mean "so close to Gaussian that it really doesn't matter."). This means that an MLP augmented with the squares of it's inputs can easily form local features with a single hidden node but can also span vast regions of the input-space, thereby effectively ignoring inputs when needed. Thus, the best of both architectural approaches is retained by using this amazingly simple trick.

The remainder of this chapter is divided into 5 more sections. Section 7.2 contains a description of the trick and briefly gives a comparison of the proposed architecture to other classes of well-known models. In Section 7.3, a function approximation problem and a pattern classification problem are used as examples to demonstrate the effectiveness of the trick. Afterwards, a well-known and challenging vowel classification problem is studied in greater detail. Section 7.4 theoretically justifies the trick by showing the equivalence of the resulting architecture and RBFNs, while Section 7.5 gives a more intuitive justification for the proposed trick by illustrating the types of surfaces and boundaries that can be formed by a single node with auxiliary square inputs. Finally, in Section 7.6, I give my conclusions.

7.2 The Trick: A SQUARE-MLP

The proposed trick involves only a simple modification to the standard MLP architecture: the input layer of an MLP is augmented with an extra set of inputs that are coupled to the squares of the original inputs. This trick can be implemented in at least two different ways. The first technique is to simply augment a data set with the extra inputs. Thus, if one had a set with each input pattern having d components, then a new data set can be made from the original that has $2d$ inputs with the extra d inputs set equal to the squares of the original inputs. Implementing the trick in this way is expensive from a memory point of view but has the advantage that not a single line of new source code need be written to try it out. Moreover, this allows the trick to be tried even on a commercial simulator where one may not have access to source code.

The second way to implement the trick is to explicitly code the actual changes into the architecture:

$$y = \sum_i w_i g \left(\sum_j u_{ij} x_j + \sum_k v_{ik} x_k^2 + a_i \right) + b, \quad (7.1)$$

with $g(x) = \tanh(x)$ or $1/(1+\exp(-x))$. I call such a network a SQuare Unit Augmented, Radially Extended, MultiLayer Perceptron (SQUARE-MLP or SMLP). The "square unit augmented" portion of the name comes from the newly added $v_{ik} x_k^2$ terms. The reason behind the "radially extended" portion of the name will become clear in Sections 7.4 and 7.5. All experiments in Section 7.3 use the architecture described by Equation 7.1. The history of this trick is rather difficult to trace primarily because it is such a trivial trick; however, a brief list of some related ideas is presented below.

Engineering and Statistics Very early related ideas have been pursued in the statistics community in the form of polynomial regression and Volterra filters in the engineering community [24, 22]. However, in both of these related approaches the model output is always linear in the polynomial terms, which is not the case with the SMLP architecture or in the other neural architectures discussed below.

Sigma-Pi Networks Some neural network architectures which are much more complicated than Equation 7.1 have the SMLP as a special case. Perhaps the earliest reference to a similar idea in the neural network literature can be traced back to Sigma-Pi networks [20], which extends an MLP's linear net input function with a summation of products, $\sum_i w_{ji} \prod_k x_{ik}$. One could imagine a multi-layer Sigma-Pi network that manages to compute the squares of the x_{ik} terms prior to them being passed through to a sigmoidal activation function. This would be a rather clumsy way of calculating the squares of the inputs, but it is possible to do it, nonetheless.

Higher-Order Networks Perhaps the closest example is the higher-order network, proposed by Lee *et al.* [14], which is similar to Equation 7.1 but uses the full quadratic net input function:

$$y = \sum_i w_i g \left(\sum_j u_{ij} x_j + \sum_k \sum_l v_{ikl} x_k x_l + a_i \right) + b. \quad (7.2)$$

With v_{ikl} set to zero when $k \neq l$ a SMLP is recovered. Thus, a SMLP is actually a higher-order network with a diagonal quadratic term. Higher-order networks have been shown to be very powerful extensions of MLPs. They can form both local and global features but only at the cost of squaring the number of weights for each hidden node. The memory requirements become an even greater issue when more sophisticated optimization routines are applied to an architecture such as Newton's or quasi-Newton methods which require memory proportional to the square of the number of weights in the network.

Functional Link Networks Another related architecture is the functional-link network [18], which is similar to a standard MLP but explicitly augments the network with the results of scalar functions applied to the inputs. For example, in some applications it may be known in advance that the desired output of the network is a function of the sine and cosine of one or more inputs (e.g., the inputs may correspond to angles of a robot arm). In this case, one would do well to include these values explicitly as inputs into the network instead of forcing the network to learn a potentially difficult-to-model concept. Functional-link networks may use any scalar function that, in the end, essentially performs a type of preprocessing on the data. Usually, expert knowledge is used to determine which extra scalar functions are to be incorporated into the network; that is, there is no general technique for choosing the best preprocessor functions *a priori*. However, given a set of nonlinear transformations on the input data one can perform principal component analysis (PCA) on the nonlinear feature space to select a subset that carries the most variance. A computationally feasible version of this technique has been proposed in [23], which they refer to as kernel PCA. In any event, by using the square function to augment a functional-link network, the SMLP is once again recovered.

While I have tried to assign proper credit, it is generally accepted that the basic idea of adding the squares of the inputs to a model is at least as old as the sage advice "preprocessing is everything."

7.3 Example Applications

The following three examples demonstrate problem domains in which an SMLP can conceivably outperform an MLP or an RBFN. All of the examples are well-known benchmarks. In each case, the output response of the models must form local features while simultaneously spanning a large region of the input-space. In general, the MLPs will have difficulty forming the local features, while the RBFNs will have trouble spanning the flat regions of the input space.

7.3.1 Hill-Plateau Function Approximation

The first problem is an admittedly contrived example that was chosen precisely because it is difficult for both MLPs and RBFNs. The "Hill-Plateau" surface [21], displayed in Figure 7.1, has a single local bump on a sigmoidal ridge. Training data for this problem consists of a two-dimensional uniform sampling on a 21×21 grid of the region shown in the figure while the testing data comes from a finer 41×41 grid.

7. Square Unit Augmented, Radially Extended, Multilayer Perceptrons

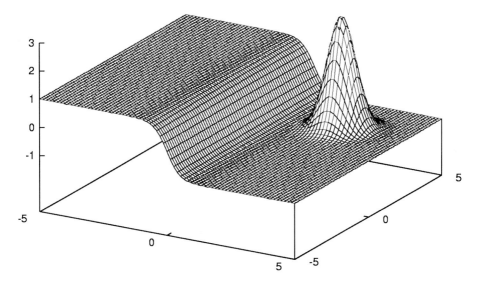

Fig. 7.1. A hill-plateau exemplifies the differences between local and global architectures. MLPs can easily form the plateau but have a hard time on the hill, while RBFNs trivially form the hill but are troubled by the plateau.

Model	# of Nodes	# of Weights	Test RMSE
RBFN	2	9	0.333406
	3	13	0.071413
	4	17	0.042067
	5	21	0.002409
MLP	2	9	0.304800
	3	13	0.015820
	4	17	0.001201
SMLP	2	13	0.000025

Table 7.1. Best of twenty runs for the Hill-Plateau surface

Besides a standard MLP, a normalized RBFN (NRBFN) was used for this problem, which is described by the two equations:

$$y = \frac{\sum_i w_i r_i(\mathbf{x})}{\sum_j r_j(\mathbf{x})} + a \quad \text{and} \tag{7.3}$$

$$r_i(\mathbf{x}) = \exp(-||\mathbf{x} - \mathbf{c}_i||^2/\sigma_i^2), \tag{7.4}$$

with $\mathbf{c}_i$ and σ_i being the ith basis center and width, respectively. To train the NRBFNs the centers were first clustered in the input-space of the training patterns with the k-means clustering algorithm. The width of each basis function was then set proportional to the distance to the nearest neighbor. Afterwards, the least-mean-square solution of the linear terms, w_i and a, were solved for

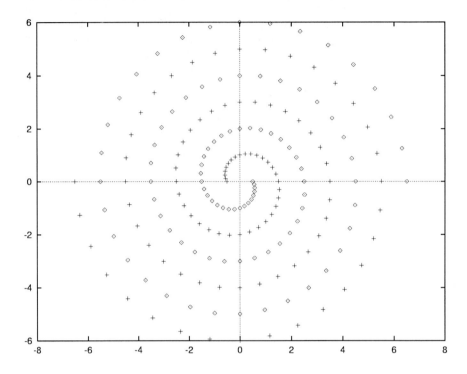

Fig. 7.2. Data for the Two-Spiral problem

exactly using a singular value decomposition to compute the pseudo-inverse. All of this formed the initial set of weights for the quasi-Newton's method (BFGS) optimization routine which was used on all weights simultaneously for up to 200 epochs.

For the MLPs, a single hidden layer with the tanh(x) activation function and a linear output was used. All weights were initially set to uniform random values in a -0.1 to 0.1 range. All weights were then trained with quasi-Newton method (BFGS) for up to 200 epochs. The SMLPs were setup and trained exactly as the MLPs.

Table 7.1 shows the results for all three architectures. As can be seen, both the RBFNs and the MLPs have a fair amount of difficulty with this task, even though the training data is noise free and the training procedures are fairly sophisticated. Contrary to this, the SMLP manages to nail the surface with only two hidden nodes. Moreover, the testing error is orders of magnitude better than the best results from the other two architectures.

7.3.2 Two-Spirals Classification

The two-spirals classification problem is a well-known benchmark that is extremely challenging for all neural network architectures; additionally, virtually no results have been reported for RBFNs as the problem is such that local models would have to memorize the training data with many basis functions (> 100) in order to come even close to solving it. Thus, this is an example of a problem that RBFNs are not even viable candidates, which is why they are not considered further. Figure 7.2 shows the data for the two-spirals problem, which consists of 194 points on the x-y-plane that belong to one of two spirals, each of which rotates around the origin three times.

Previous results by other researchers for this problem have mostly focused on traditional MLPs and MLPs with shortcut connections. The best reported results for 100% classification accuracy are summarized below:

- (Lang & Witbrock [9]): 2-5-5-5-1 MLP with shortcut connections and 138 total weights. Average convergence time of 20,000 batched backpropagation epochs.
- (Lang & Witbrock [9]): 2-5-5-5-1 MLP with shortcut connections, 138 total weights, and cross-entropy error function. Average convergence time of 12,000 batched backpropagation epochs.
- (Lang & Witbrock [9]): 2-5-5-5-1 MLP with shortcut connections and 138 total weights. Average of 7,900 quickprop [3] epochs.
- (Frostrom [unpublished]): 2-20-10-1 MLP with no shortcut connections and 281 weights. Required 13,900 batched backpropagation with momentum epochs.
- (Fahlman and Lebiere [4]): Cascade-Correlation MLP using 12 to 19 hidden units (15 average) and an average of 1700 quickprop epochs. Because of the cascade correlation topology, these networks used between 117 and 250 weights.

As these results show, the two-spiral problem is exceptionally difficult, requiring both complicated network topologies and long training times.

Compared to the results above, the SMLP architecture seems to be very well-suited to this problem. An SMLP with 15 hidden hyperbolic tangent units (for a total of only 91 weights) was trained with a conjugate gradient optimization routine. In ten out of ten trials, the SMLP solved the two-spirals problem with an average of 2500 training epochs (but as few as 800). Notice that the architecture for the SMLP, both topologically and in the number of weights, is much simpler than those used in the studies with the MLPs. As a result, the optimization algorithms can be much more efficient. This is a case of the representation power of an architecture simplifying the learning, thereby making weight optimization a faster process.

Although it was not always possible to consistently train a simpler SMLP to solve this problem, an SMLP with 10 hidden nodes (and only 61 weights) succeeded on three separate trials, taking an average of 1500 epochs. The output response surface of this SMLP is shown in Figure 7.3. In Section 7.5 we will examine the different types of surfaces that can be formed by a single SMLP hidden node. We shall see that the the SMLP's ability to easily form local and global features is crucial to its ability to rapidly solve the two-spiral problem.

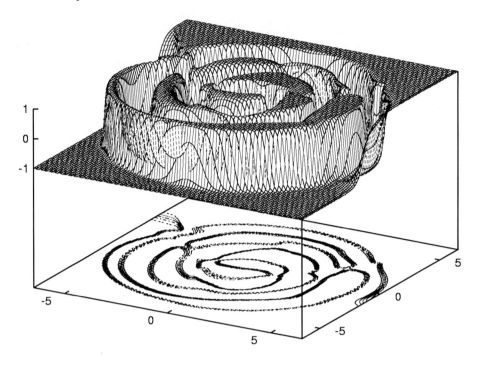

Fig. 7.3. SMLP reconstructed Two-Spiral surface from only ten hidden nodes

7.3.3 Vowel Classification

The Deterding vowel recognition data set [2] is another widely studied benchmark that is much more difficult than the two earlier problems and is also more indicative of the type of problem that a neural network practitioner could be faced with. The data consists of auditory features of steady state vowels spoken by British English speakers. There are 528 training patterns and 462 test patterns with each pattern consisting of 10 features and belonging to exactly one of 11 classes that correspond to the spoken vowel. The speakers are of both genders, making this a very interesting problem.

All results for this section use an architecture with 10 inputs, a varying number of hidden units, and 11 outputs. Some results from previous studies are summarized in Table 7.2. Some of the earlier studies are somewhat anecdotal in that they used either a single experiment or only a few experiments but they are informative as they demonstrate what the sophisticated neural network practitioner could expect to achieve on this problem with a wide number of architectures. Interestingly, Robinson's results show that a nearest neighbor classifier is very difficult to beat for this problem. With a 56% correct classification rate, a nearest neighbor classifier outperforms all of Robinson's neural solutions. However, nearest neighbor approaches require vast amounts of data to be stored as a look-up table, so this is not a particularly encouraging result.

The best score, reported by Hastie and Tibshirani [7], was achieved with a Discriminant Adaptive Nearest Neighbor (DANN) classifier. The score of 61.7% was from the best classifier found in a number of simulation studies; hence, this score represents the best known prior result as found by an expert attempting multiple solutions.

Table 7.3 lists the results of the experiments done specifically for this work. Five different model/optimization combinations are shown in the table, and each row in the table corresponds to fifty separate trials started with different random initial conditions. For the major entries labeled as "trained" the weights of the model were determined by conjugate gradient for models with more than 1,000 weights and quasi-Newton's method for models with fewer than 1,000 weights. The optimization routines were set to minimize an error function of the form $E = e^2 + \lambda ||\mathbf{w}||^2$ where e is the difference between the actual and desired outputs and λ is a weight decay term that penalizes large weights. λ was set to 10^{-4} for all experiments. Values of λ equal to 10^{-3} and 10^{-5} consistently gave worse results for all architectures, so 10^{-4} was a fair compromise.[1] All MLP and SMLP architectures used a tanh(x) activation function, while the RBFN is the same as described in Section 7.3.1 but is unnormalized (The normalized RBFN gave consistently inferior results). The optimization routines were always allowed to run until convergence (change in error measure is less 10^{-20}) unless otherwise noted.

The weights in the RBFN and NRBFN models were "solved" with a three step process: 1) set the centers to the cluster centers generated by the k-means clustering algorithm applied to the input vectors of the training data; 2) set the widths proportional to the distance to the nearest neighbor of each center; and 3) solve the remaining linear weights as a least mean square problem with a matrix pseudo-inverse.

The SMLP architecture can be solved for in a manner similar to how the RBFN and NRBFN networks are solved. The details of this procedure are covered in Section 7.4, but it suffices to say at this point that the procedure is nearly identical with the exception that the weights corresponding to the centers and the widths must be slightly transformed

Interestingly, the SMLP can use this "solve" procedure to compute an initial set of weights for an SMLP that is then trained with a gradient-based method. This has the effect of predisposing the SMLP to a very good initial solution that can be refined by the gradient-based optimization routines. The row in Table 7.3 with the parenthetical label "hybrid" corresponds to SMLPs trained in this way.

Three of the columns in Table 7.3 show three different ways of measuring success for the various models, of which the only statistically significant measure is the column labeled "% Correct (Average)," which is the average test set score achieved after the optimization procedure halted on the training data. The scores reported under the column heading "% Correct (Best)" correspond to the best final test score achieved from the 50 runs, while the "% Correct (Cheating)" is

---

[1] Note that since the SMLP has an extra set of weights, care must be taken to control the capacity and avoid over-fitting the data.

Model	Number of Hidden	Number of Weights	Percent Correct
Single-Layer Perceptron	—	11	33
Multilayer Perceptron [19]	11	253	44
	22	495	45
	88	1,947	51
Multilayer Perceptron [12]	5	121	50.1*
(with renormalization)	10	231	57.5*
	20	451	50.6*
Stochastic Network [5]	8	297	54*
(FF-R classifier)	16	473	56*
	32	825	57.9*
Radial Basis Function	88	1,936	48
	528	11,616	53
Gaussian Node Network	11	253	47
	22	495	54
	88	1,947	53
	528	11,627	55
Square Node Network	11	253	50
(not an SMLP)	22	495	51
	88	1,947	55
Modified Kanerva Model	88	968	43
	528	5808	50
Local Approximation	2	5808	50.0
	3	5808	52.8
	5	5808	53.0
	10	5808	48.3
	20	5808	45.0
Nearest Neighbor	—	(5,808)	56
Linear Discriminant Analysis	—	715	44
Softmax	—	-?-	33
Quadratic Discriminant Analysis	—	-?-	47
CART	—	-?-	44
CART (linear comb. splits)	—	-?-	46
FDA / BRUTO	—	-?-	56
Softmax / BRUTO	—	-?-	50
FDA / MARS (degree 1)	—	-?-	55
FDA / MARS (degree 2)	—	-?-	58
Softmax / MARS (degree 1)	—	-?-	52
Softmax / MARS (degree 2)	—	-?-	50
LOCOCODE / Backprop	11	473	58*
(30 inputs)			
DANN	—	-?-	61.7

Table 7.2. Previous result on the vowel data as summarized in [19], [12], [5] [8], [6], and [7]. All entries are either deterministic techniques or are the best scores reported, unless the score appears with a "*," in which case the score represents an average over multiple runs.

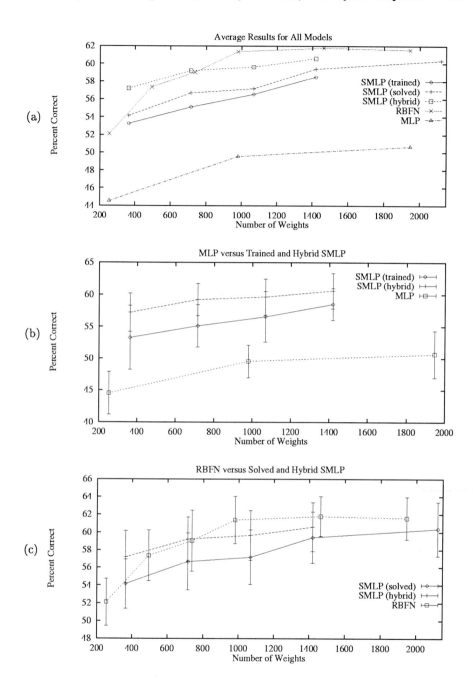

Fig. 7.4. The results from Table 7.3 shown in a graphical format: (a) the average results shown for all models types; (b) the average of the SMLP and MLP models with error bars shown; (c) the average of the SMLP and RBFN models with error bars shown.

Model	# of Hidden	# of Weights	% Correct (Cheating)	% Correct (Best)	% Correct) (Average)	Standard Deviation
MLP (trained)	11	253	53.24	51.08	44.53	3.34
	44	979	58.00	54.76	49.54	2.55
	88	1947	57.57	57.14	50.67	3.69
SMLP (trained)	11	363	63.63	60.82	53.25	5.02
	22	715	64.93	63.63	55.11	3.31
	33	1067	65.15	65.15	56.54	3.92
	44	1419	66.66	65.15	58.50	2.51
RBFN (solved)	11	253	———	56.92	52.11	2.64
	22	495	———	63.20	57.36	2.89
	33	737	———	66.88	59.03	3.45
	44	979	———	67.53	61.38	2.66
	66	1463	———	65.80	61.79	2.23
	88	1947	———	67.09	61.58	2.38
SMLP (solved)	11	363	———	58.87	54.14	2.79
	22	715	———	63.63	56.68	3.23
	33	1067	———	63.85	57.17	3.09
	44	1419	———	67.31	59.41	2.89
	66	2123	———	68.39	60.36	3.04
	88	2827	———	67.09	60.30	2.57
SMLP (hybrid)	11	363	66.66	63.85	57.19	2.98
	22	715	66.88	63.41	59.22	2.50
	33	1067	66.45	64.71	59.64	2.81
	44	1419	68.18	66.88	60.59	2.74

Table 7.3. Results from this study: All averages are computed from 50 trials with any result less than 33% (the score of a perceptron) being discarded as non-convergent. See the text for an explanation of the terms "Cheating," "Best," "and "Average."

the best test score achieved at any time during the training by any of the models in the fifty runs. Since the "solved" models have their weights computed in a single step, the "cheating" score only has meaning for the iterative techniques. One way of interpreting the "cheating" score is that it is the best score that could be achieved if one had a perfect cross validation set to use for the purpose of early stopping.

Some of the information in Table 7.3 is graphically summarized in Figure 7.3.3 and can, therefore, be better appreciated. In every case the SMLPs and the RBFNs outperform the MLPs by a statistically significant margin. However, the difference between the SMLPs and the RBFNs is much narrower, with the RBFNs and the hybrid SMLPs being nearly identical performance-wise. Also note that the hybrid training scheme appears to offer some improvement over both the trained and the solved SMLPs.

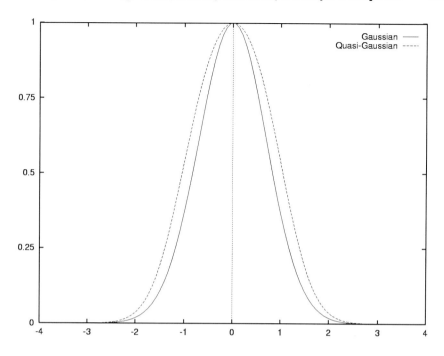

Fig. 7.5. A "quasi-Gaussian" activation function as a affine transformed sigmoidal

7.4 Theoretical Justification

Given the context of the numerical experiments from the previous section, we are now ready to see how an SMLP can be thought of as a "radially extended" version of an MLP. In this section, I will rewrite Equation 7.1 into a form that is equivalent to an RBFN; thus, we will see how it is possible for an SMLP to almost perfectly approximate an RBFN.

The first step is to more closely examine the sigmoidal activation function. Let $\text{sigmoid}(x) = 1/(1 + \exp(-x))$ and $\text{gauss}(x) = \exp(-x^2)$. We can define a quasi-Gaussian function as:

$$q(x) = 2 - 2/(\text{gauss}(x) + 1) = 2 - 2\,\text{sigmoid}(x^2). \tag{7.5}$$

This means that a local kernel function can be formed from an affine transformation of a sigmoid whose input has been squared.

Figure 7.5 shows how the quasi-Gaussian function relates to the true Gaussian function. Both functions are unimodal and exponentially decay in both directions. Moreover, a similar transformation can be applied to a hyperbolic tangent function; hence, it really doesn't matter which of the two common sigmoid functions are used as either can be transformed into a basis function.

Since a basis function has a center and a width, we want to be able to form local features of arbitrary size at an arbitrary location. Typically, this means that a basis function incorporates a distance measure such as Euclidean distance. With a center denoted by c_i and a width proportional to σ, we can rewrite a normalized Euclidean distance function as follows:

$$\frac{1}{\sigma_i^2}\|\mathbf{x} - \mathbf{c_i}\|^2 = \frac{1}{\sigma_i^2}\left(\mathbf{x}\cdot\mathbf{x} - 2\mathbf{c_i}\cdot\mathbf{x} + \mathbf{c_i}\cdot\mathbf{c_i}\right)$$

$$= \left(-\frac{2}{\sigma_i^2}\mathbf{c_i}\right)\cdot\mathbf{x} + \left(\frac{1}{\sigma_i^2}\right)\mathbf{x}\cdot\mathbf{x} + \left(\frac{1}{\sigma_i^2}\mathbf{c_i}\cdot\mathbf{c_i}\right)$$

$$= \sum_j\left(-\frac{2}{\sigma_i^2}c_{ij}\right)x_j + \sum_k\left(\frac{1}{\sigma_i^2}\right)x_k^2 + \left(\frac{1}{\sigma_i^2}\mathbf{c_i}\cdot\mathbf{c_i}\right) \quad (7.6)$$

Thus, the equation

$$2 - 2\,\mathrm{sigmoid}\left(\sum_j u_{ij}x_j + \sum_k v_{ik}x_k^2 + a_i,\right) \quad (7.7)$$

looks a lot like a radial basis function. By comparing Equation 7.6 to Equation 7.7 it is trivial to set the u_{ij}, v_{ij}, and a_i terms in such a way that a local "bump" is placed at a specific location with a specific width. This means that a single hidden node in an SMLP network can form a local feature in an input of any dimension. By way of comparison, Lapedes and Farber [10, 11] similarly constructed local features with standard MLPs. However, in a d-dimensional input space, one would need an MLP with two hidden layers, $2d$ hidden nodes in the first hidden layers, and another hidden node in the second hidden layer, just to form a single local "bump".

This simple analysis shows that local features are exceptionally easy to form in an SMLP but are potentially very difficult to form in an MLP. As mentioned in Section 7.3.3, it is possible to exploit the similarity between SMLPs and RBFNs and "solve" the weights in an SMLP with a non-iterative procedure. The first step is to choose a set of basis centers that can be determined by sub-sampling or clustering the input-space of the training data. After the centers are chosen, the nearest neighbor of each center with respect to the other centers can be calculated. These distances can be used as the widths of the basis centers. Next, the centers c_i and widths σ_i can be plugged into Equation 7.6 to determine the values of the u_{ij}, v_{ik} and a_i weights. Finally, the linear weights in the SMLP, w_i and b from Equation 7.1, can be solved for exactly by using a matrix pseudo-inverse procedure.

Thus, one can train an SMLP as one would an MLP or one could solve an SMLP as one would an RBFN. It is also possible to combine the approaches and let the solved weights be the initial weights for a training procedure. Using the procedure to solve the weights can sometimes cut the computational overhead for computing the weight by orders of magnitude compared to typical training methods. Moreover, as was found in the numerical experiments in Section 7.3.3, solutions found with this hybrid scheme may easily exceed the quality of solutions found with more traditional approaches.

7. Square Unit Augmented, Radially Extended, Multilayer Perceptrons 159

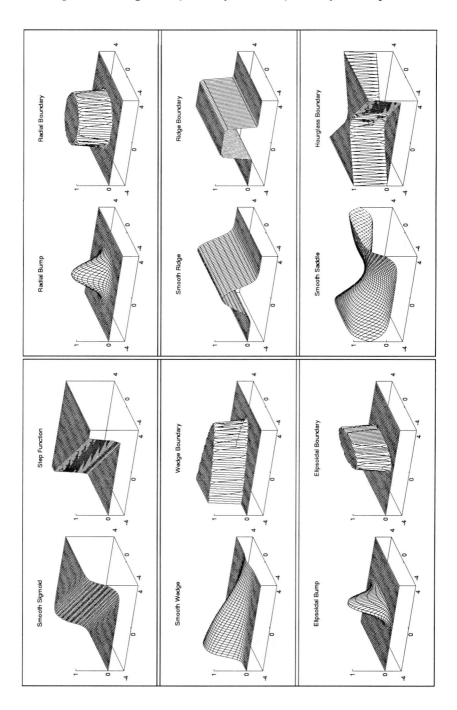

Fig. 7.6. Several difference types of surfaces and decision boundaries that can be formed by a single SMLP node

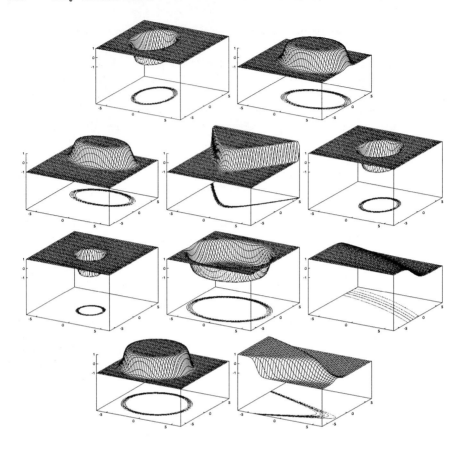

Fig. 7.7. Output response surfaces of the 10 hidden nodes in the SMLP network that solved the two spiral problem

7.5 Intuitive and Topological Justification

While the previous analysis shows that an SMLP can efficiently approximate an RBFN, the transformation from an SMLP into an RBFN examined only a single special case of the type of features that can be formed by an SMLP node. In fact, a single SMLP node can form many other types of features besides hyper-sigmoids and local bumps. To show that this is indeed the case, one only needs to compose a two-dimensional diagonal quadratic equation into a sigmoidal function to see what type of surfaces are possible.

Figure 7.6 shows some familiar surfaces and decision boundaries that can be formed with a single SMLP node. As expected, hyper-sigmoids and bumps can be formed. What is interesting, however, is that there are many types of SMLP features that are neither local nor global. For example, a ridge, wedge or saddle may look like a local or global feature if it is projected onto one lower dimension;

however, whether this projection is local or global depends on the subspace that the projection is formed.

How important is it for an architecture to be able to form these features? The very fact that these types of boundaries and surfaces have names means that they are important enough that one may need a type of model that can efficiently form them. However, if we reexamine the two-spirals problem from Section 7.3.2 it is possible to dissect the decision boundary formed by the SMLP (shown in Figure 7.3) to see how the surface was formed. What it truly interesting is that radial boundaries, wedges, and a sigmoid were all crucial to forming the entire decision surface. If the SMLP lacked the ability to form any of these features, then it is easily possible that the hidden node requirements for this problem would explode.

There is, however, one caveat with the variety of features that can be formed with a single SMLP node. The wedge, ridge, ellipse and saddle structures shown in Figure 7.6 must always be aligned with one of the input axes. In other words, it is impossible to make a ridge that would run parallel to the line defined by $x = y$. We can see that this is true by noting that $(x-y)^2$ has the term $-2xy$ in its expansion, which means that the quadratic form is non-diagonal. In general, in order to have the ability to rotate all of the features in Figure 7.6, one would need the full quadratic form, thus requiring a higher-order network instead of the SMLP. Hence, while eliminating the off-diagonal terms in a higher-order network saves a considerable number of weights, there is a cost in the types of features that can be formed.

7.6 Conclusions

We have seen that there are problems whose solutions require features of both local and global scope. MLPs excel at forming global features but have a difficult time forming local features. RBFNs are exactly the opposite. The SMLP architecture can efficiently form both types of features with only a small penalty in the number of weights. However, the increase in the number of weights is more than compensated for by improvements in the network's ability to form features. This often results in simpler networks with fewer weights that can learn much faster and approximate more accurately.

For the two main numerical studies in this work, it was found that the SMLP architecture performed as well or better than the best known techniques for the two-spirals problem and the Deterding vowel recognition data. Moreover, these results are strengthened by the fact that the average performance of the SMLP was often superior to the best known results for the other techniques.

It was also found that the dual nature of the SMLP can be exploited in the form of hybrid algorithms. SMLPs can be "solved" like an RBFN, trained like an MLP, or both. It is also noteworthy that the nonlinear weights in an SMLP are only "moderately" nonlinear. For example, gradient-based learning algorithms can be used on RBFNs but it is known that the high degree of nonlinearity found in the weights which correspond to centers and widths can often make gradient training very slow for RBFNs when the nonlinear weight are included. Similarly, since MLPs require two hidden layers of nodes to form local features efficiently, the first layer of weights in an MLP are exceedingly nonlinear because they are eventually passed through two layers of nonlinear nodes. Counter to this, the nonlinear nodes in an SMLP are only passed through a single layer of nonlinear nodes. Although it is unproven at this time, it seems like a reasonable conjecture that SMLP networks may be intrinsically better conditioned for gradient-based learning of local "bumps" than MLPs with two hidden layers.

Acknowledgements

I thank Frans Coetzee, Chris Darken, Lee Giles, Jenny Orr, Ray Watrous, and the anonymous reviewers for many helpful comments and discussions.

References

1. M. Casdagli. Nonlinear prediction of chaotic time series. *Physica D*, 35:335–356, 1989.
2. D. H. Deterding. *Speaker Normalisation for Automatic Speech Recognition.* PhD thesis, University of Cambridge, 1989.
3. S.E. Fahlman. Faster-learning variations on back-propagation: An empirical study. In *Proceedings of the 1988 Connectionist Models Summer School.* Morgan Kaufmann, 1988.
4. S.E. Fahlman and C. Lebiere. The cascade-correlation learning architecture. In S. Touretzky, editor, *Advances in Neural Information Processing Systems 2.* Morgan Kaufmann, 1990.
5. M. Finke and K.-R. Müller. Estimating a-posteriori probabilities using stochastic network models. In M. Mozer, P. Smolensky, D.S. Touretzky, J.L. Elman, and A.S. Weigend, editors, *Proceedings of the 1993 Connectionist Models summer school,* pages 324–331, Hillsdale, NJ, 1994. Erlenbaum Associates.
6. T. Hastie and R. Tibshirani. Flexible discriminant analysis by optimal scoring. Technical report, AT&T Bell Labs, Murray Hill, New Jersey, 1993.
7. T. Hastie and R. Tibshirani. Discriminant adaptive nearest neighbor classification. *IEEE Transactions on Pattern Analysis and Machine Intelligence,* 18(6):607–616, June 1996.
8. S. Hochreiter and J. Schmidhuber. Lococode. Technical Report FKI-222-97, Fakultät für Informatik, Technische Universität München, 1997.
9. K. J. Lang and M. J. Witbrock. Learning to tell two spirals apart. In *Proceedings of the 1988 Connectionist Models Summer School.* Morgan Kaufmann, 1988.
10. A. Lapedes and R. Farber. Nonlinear signal processing using neural networks: Prediction and system modelling. Technical Report LA-UR-87-2662, Los Alamos National Laboratory, Los Alamos, NM, 1987.
11. A. Lapedes and R. Farber. How neural nets work. In D.Z. Anderson, editor, *Neural Information Processing Sysytems,* pages 442–456. American Institute of Physics, New York, 1988.
12. S. Lawrence, A. C. Tsoi, and A. D. Back. Function approximation with neural networks and local methods: Bias, variance and smoothness. In Peter Bartlett, Anthony Burkitt, and Robert Williamson, editors, *Australian Conference on Neural Networks,* pages 16–21. Australian National University, 1996.

13. S. Lee and R.M. Kil. Multilayer feedforward potential function networks. In *IEEE international Conference on Neural Networks*, pages I:161–171. San Diego: SOS Printing, 1988.
14. Y.C. Lee, G. Doolen, H.H. Chen, G.Z. Sun, T. Maxwell, H.Y. Lee, and C.L. Giles. Machine learning using higher order correlation networks. *Physica D*, 22-D:276–306, 1986.
15. J. Moody and C. Darken. Learning with localized receptive fields. In D. Touretsky, G. Hinton, and T. Sejnowski, editors, *Proceedings of the 1988 Connectionist Models Summer School*. Morgan-Kaufmann, 1988.
16. J. Moody and C. Darken. Fast learning in networks of locally-tuned processing units. *Neural Computation*, 1:281–294, 1989.
17. M. Niranjan and F. Fallside. Neural networks and radial basis functions in classifying static speech patterns. *Computer Speech and Language*, 4(275-289), 1990.
18. Y.H. Pao. *Adaptive Pattern Recognition and Neural Networks*. Addison-Wesley Publishing Company, Inc., Reading, Massachusetts, 1989.
19. A. J. Robinson. *Dynamic Error Propagation Networks*. PhD thesis, Cambridge University, 1989.
20. D.E. Rumelhart, J.L. McClelland, and the PDP Research Group. *Parallel Distributed Processing: Explorations in the Microstructure of Cognition*. The MIT Press., 1986. 2 vols.
21. W. Sarle. The `comp.ai.neural-nets` Frequently Asked Questions List, 1997.
22. M. Schetzen. *The Volterra and Wiener Theories of Nonlinear Systems*. John Wiley and Sons, New York, 1980.
23. B. Schölkopf, A. Smola, and K.-R. Müller. Nonlinear component analysis as a kernel eigenvalue problem. Technical report, Max-Planck-Institut für biologische Kybernetik, 1996. and *Neural Computation*, 10, 5, 1299–1319. 1998.
24. V. Volterra. *Theory of Functionals and of Integro-differential Equations*. Dover, 1959.
25. P. Werbos. *Beyond Regression: New Tools for Prediction and Analysis in the Behavioral Sciences*. PhD thesis, Harvard University, 1974.

8
A Dozen Tricks with Multitask Learning

Rich Caruana

Just Research and Carnegie Mellon University,
4616 Henry Street, Pittsburgh, PA 15213
caruana@cs.cmu.edu
http://www.cs.cmu.edu/ caruana/

Abstract. Multitask Learning is an inductive transfer method that improves generalization accuracy on a main task by using the information contained in the training signals of other *related* tasks. It does this by learning the extra tasks in parallel with the main task while using a shared representation; what is learned for each task can help other tasks be learned better. This chapter describes a dozen opportunities for applying multitask learning in real problems. At the end of the chapter we also make several suggestions for how to get the most our of multitask learning on real-world problems.

When tackling real problems, one often encounters valuable information that is not easily incorporated in the learning process. This chapter shows a dozen ways to benefit from the information that often gets ignored. The basic trick is to create extra tasks that get trained on the same net with the main task. This *Multitask Learning* is a form of inductive transfer[1] that improves performance on the main task by using the information contained in the training signals of other *related* tasks. It does this by learning the main task in parallel with the extra tasks while using a shared representation; what is learned for each task can help other tasks be learned better.

We use the term "task" to refer to a function that will be learned from a training set. We call the important task that we wish to learn better the **main task**. Other tasks whose training signals will be used by multitask learning to learn the main task better are the **extra tasks**. Often, we do not care how well extra tasks are learned. Their sole purpose is to help the main task be learned better. We call the union of the main task and the extra tasks a **domain**. Here we restrict ourselves to domains where the tasks are defined on a common set of input features, though some of the extra tasks may be functions of only a subset of these input features.

This chapter shows that most real-world domains present a number of opportunities for multitask learning (MTL). Because these opportunities are not always obvious, most of the chapter is dedicated to showing different ways useful

---
[1] Inductive transfer is the process of transferring anything learned for one problem to help learning of other related problems.

extra tasks arise in real problems. We demonstrate several of these opportunities using real data. The chapter ends with a few suggestions that help you get the most out of multitask learning. Some of these suggestions are so important that if you don't follow them, MTL can easily hurt performance on the main task instead of helping it.

8.1 Introduction to Multitask Learning in Backprop Nets

Consider the following boolean functions defined on eight bits, $B_1 \cdots B_8$:

$$Task1 = B_1 \vee Parity(B_2 \cdots B_6)$$
$$Task2 = \neg B_1 \vee Parity(B_2 \cdots B_6)$$
$$Task3 = B_1 \wedge Parity(B_2 \cdots B_6)$$
$$Task4 = \neg B_1 \wedge Parity(B_2 \cdots B_6)$$

where "B_i" represents the ith bit, "$\neg$" is logical negation, "$\vee$" is disjunction, "$\wedge$" is conjunction, and "$Parity(B_2 \cdots B_6)$" is the parity of bits 2–6. Bits B_7 and B_8 are not used by the functions. These four tasks are related in several ways:

- they are all defined on the same inputs, bits $B_1 \cdots B_8$;
- they all ignore the same inputs, bits B_7 and B_8;
- each uses a common computed subfeature, $Parity(B_2 \cdots B_6)$;
- when $B_1 = 0$, Task 1 needs $Parity(B_2 \cdots B_6)$, but Task 2 does not, and vice versa;
- as with Tasks 1 and 2, when Task 3 needs $Parity(B_2 \cdots B_6)$, Task 4 does not need it, and vice versa.

We can train artificial neural nets on these tasks with backprop. Bits $B_1 \cdots B_8$ are the inputs to the net. The task values computed by the four functions are the target outputs. We create a data set by enumerating all 256 combinations of the eight input bits, and computing for each setting of the bits the task signals for Tasks 1, 2, 3, and 4 using the definitions above. This yields 256 different cases, with four different training signals for each case.

8.1.1 Single and Multitask Learning of Task 1

Consider Task 1 the main task. Tasks 2, 3, and 4 are the extra tasks. That is, we are interested only in improving the accuracy of models trained for Task 1. We've done an experiment where we train Task 1 on the three nets shown in Figure 8.1. All the nets are fully connected feed-forward nets with 8 inputs, 100 hidden units, and 1–4 outputs. Where there are multiple outputs, each output is fully connected to the hidden units. Nets were trained in batch mode using backprop with MITRE's Aspirin/MIGRAINES 6.0 with learning rate = 0.1 and momentum = 0.9.

Task 1 is trained alone on the net on the left of Figure 8.1. This is a backprop net trained on a single task. We refer to this as single task learning (STL) or single task backprop (STL-backprop). The net in the center of Figure 8.1 trains Task 1 on a net that is also trained on Task 2. The hidden layer of this net is shared by Tasks 1 and 2. This is multitask backprop (MTL-backprop) with two tasks. The net on the right side of Figure 8.1 trains Task 1 with Tasks 2, 3, and 4. The hidden layer of this net is shared by all four tasks. This is MTL-backprop with four tasks. How well will Task 1 be learned by the different nets?

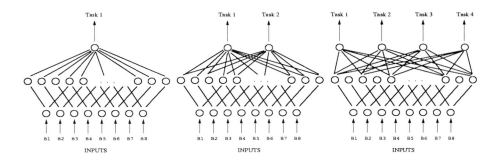

Fig. 8.1. Three Neural Net Architectures for Learning Task 1

We performed 25 independent trials by resampling training and test sets from the 256 cases. From the 256 cases, we randomly sample 128 cases for the training set, and use the remaining 128 cases as a test set. (For now we ignore the complexity of early stopping, which can be tricky with MTL nets. See section 8.3.2 for a thorough discussion of early stopping in MTL nets.)

For each trial, we trained three nets: an STL net for Task 1, an MTL net for Tasks 1 and 2, and an MTL net for Tasks 1–4. We measure performance only on the output for Task 1. When there are extra outputs for Task 2 or Tasks 2–4, these are trained with backprop, but ignored when the net is evaluated. The sole purpose of the extra outputs is to affect what is learned in the hidden layer these outputs share with Task 1.

8.1.2 Results

Every 5000 epochs we evaluated the performance of the nets on the test set. We measured the RMS error of the output with respect to the target values, the criterion being optimized by backprop. We also measured the accuracy of the output in predicting the boolean function values. If the net output is less than 0.5, it was treated as a prediction of 0, otherwise it was treated as a prediction of 1.

Figure 8.2 shows the RMSE for Task 1 on the test set during training. The three curves in the graph are each the average of 25 trials.[2] RMSE on the main task, Task 1, is reduced when Task 1 is trained on a net simultaneously trained on other related tasks. RMSE is reduced when Task 1 is trained with extra Task 2, and is further reduced when extra Tasks 3 and 4 are added. *Training multiple tasks on one net does not increase the number of training patterns seen by the net. Each net sees exactly the same training cases. The MTL nets do not see more training cases; they receive more training signals with each case.*

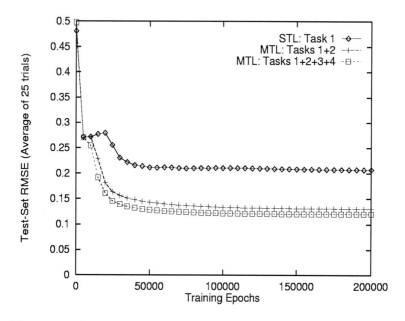

Fig. 8.2. RMSE Test-set Performance of Three Different Nets on Task 1.

Figure 8.3 shows the average test-set accuracy on Task 1 for 25 trials with the three different nets. Task 1 has boolean value 1 about 75% of the time. A simple learner that learned to predict 1 all the time should achieve about 75% accuracy on Task 1. When trained alone (STL), performance on Task 1 is about 80%. When Task 1 is trained with Task 2, performance increases to about 88%. When Task1 is trained with Tasks 2, 3, and 4, performance increases further to

---

[2] Average training curves can be misleading, particularly if training curves are not monotonic. For example, it is possible for method A to always achieve better error than method B, but for the average of method A to be everywhere worse than the average of method B because the regions where performance on method A is best do not align, but do align for method B. Before presenting average training curves, we always examine the individual curves to make sure the average curve is not misleading.

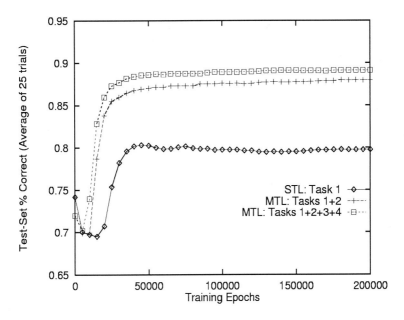

Fig. 8.3. Test-set Percent Correct of Three Different Nets on Task 1.

about 90%. Table 8.1 summarizes the results of examining the training curve from each trial.

Table 8.1. Test-set performance on Task 1 of STL of Task 1, MTL of Tasks 1 and 2, and MTL of Tasks 1, 2, 3, and 4. *** indicates performance is statistically better than STL at .001 or better, respectively.

NET	STL: 1	MTL: 1+2	MTL: 1+2+3+4
Root-Mean-Squared-Error	0.211	0.134 ***	0.122 ***
Percent Correct	79.7%	87.5% ***	88.9% ***

8.1.3 Discussion

Why is the main task learned better if it is trained on a net learning other related tasks at the same time? We ran a number of experiments to verify that the performance increase with MTL is due to the fact that the tasks are related, and not just a side effect of training multiple outputs on one net.

Adding noise to neural nets sometimes improves their generalization performance [22]. To the extent that MTL tasks are *uncorrelated*, their contribution to the aggregate gradient may appear as noise to other tasks and this might improve generalization. To see if this explains the benefits we see from MTL, in one experiment we train Task 1 on a net with three *random* tasks.

A second effect to be concerned about is that adding tasks tends to increase the effective learning rate on the input-to-hidden layer weights because the gradients from the multiple outputs add at the hidden layer, and this might favor nets with multiple outputs. To test this, we train an MTL net with four copies of Task 1. Each of the four outputs receives exactly the same training signal. This is a degenerate form of MTL where no extra information is given to the net by the extra tasks.

A third effect that needs to be ruled out is net capacity. 100 hidden units is a lot for these tasks. Does the MTL net, which has to share the 100 hidden units among four tasks, generalize better because each task has fewer hidden units? To test for this, we train Task 1 on STL nets with 200 hidden units and with 25 hidden units. This will tell us if generalization would be better with more or less capacity.

Finally, we ran a fourth experiment based on the heuristic used in [37]. We shuffle the training signals (the target output values) for Tasks 2, 3, and 4 before training an MTL net on the four tasks. Shuffling reassigns the target values to the input vectors in the training set for Tasks 2, 3, and 4. The main task, Task 1, is not affected. The distributions of the training signals for outputs 2–4 have not changed, but the training signals are no longer related to Task 1. This is a powerful test that has the potential to rule-out many mechanisms that do not depend on relationships between the tasks.

We ran each experiment 25 times using exactly the same data sets used in the previous section. Figure 8.4 shows the generalization performance on Task 1 in the four experiments. For comparison, the performance of of STL, MTL with Tasks 1 and 2, and MTL with Tasks 1–4 from the previous section are also shown.

When Task 1 is trained with random extra tasks, performance on Task 1 drops below the performance on Task 1 when it is trained alone on an STL net. We conclude MTL of Tasks 1–4 probably does not learn Task 1 better by adding noise to the learning process through the extra outputs.

When Task 1 is trained with three additional copies of Task 1, the performance is comparable to that when Task 1 is trained alone with STL.[3] We conclude that MTL does not learn Task 1 better just because backprop works better with multiple outputs.

When Task 1 is trained on an STL net with 25 hidden units, performance is comparable to the performance with 100 hidden units. Moreover, when Task

---

[3] We sometimes observe that training multiple copies of a task on one net does improve performance. When we have observed this, the benefit is never large enough to explain away the benefits observed with MTL. But it is interesting and surprising, as the improvement is gained without any additional information being given to the net. The most likely explanation is that the multiple connections to the hidden layer allow different hidden layer predictions to be averaged and thus act as a weak boosting mechanism.

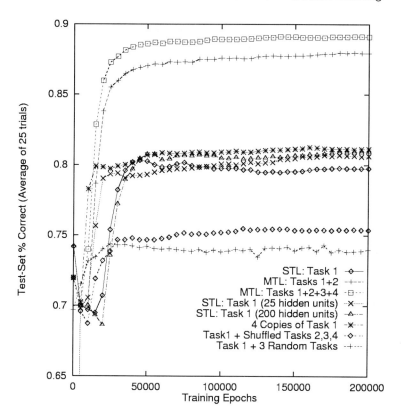

Fig. 8.4. RMSE test-set performance of Task 1 when trained with: MTL with three random tasks; MTL with three more copies of Task 1; MTL with shuffled training signals for Tasks 2–4; STL on nets with 25 or 200 hidden units.

1 is trained on an STL net with 200 hidden units, it is slightly better. (The differences between STL with 25, 100, and 200 hidden units are not statistically significant.) We conclude that performance on Task 1 is relatively insensitive to net size for nets between 25 and 200 hidden units, and, if anything, Task 1 would benefit from a net with more capacity, not one with less capacity. Thus it is unlikely that MTL on Tasks 1–4 performs better on Task 1 because Tasks 2–4 are using up extra capacity that is hurting Task 1.

When Task 1 is trained with training signals for Tasks 2–4 that have been shuffled, the performance of MTL drops below the performance of Task 1 trained alone on an STL net. Clearly the benefit we see with MTL on these problems is not due to some accident caused by the distribution of the extra outputs. The extra outputs must be *related* to the main task to help it.

These experiments rule out most explanations for why MTL outperforms STL on Task 1 that do not require Tasks 2–4 be related to Task 1. So why is the main task learned better when trained in parallel with Tasks 2–4?

One reason is that Task 1 needs to learn the subfeature $Parity(B_2 \cdots B_6)$ that it shares with Tasks 2–4. Tasks 2–4 give the net information about this subfeature that it would not get from Task 1 alone. For example, when $B_1 = 1$, the training signal for Task 1 contains no information about $Parity(B_2 \cdots B_6)$. We say B_1 *masks* $Parity(B_2 \cdots B_6)$ when $B_1 = 1$. But the training signals for Task 2 provide information about the Parity subfeature in exactly those cases where Task 1 is masked. Thus the hidden layer in a net trained on both Tasks 1 and 2 gets twice as much information about the Parity subfeature as a net trained on one of these tasks, despite the fact that they see exactly the same training cases. The MTL net is getting more information with each training case.

Another reason why MTL helps Task 1 is that all the tasks are functions of the same inputs, bits $B_1 \cdots B_6$, and ignore the same inputs, B_7 and B_8. Because the tasks overlap on the features they use and don't use, the MTL net is better able select which input features to use.

A third reason why MTL helps Task 1 is that there are relationships between the way the different tasks use the inputs that promote learning good internal representations. For example, all the tasks logically combine input B_1 with a function of inputs $B_2 \cdots B_6$. This similarity tends to prevent the net from learning internal representations that, for example, directly combine bits B_1 and B_2. A net trained on all the tasks together is biased to learn more modular, in this case more correct, internal representations that support the multiple tasks. This bias towards modular internal representations reduces the net's tendency to learn spurious correlations that occur in any finite training sample: there may be a random correlation between bit B_3 and the output for Task 1 that looks fairly strong in this one training set, but if that spurious correlation does not also help the other tasks, it is less likely to be learned.

8.2 Tricks for Using Multitask Learning in the Real World

The previous section introduced multitask learning (MTL) in backprop nets using four tasks carefully designed to have relationships that make learning them in parallel work better than learning them in isolation. How often will real problems present extra tasks that allow multitask learning to improve performance on the main task?

This section shows that many real world problems yield opportunities for multitask learning. We present a dozen prototypical real-world applications where the training signals for related extra tasks are available and can be leveraged. We believe most real-world problems fall into one or more of these prototypical classes. This claim might sound surprising given that few of the test problems traditionally used in machine learning are multitask problems. We believe most of the problems used in machine learning so far have been heavily preprocessed to fit the single task learning mold. Most of the opportunities for MTL in these problems were eliminated as the problems were defined.

8.2.1 Using the Future to Predict the Present

Often valuable features become available *after* predictions must be made. If learning is done offline, these features can be collected for the training set and used for learning. These features can't be used as inputs, because they will not be available when making predictions for future test cases. They can, however, be used as extra outputs for multitask learning. The predictions the learner makes for these extra tasks will be ignored when the system is used to make predictions for the main task. Their sole function is to provide extra information to the learner during training so that it can learn the main task better.

One source of applications of learning from the future is sequential decision making in medicine. Given the initial symptoms, decisions are made about what tests to make and what treatment to begin. New information becomes available when the tests are completed and as the patient responds (or fails to respond) to the treatment. From this new information, new decisions are made. Should more tests be made? Should the treatment be changed? Has the patient's condition changed? Is this patient now high risk, or low risk? Does the patient need to be hospitalized? Etc.

When machine learning is applied to early stages in the decision making process, only those input features that typically would be available for patients at this stage of the process are usually used. This is unfortunate. In an historical database, all of the patients may have run the full course of medical testing and treatment and their final outcome may be known. Must we ignore the results of lab tests and other valuable features in the database just because these will not be available for patients at the stage of medical decision making for which we wish to learn a model?

The Pneumonia Risk Prediction Problem Consider pneumonia. There are 3,000,000 cases of pneumonia each year in the U.S., 900,000 of which get hospitalized. Most pneumonia patients recover given appropriate treatment, and many can be treated effectively without hospitalization. Nonetheless, pneumonia is serious: 100,000 of those hospitalized for pneumonia die from it, and many more are at elevated risk if not hospitalized.

Consider the problem of predicting a patient's risk from pneumonia before they are hospitalized. (The problem is not to diagnose if the patient has pneumonia, but to determine how much risk the pneumonia poses to the patient.) A primary goal in medical decision making is to accurately, swiftly, and economically identify patients at high risk from diseases like pneumonia so that they may be hospitalized to receive aggressive testing and treatment; patients at low risk may be more comfortably, safely, and economically treated at home.

Some of the most useful features for assessing risk are the lab tests that become available only after a patient is hospitalized. It is the *extra* lab tests made after patients are admitted to the hospital that we use as extra tasks for

MTL; they cannot be used as inputs because they will not be available for most future patients when making the decision to hospitalize.[4]

The most useful decision aid for this problem would be to predict which patients will live or die. This is too difficult. In practice, the best that can be achieved is to estimate a probability of death (POD) from the observed symptoms. In fact, it is sufficient to learn to order patients by POD so lower-risk patients can be discriminated from higher risk patients; patients at least risk may then be considered for outpatient care.

The performance criteria used by others working with this database [15] is the accuracy with which one can select prespecified fractions of a patient population who will live. For example, given a population of 10,000 patients, find the 20% of this population at *least* risk. To do this we learn a risk model, and a threshold for this risk model, that allows 20% of the population (2000 patients) to fall below it. If 30 of the 2000 patients below this threshold die, the error rate is $30/2000 = 0.015$. We say that the error rate for FOP 0.20 is 0.015 (FOP stands for "fraction of population"). Here we consider FOPs 0.1, 0.2, 0.3, 0.4, and 0.5. Our goal is to learn models and thresholds such that the error rate at each FOP is minimized.

Multitask Learning and Pneumonia Risk Prediction The straightforward approach to this problem is to use backprop to train an STL net to learn to predict which patients live or die, and then use the real-valued predictions of this net to sort patients by risk. This STL net has 30 inputs for the 30 basic prehospitalization measurements, a single hidden layer, and a single output trained with targets 0=lived, 1=died.[5] Given a large training set, a net trained this way should learn to predict the probability of death for each patient, not which patients live or die. If the training sample is small, the net will overfit and learn a very nonlinear function that outputs values near 0/1 for cases in the training set, but which does not generalize well. It is critical to use early stopping to halt training before this happens.

We developed a method called *Rankprop* specifically for this domain. Rankprop learns to rank patients without learning to predict mortality (0=lived,1=died). Figure 8.5 compares the performance of squared error on 0/1 targets with rankprop on this problem. Rankprop outperforms traditional backprop using squared error on targets 0=lived,1=died by 10%-40% on this domain, depending on which FOP is used for comparison. See [9] for details about rankprop.[6]

---

[4] Other researchers who tackled this problem ignored the the lab tests because they knew they would not be available at run time and did not see ways to use them other than as inputs.

[5] We tried both squared error and cross entropy. The difference between the two was small. Squared error performed slightly better.

[6] We use rankprop for the rest of our experiments on this domain because it is the best performer we know of on this problem. We want to see if MTL can make the best method better.

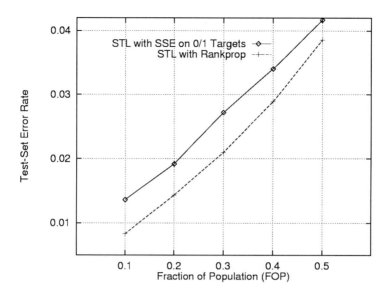

Fig. 8.5. The performance of SSE (0/1 targets) and rankprop on the 5 FOPs in the pneumonia domain. Lower error indicates better performance.

There are 35 future lab values that we use as extra backprop *outputs,* as shown in Figure 8.6. The expectation is that these extra outputs will bias the shared hidden layer toward representations that better capture important features of each patient's condition, and that this will lead to more accurate predictions of patient risk at the main task output.

The STL net has 30 inputs, 8 hidden units, and one output trained to predict risk with rankprop. The MTL net has the same 30 inputs, 64 hidden units, one output for rankprop, and 35 extra outputs trained with squared error. (Preliminary experiments suggested 8–32 hidden units was optimal for STL, and that MTL performs best with nets as large as 512 hidden units. We used 8 and 64 hidden units so that we could run many experiments.) The 35 extra outputs on the MTL net (see Figure 8.6) are trained at the same time the net is trained to predict risk.

We train the net using training and validation sets containing 1000 patients randomly drawn from the database. Training is halted on both the STL and MTL nets when overfitting is observed on the main rankprop risk task. On the MTL net, the performance of the extra tasks is not taken into account for early stopping. Only the performance of the output for the main task is considered when deciding when to stop training. (See section 8.3.2 for more discussion of early stopping with MTL nets.) Once training is halted, the net is tested on the remaining unused patients in the database.

Results Table 8.2 shows the mean performance of ten runs of rankprop using STL and MTL. The bottom row shows the percent improvement in performance

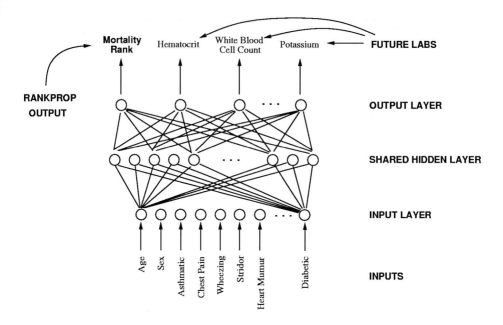

Fig. 8.6. Using future lab results as extra outputs to bias learning for the main risk prediction task. The lab tests would help most if they could be used as inputs, but will not yet have been measured when risk must be predicted, so we use them as extra outputs for MTL instead.

obtained on this problem by using the future lab measurements as extra MTL outputs. Negative percentages indicate MTL reduces error. Although MTL lowers the error at each FOP compared with STL, only the differences at FOP 0.3, 0.4, and 0.5 are statistically significant with ten trials using a standard t-test.

Table 8.2. Error Rates (fraction deaths) for STL with Rankprop and MTL with Rankprop on Fractions of the Population predicted to be at low risk (FOP) between 0.0 and 0.5. MTL makes 5–10% fewer errors than STL.

FOP	0.1	0.2	0.3	0.4	0.5
STL Rankprop	.0083	.0144	.0210	.0289	.0386
MTL Rankprop	.0074	.0127	.0197	.0269	.0364
% Change	-10.8%	-11.8%	-6.2% *	-6.9% *	-5.7% *

The improvement from MTL is 5–10%. This improvement can be of considerable consequence in medical domains. To verify that the benefit from MTL is due to relationships between what is learned for the future labs and the main task, we ran the shuffle test (see section 8.1.3). We shuffled the training signals for the extra tasks in the training sets before training the nets with MTL.

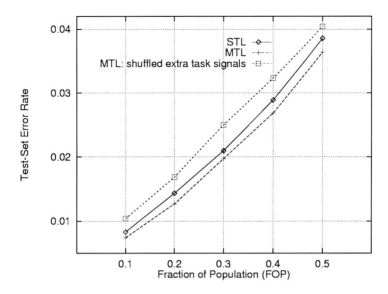

Fig. 8.7. Performance of STL, MTL, and MTL with shuffled extra task signals on pneumonia risk prediction at the five FOPs.

Figure 8.7 shows the results of MTL with shuffled training signals for the extra tasks. The results of STL and MTL with unshuffled extra tasks are also shown. Shuffling the training signals for the extra tasks reduces the performance of MTL below that of STL. We conclude that it is the relationship between the main task and the extra tasks that lets MTL perform better on the main task; the benefit disappears when these relationships are broken by shuffling the extra task signals.

We have also run experiments where we use the future lab tests as inputs to a net trained to predict risk, and impute the values for the lab tests when they are missing on future test cases. Imputing missing values for the lab tests did not yield performance comparable to MTL on this problem. Similar experiments with feature nets [17] also failed to yield improvements comparable to MTL.

Future measurements are available in many *offline* learning problems. As just one very different example, a robot or autonomous vehicle can more accurately measure the size, location, and identity of objects if it passes nearer them in the future. For example, road stripes and the edge of the road can be detected reliably as a vehicle passes alongside them, but detecting them far ahead of the vehicle is hard. Since driving brings future road closer to the car, stripes and road borders can be measured accurately as the car passes them. Dead reckoning allows these future measurements to be added to the training set. They can't be used as *inputs*; They won't be available in time while driving. But they can be used to augment a training set. We suspect that using future measurements as extra outputs will be a frequent source of extra tasks in real problems.

8.2.2 Multiple Metrics

Sometimes it is hard to capture everything that is important in one error metric. When alternate metrics capture different, but useful, aspects of a problem, MTL can be used to benefit from the multiple metrics. One example of this is the pneumonia problem in the previous section. Rankprop outperforms backprop using traditional squared error on this problem, but has trouble learning to rank cases at such low risk that virtually all patients survive because these cases provide little ordering information. Interestingly, squared error performs best when cases have high purity, such as in regions of feature space where most cases have low risk. *Squared error is at its best where rankprop is weakest.* Adding an extra output trained with squared error to a net learning to predict pneumonia risk with rankprop improves the accuracy of the rankprop output an additional 5-10% for the least-risk cases. The earliest example of using multiple output representations we know of is [38] which uses both SSE and cross-entropy outputs for the same task.

8.2.3 Multiple Output Representations

Sometimes it is not apparent what output encoding is best for a problem. Distributed output representations often help *parts* of a problem be learned well because the parts have separate error gradients. But non-distributed output representations are sometimes more accurate. Consider the problem of learning to classify a face as one of twenty faces. One output representation is to have one output code for each face. Another representation is to have outputs code for features such as beard/no_beard, glasses/no_glasses, long_hair/short, eye_color(blue, brown), male/female, that are sufficient to distinguish the faces. Correct classification, however, may require that each feature be correctly predicted. The non-distributed output coding that uses one output for each individual may be more reliable. But training the net to recognize specific traits should help, too. MTL is one way to merge these conflicting requirements in one net by using both output representations, even if only one representation will be used for prediction.

A related approach to multiple output encodings is error correcting codes [18]. Here, multiple encodings for the outputs are designed so that the combined prediction is less sensitive to occasional errors in some of the outputs. It is not clear how much ECOC benefits from MTL-like mechanisms. In fact, ECOC may benefit from being trained on STL nets (instead of MTL nets) so that different outputs do not share the same hidden layer and thus are less correlated. But see [27] for ways of using MTL to *decorrelate* errors in multiple outputs to boost committee machine performance.

8.2.4 Time Series Prediction

The simplest way to use MTL for time series prediction is to use a single net with multiple outputs, each output corresponding to the same task at a different time. This net makes predictions for the same task at different times. We tested this on a robot domain where the goal is to predict what the robot will sense 1, 2, 4, and 8 meters in the future as it moves forward. Training all four of these distances on one MTL net improved the accuracy of the long range predictions about 10% (see chapter 17 where MTL is used in a time series application).

8.2.5 Using Non-operational Features

Some features are impractical to use at run time, either because they are too expensive to compute, or because they need human expertise that won't be available. We usually have more time, however, to prepare our training sets. When it is impractical to compute some features on the fly at run time, but practical to compute them for the training set, these features can be used as extra outputs to help learning. Pattern recognition provides a good example of this. We tested MTL on a door recognition problem where the goal is to recognize doorways and doorknobs. The extra tasks were features such as the location of door edges and doorway centers that required laborious hand labelling that would not be applied to the test set. The MTL nets that were trained to predict these additional hand-labelled features were 25% more accurate at locating doors and doorknobs. Other domains where hand-labelling can be used to augment training sets this way include text domains, medical domains, acoustic domains, and speech domains.

8.2.6 Using Extra Tasks to Focus Attention

Learning often uses large, ubiquitous patterns in the inputs, while ignoring small or less common inputs that might also be useful. MTL can be used to coerce the learner to attend to patterns in the input it would otherwise ignore. This is done by forcing it to learn internal representations to support related tasks that depend on these patterns.

A good example is road following. Here, STL nets often ignore lane markings when learning to steer because lane markings are usually a small part of the image, are constantly changing, and are often difficult to see (even for humans). If a net learning to steer is also required to learn to recognize road stripes, the net will learn to attend to those parts of the image where stripes occur. To the extent that the stripe tasks are learnable, the net will develop internal representations to support them. Since the net is also learning to steer using the same hidden layer, the steering task can use the parts of the stripe hidden representation that are useful for steering.

We tested this idea using a road image simulator developed by Pomerleau to permit rapid testing of learning methods for road-following domains [28]. Figure 8.8 shows several 2-D road images.

The principal task is to predict steering direction. For the MTL experiments, we used eight additional tasks:

- whether the road is one or two lanes
- location of centerline (if any)

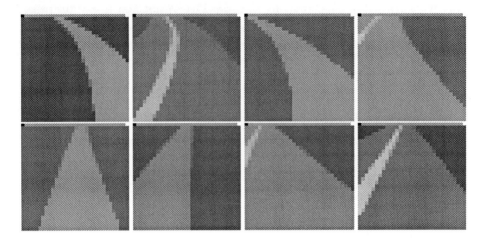

Fig. 8.8. Sample single and two lane roads generated with Pomerleau's road simulator.

- location of left edge of road
- location of road center
- intensity of region bordering road
- location of right edge of road
- intensity of road surface
- intensity of centerline (if any)

These additional tasks are all computable from the internal variables in the simulator. Table 8.3 shows the average performance of ten runs of single and multitask learning on each of these tasks. The MTL net has 32 inputs, 16 hidden units, and 9 outputs. The 36 STL nets have 32 inputs, 2, 4, 8 or 16 hidden units, and 1 output each.

Table 8.3. Performance of STL and MTL on the road following domain. The underlined entries in the STL columns are the STL runs that performed best. Differences statistically significant at .05 or better are marked with an *.

TASK	ROOT-MEAN SQUARED ERROR ON TEST SET						
	Single Task Backprop (STL)				MTL	Change MTL	Change MTL
	2HU	4HU	8HU	16HU	16HU	to Best STL	to Mean STL
1 or 2 Lanes	.201	.209	.207	.178	.156	-12.4% *	-21.5% *
Left Edge	.069	.071	.073	.073	.062	-10.1% *	-13.3% *
Right Edge	.076	.062	.058	.056	.051	-8.9% *	-19.0% *
Line Center	.153	.152	.152	.152	.151	-0.7%	-0.8%
Road Center	.038	.037	.039	.042	.034	-8.1% *	-12.8% *
Road Greylevel	.054	.055	.055	.054	.038	-29.6% *	-30.3% *
Edge Greylevel	.037	.038	.039	.038	.038	2.7%	0.0%
Line Greylevel	.054	.054	.054	.054	.054	0.0%	0.0%
Steering	.093	.069	.087	.072	.058	-15.9% *	-27.7% *

The last two columns compare STL and MTL. The first is the percent reduction in error of MTL over the best STL run. Negative percentages indicate MTL performs better. The last column is the percent improvement of MTL over the average STL performance. On the important steering task, MTL outperforms STL 15–30%.

We ran a follow-up experiment to test how important centerstripes are to the STL and MTL nets. We eliminated the stripes from the images in a test set. If MTL learned more about centerstripes than STL, and uses what it learned about centerstripes for the main steering task, we expect to see steering performance degrade more for MTL than for STL when we remove the centerstripes from the images. Error increased more for the MTL nets than for the STL nets, suggesting the MTL nets are making more use of the stripes in the images.

8.2.7 *Hints:* Tasks Hand-Crafted by a Domain Expert

Extra outputs can be used to *inject rule hints* into nets about what they should learn [32, 33]. This is MTL where the extra tasks are carefully engineered to coerce the net to learn specific internal representations. Hints can also be provided to backprop nets via extra terms in the error signal backpropagated for the main task output [1, 2]. The extra error terms constrain what is learned to satisfy desired properties of main task such as monotonicity [31], symmetry, or transitivity with respect to certain sets of inputs. MTL, which does not use extra error terms on task outputs, could be used in concert with these techniques.

8.2.8 Handling *Other* Categories in Classification

In real-world applications of digit recognition, some of the images given to the classifier may be alphabetic characters or punctuation marks instead of digits. One way to prevent accidentally classifying a "t" as a one or seven is to create an "other" category that is the correct classification for non-digit images. The large variety of characters mapped to this "other" class makes learning this class potentially very difficult. MTL suggests an alternate way to do this. Split the "other" class into separate classes for the individual characters that are trained in parallel with the main digit tasks. A single output coding for the "other" class can be used, as well. Breaking the "other" category into multiple tasks gives the net more learnable error signal for these cases [26].

8.2.9 Sequential Transfer

MTL is parallel transfer. Often tasks arise serially and we can't wait for all of them to begin learning. In these cases we can use parallel transfer to perform sequential transfer. If the training data can be stored, do MTL using whatever tasks are available when it is time to start learning, and re-train as new tasks or new data arise. If training data cannot be stored, or if we already have models for which data is not available, we can still use MTL. Use the models to generate synthetic data that is then used as extra training signals. This approach to sequential transfer avoids catastrophic interference (forgetting old tasks while learning new ones). Moreover, it is applicable where the analytical methods of evaluating domain theories required by some serial transfer methods [29, 34] are not available. For example, the domain theory need not be differentiable, it only needs to make predictions. One issue that arises when synthesizing data from prior models is what distribution to sample from. See [16] for a discussion of synthetic sampling.

8.2.10 Similar Tasks with Different Data Distributions

Sometimes there are multiple instances of the same problem, but the distribution of samples differs for each instantiation. For example, most hospitals diagnose and treat the same diseases, but the demographics of the patients each hospital serves is different. Hospitals in Florida see older patients, urban hospitals see poorer patients, etc. Models trained separately for each hospital would perform best, but often there is insufficient data to train a separate model for each hospital. Pooling the data, however, may not lead to models that are accurate for each hospital. MTL provides one solution to this problem. Use one net to make predictions for each hospital, using a different output on the net for each hospital. Because each patient is a training case for only one hospital, error can be backpropagated only through the one output that has a target value for each input vector.

8.2.11 Learning with Hierarchical Data

In many domains, the data falls in a hierarchy of classes. Most applications of machine learning to hierarchical data make little use of the hierarchy. MTL provides one way of exploiting hierarchical information. When training a model to classify data at one level in the hierarchy, include as extra tasks the classification tasks that arise for ancestors, descendants, and siblings of the current classification task. The easiest way to to accomplish this is to train one MTL net to predict all class distinctions in the hierarchy at the same time.

8.2.12 Some Inputs Work Better as Outputs

The common practice in backprop nets is to use all features that will be available for test cases as inputs, and have outputs only for tasks that need to be predicted. On real problems, however, learning often works better given a carefully selected subset of the features to use inputs[7, 23, 24]. One way to benefit from features not used as inputs is to use them as extra outputs for MTL. We've done experiments with both synthetic and real problems where moving some features from the input side of the net to the output side of the net improves performance on the main task. We use feature selection to select those features that should be used as inputs, and then treat some of the remaining features as extra tasks.

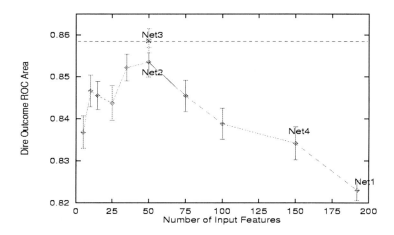

Fig. 8.9. ROC Area on the Pneumonia Risk Prediction Task vs. the number of input features used by the backprop net.

Figure 8.9 shows the ROC Area on a pneumonia problem as the number of input features on the backprop net varies.[7] ROC Areas closer to 1 indicate better performance. There are 192 features available for most patients. Using all 192 features as inputs (Net1) is suboptimal. Better performance is obtained by using the first 50 features selected with feature selection (Net2). The horizontal line at the top of the graph (Net3) shows the ROC Area obtained by using the first 50 features as inputs, and the *next* 100 features as extra *outputs*. Using these same 150 features all as inputs (Net4) yields worse performance.[8]

8.3 Getting the Most out of MTL

The basic machinery for doing multitask learning in neural nets is present in backprop. Backprop, however, was not designed to do MTL well. This chapter presents suggestions for how to make MTL in backprop nets work better. Some of the suggestions are counterintuitive, but if not used, can cause MTL to hurt generalization on the main task instead of helping it.

8.3.1 Use Large Hidden Layers

The basic idea behind MTL in backprop nets is that what is learned in the hidden layer for one task can be useful to other tasks. MTL works when tasks share hidden units. One might think that small hidden layers would help MTL by promoting sharing between tasks.

---

[7] This is not the same pneumonia problem used in section 8.2.1.
[8] Although the 95% confidence intervals for Net2 and Net3 overlap with ten trials, a paired t-test shows the results are significant at .01.

For the kinds of problems we've examined here, this usually does not work. Usually, tasks are different enough that much of what each task needs to learn does not transfer to many (or any) other tasks. Using a large hidden layer insures that there are enough hidden units for tasks to learn independent hidden layer representations when they need to. Sharing can still occur, but only when the overlap between the hidden layer representations for different tasks is strong. In many real-world problems, the loss in accuracy that results from forcing tasks to share by keeping the hidden layer small is larger than the benefit that arises from the sharing. Usually it is important to use large hidden layers with MTL.

8.3.2 Do Early Stopping for Each Task Separately

The classic NETtalk application [30] used one trained both phonemes and stresses on one backprop net. NETtalk is an early example of MTL. But the builders of NETtalk viewed the multiple outputs as codings for a single problem, not as independent tasks that benefited each other by being trained together.

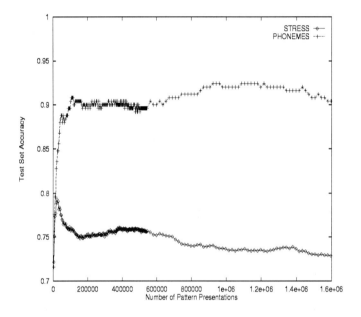

Fig. 8.10. On NETtalk, the Stress task trains very quickly and overfits long before the Phoneme task reaches peak performance.

Figure 8.10 shows the learning curves for the phoneme and stress subtasks separately. It is clear that the stress tasks begin to overfit before the phoneme tasks reach peak performance. Better performance could be obtained on NETtalk by doing early stopping on the stress and phoneme tasks individually, or by balancing their learning rates so they reach peak performance at the same time.

Early stopping prevents overfitting by halting the training of error-driven procedures like backprop before they achieve minimum error on the training set (see chapter 2). Recall the steering prediction problem from section 8.2.6. We applied MTL to this problem by training a net on eight extra tasks in addition to the main steering task. Figure 8.11 shows nine learning curves, one for each of the tasks on this MTL net. Each graph is the validation set error during training.

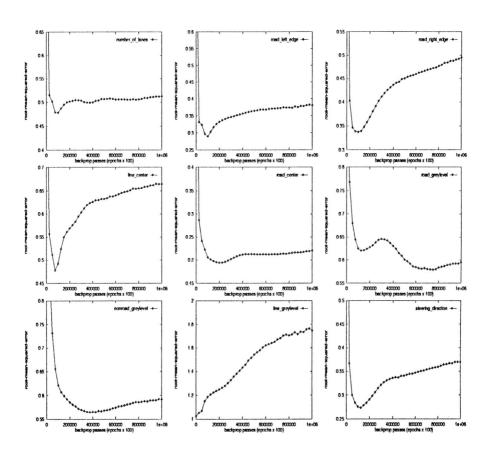

Fig. 8.11. Test-Set Performance of MTL Net Trained on Nine Tasks.

Table 8.4 shows the best place to halt each task. There is no one epoch where training can be stopped so as to achieve maximum performance on all tasks. If all tasks are important, and one net is used to predict all the tasks, halting training where the error summed across all outputs is minimized is the best you can do. Figure 8.12 shows the combined RMS error of the nine tasks. The best average RMSE occurs at 75,000 backprop passes.

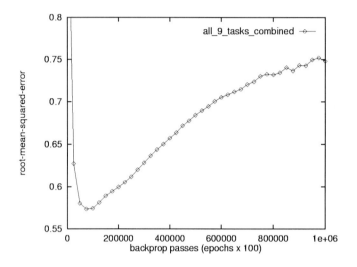

Fig. 8.12. Combined Test-Set Performance on all Nine Tasks.

But using one net to make predictions for all the tasks is suboptimal. Better performance is achieved by using the validation set to do early stopping on each output individually. The trick is to make a copy of the net at the epoch where performance on each task is best, and use this copy to make predictions for that task. After making each copy, continue training the net until the other tasks reach peak performance. Sometimes, it is best to continue training all outputs on the net, including those that have begun to overfit. Sometimes, it is better to stop training (or use a lower learning rate) for outputs that have begun to overfit. Keep in mind that once an output has begun to overfit, we no longer care how well the net performs on that task because we have a copy of the net from an earlier epoch when performance on that task was best. The only reason to continue training the task is because it may benefit other tasks that have not reached peak performance yet.

Table 8.4 compares the performance of early stopping done per task with the performance one obtains by halting training for the entire MTL net at one place using the combined error. On average, early stopping for tasks individually reduces error 9.0%. This is a large difference. For some tasks, the performance of the MTL net is worse than the performance of STL on this task if the MTL net is not halted on that task individually.

Before leaving this topic, it should be mentioned that the training curves for the individual outputs on an MTL net are not necessarily monotonic. While

8. A Dozen Tricks with Multitask Learning

Table 8.4. Performance of MTL on the nine tasks in the steering domain when training is halted on each task individually compared with halting using the combined error across all tasks.

TASK	Halted Individually		Halted Combined		Difference
	BP Pass	Performance	BP Pass	Performance	
1: 1 or 2 Lanes	100000	0.444	75000	0.456	2.7%
2: Left Edge	100000	0.309	75000	0.321	3.9%
3: Right Edge	100000	0.376	75000	0.381	1.3%
4: Line Center	75000	0.486	75000	0.486	0.0%
5: Road Center	200000	0.208	75000	0.239	14.9%
6: Road Greylevel	750000	0.552	75000	0.680	23.2%
7: Edge Greylevel	375000	0.518	75000	0.597	15.3%
8: Line Greylevel	1	1.010	75000	1.158	14.7%
9: Steering	125000	0.276	75000	0.292	5.8%

it is not unheard of for the test-set error of an STL net to be multimodal, the training-set error for an STL net should descend monotonically or become flat. This is not true for the errors of individual outputs on an MTL net. The training-set error summed across all outputs should never increase, but any one output may exhibit more complex behavior. The graph for road_greylevel (graph number 6) in Figure 8.11 shows a multimodal test-set curve. The training set curve for this output is similar. This makes judging when to halt training more difficult with MTL nets. Because of this, we always do early stopping on MTL nets by training past the epoch where performance on each task appears to be best, and either retrain the net a second time (with the same random seed) to get the copies, or are careful to keep enough copies during the first run that we have whatever copies we will need.

8.3.3 Use Different Learning Rates for Different Tasks

Is it possible to control the rates at which different tasks train so they each reach their best performance at the same time? Would best performance on each task be achieved if each task reached peak performance at the same time? If not, is it better for extra tasks to learn slower or faster than the main task?

The rate at which different tasks learn using vanilla backprop is rarely optimal for MTL. Task that train slower than the main task will not have learned enough to help the main task when training on the main task is stopped. Tasks that train faster than the main task may overfit so much before the main task is learned well that either what they have learned is no longer useful to the main task, or they may drive the main task into premature overfitting.

The most direct method of controlling the rate at which different tasks learn is to use a different learning rate for each task, i.e., for each output. We have experimented with using gradient descent to find learning rates for each extra output to maximize the generalization performance on the main task. Table 8.5 shows the performance on the main steering task before and after optimizing the

188   Rich Caruana

learning rates of the other eight extra tasks. Optimizing the learning rates for the extra MTL tasks improved performance on the main task an additional 11.5%. This improvement is over and above the original improvement of 15%–25% for MTL over STL.

Table 8.5. Performance of MTL on the main Steering Direction task before and after optimizing the learning rates of the other eight extra tasks.

TRIAL	Before Optimization	After Optimization	Difference
Trial 1	0.227	0.213	-6.2%
Trial 2	0.276	0.241	-12.7%
Trial 3	0.249	0.236	-5.2%
Trial 4	0.276	0.231	-16.3%
Trial 5	0.276	0.234	-15.2%
Average	0.261	0.231	-11.5% *

Examining the training curves for all the tasks as the learning rates are optimized shows that the changes in the learning rates of the extra tasks has a significant effect on the rate at which the extra tasks are learned. Interestingly, it also has a significant effect on the rate at which the main task is learned. This is surprising because we keep the learning rate for the main task fixed during optimization. Perhaps even more interesting is the fact that optimizing the learning rate to maximize the generalization accuracy of the main task also improved generalization on the extra tasks nearly as much as it helped the main task. What is good for the goose appears to be good for the gander.

8.3.4 Use a Private Hidden Layer for the Main Task

Sometimes the optimal number of hidden units is 100 hidden units or more per output. If there are hundreds of extra tasks this translates to thousands of hidden units. This not only creates computational difficulties, but degrades performance on the main task because most of the hidden layer repersentation is constructed for other tasks. The main task output unit has a massive hidden unit selection problem as it tries to use only those few hidden units that are useful to it.

Figure 8.13 shows a net architecture that solves this problem. Instead of one hidden layer shared equally by all tasks, there are two disjoint hidden layers. Hidden layer 1 is a private hidden layer used only by the main task(s). Hidden layer 2 is shared by the main task(s) and the extra tasks. This is the hidden layer that supports MTL transfer. Because the main task sees and affects the shared hidden layer, but the extra tasks do not affect the hidden layer reserved for the main tasks(s), hidden layer 2 can be kept small without hurting the main task.

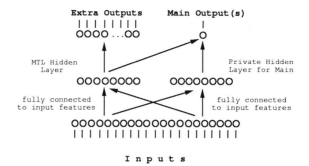

Fig. 8.13. MTL Architecture With a Private Hidden Layer for the Main Task(s), and a Shared Hidden Layer Used by the Main Task(s) and the Extra Tasks.

8.4 Chapter Summary

We usually think of the inputs to a backprop net as the place where information is given to the net, and the outputs as the place where the net outputs predictions. Backprop, however, pushes information into the net through the *outputs* during training. The information fed into a net through its outputs is as important as the information fed into it through its inputs.

Multitask Learning is a way of using the outputs of a backprop net to push additional information into the net during training. If the net architecture allows sharing of what is learned for different outputs, this additional information can help the main task be learned better. (See [5, 6, 4, 8, 3, 9, 11, 10, 20, 35, 36, 12, 21, 13, 14] for additional discussion about multitask learning.)

MTL trains multiple tasks in parallel not because this is a more efficient way to learn multiple tasks, but because the information in the training signals for the extra tasks can help the main task be learned better. Sometimes what is optimal for the main task is not optimal for the extra tasks. It is important to optimize the technique so that performance on the *main* task is best, even if this hurts performance on the extra tasks. If the extra tasks are important too, it may be best to rerun learning for each important extra task, with the technique optimized for each task one at a time.

This chapter presented a number of opportunities for using extra outputs to leverage information that is available in real domains. The trick in most of these applications is to view the outputs of the net as inputs that are used only during learning. Any information that is available when the net is trained, but which would not be available later when the net is used for prediction, can potentially be used as extra outputs. There are many domains where useful extra tasks will be available. The list of prototypical domains provided in this chapter is not complete. More kinds of extra tasks will be identified in the future.

Acknowledgements R. Caruana was supported by ARPA grant F33615-93-1-1330, NSF grant BES-9315428, and Agency for Health Care Policy grant

HS06468. The work to find input features that worked better when used as extra outputs is joint work with Virginia de Sa, who was supported by postdoctoral fellowship from the Sloan Foundation. We thank the University of Toronto for the Xerion Simulator, and D. Koller and M. Sahami for the use of their feature selector.

References

1. Y. S. Abu-Mostafa, "Learning from Hints in Neural Networks," *Journal of Complexity*, 1990, 6(2), pp. 192–198.
2. Y. S. Abu-Mostafa, "Hints," *Neural Computation*, 1995, **7**, pp. 639-671.
3. J. Baxter, "Learning Internal Representations," *COLT-95*, Santa Cruz, CA, 1995.
4. J. Baxter, "Learning Internal Representations," Ph.D. Thesis, The Flinders Univeristy of South Australia, Dec. 1994.
5. R. Caruana, "Multitask Learning: A Knowledge-Based Source of Inductive Bias," *Proceedings of the 10th International Conference on Machine Learning*, ML-93, University of Massachusetts, Amherst, 1993, pp. 41-48.
6. R. Caruana, "Multitask Connectionist Learning," *Proceedings of the 1993 Connectionist Models Summer School*, 1994, pp. 372-379.
7. R. Caruana and D. Freitag, "Greedy Attribute Selection," *ICML-94*, 1994, Rutgers, NJ, pp. 28-36.
8. R. Caruana, "Learning Many Related Tasks at the Same Time with Backpropagation," *NIPS-94*, 1995, pp. 656-664.
9. R. Caruana, S. Baluja, and T. Mitchell, "Using the Future to "Sort Out" the Present: Rankprop and Multitask Learning for Medical Risk Prediction," *Advances in Neural Information Processing Systems 8*, (Proceedings of NIPS-95), 1996, pp. 959-965.
10. R. Caruana, and V. R. de Sa, "Promoting Poor Features to Supervisors: Some Inputs Work Better As Outputs," *NIPS-96*, 1997.
11. R. Caruana, "Multitask Learning," *Machine Learning*, 28, pp. 41-75, 1997.
12. R. Caruana, "Multitask Learning," Ph.D. thesis, Carnegie Mellon University, CMU-CS-97-203, 1997.
13. R. Caruana and J. O'Sullivan, "Multitask Pattern Recognition for Autonomous Robots," to appear in *The Proceedings of the IEEE Intelligent Robots and Systems Conference,*, (IROS'98), Victoria, 1998.
14. R. Caruana and V. R. de Sa, "Using Feature Selection to Find Inputs that Work Better as Outputs," to appear in *The Proceedings of the International Conference on Neural Nets*, (ICANN'98), Sweden, 1998.
15. G. F. Cooper, C. F. Aliferis, R. Ambrosino, J. Aronis, B. G. Buchanan, R. Caruana, M. J. Fine, C. Glymour, G. Gordon, B. H. Hanusa, J. E. Janosky, C. Meek, T. Mitchell, T. Richardson, and P. Spirtes, "An Evaluation of Machine Learning Methods for Predicting Pneumonia Mortality," *Artificial Intelligence in Medicine 9*, 1997, pp. 107-138.
16. M. Craven and J. Shavlik, "Using Sampling and Queries to Extract Rules from Trained Neural Networks," *Proceedings of the 11th International Conference on Machine Learning*, ML-94, Rutgers University, New Jersey, 1994, pp. 37-45.
17. I. Davis and A. Stentz, "Sensor Fusion for Autonomous Outdoor Navigation Using Neural Networks," *Proceedings of IEEE's Intelligent Robots and Systems Conference*, 1995.

18. T. G. Dietterich and G. Bakiri, "Solving Multiclass Learning Problems via Error-Correcting Output Codes," *Journal of Artificial Intelligence Research*, 1995, 2, pp. 263-286.
19. M. J. Fine, D. Singer, B. H. Hanusa, J. Lave, and W. Kapoor, "Validation of a Pneumonia Prognostic Index Using the MedisGroups Comparative Hospital Database," *American Journal of Medicine*, 1993.
20. Ghosn, J. and Bengio, Y., "Multi-Task Learning for Stock Selection," *NIPS-96*, 1997.
21. T. Heskes, "Solving a Huge Number of Similar Tasks: A Combination of Multitask Learning and a Hierarchical Bayesian Approach," *Proceedings of the 15th International Conference on Machine Learning*, Madison, Wisconsin, pp./233-241, 1998.
22. L. Holmstrom and P. Koistinen, "Using Additive Noise in Back-propagation Training," *IEEE Transactions on Neural Networks*, 1992, 3(1), pp. 24-38.
23. G. John, R. Kohavi, and K. Pfleger, "Irrelevant Features and the Subset Selection Problem," *ICML-94*, 1994, Rutgers, NJ, pp. 121-129.
24. D. Koller and M. Sahami, "Towards Optimal Feature Selection," *ICML-96*, Bari, Italy, 1996, pp. 284-292.
25. Y. Le Cun, B. Boser, J. S. Denker, D. Henderson, R. E. Howard, W. Hubbard, and L. D. Jackal, "Backpropagation Applied to Handwritten Zip-Code Recognition," *Neural Computation*, 1989, 1, pp. 541-551.
26. Y. Le Cun, private communication, 1997.
27. P. W. Munro and B. Parmanto, "Competition Among Networks Improves Committee Performance," to appear in *Advances in Neural Information Processing Systems 9*, (Proceedings of NIPS-96), 1997.
28. D. A. Pomerleau, "Neural Network Perception for Mobile Robot Guidance," Doctoral Thesis, Carnegie Mellon University: *CMU-CS-92-115*, 1992.
29. L. Y. Pratt, J. Mostow, and C. A. Kamm, "Direct Transfer of Learned Information Among Neural Networks," *Proceedings of AAAI-91*, 1991.
30. T. J. Sejnowski and C. R. Rosenberg, "NETtalk: A Parallel Network that Learns to Read Aloud," John Hopkins: *JHU/EECS-86/01*, 1986.
31. J. Sill and Y. Abu-Mostafa, "Monotonicity Hints," to appear in *Neural Information Processing Systems 9*, (Proceedings of NIPS-96), 1997.
32. S. C. Suddarth and A. D. C. Holden, "Symbolic-neural Systems and the Use of Hints for Developing Complex Systems," *International Journal of Man-Machine Studies*, 1991, 35(3), pp. 291-311.
33. S. C. Suddarth and Y. L. Kergosien, "Rule-injection Hints as a Means of Improving Network Performance and Learning Time," *Proceedings of EURASIP Workshop on Neural Nets*, 1990, pp. 120-129.
34. S. Thrun, *Explanation-Based Neural Network Learning: A Lifelong Learning Approach*, 1996, Kluwer Academic Publisher.
35. S. Thrun and L. Pratt, editors, *Machine Learning. Second Special Issue on Inductive Transfer*, 1997.
36. S. Thrun and L. Pratt, editors, *Learning to Learn*, Kluwer, 1997.
37. R. Valdes-Perez and H. A. Simon, "A Powerful Heuristic for the Discovery of Complex Patterned Behavior," *Proceedings of the 11th International Conference on Machine Learning*, ML-94, Rutgers University, New Jersey, 1994, pp. 326-334.
38. A. Weigend, D. Rumelhart, and B. Huberman, "Generalization by Weight-Elimination with Application to Forecasting," *Advances in Neural Information Processing Systems 3*, (Proceedings of NIPS-90), 1991, pp. 875-882.

9
Solving the Ill-Conditioning in Neural Network Learning

Patrick van der Smagt and Gerd Hirzinger

German Aerospace Research Establishment,
Institute of Robotics and System Dynamics,
P. O. Box 1116, D–82230 Wessling, Germany,
smagt@dlr.de   http://www.robotic.dlr.de/Smagt/

Abstract. In this paper we investigate the feed-forward learning problem. The well-known ill-conditioning which is present in most feed-forward learning problems is shown to be the result of the structure of the network. Also, the well-known problem that weights between 'higher' layers in the network have to settle before 'lower' weights can converge is addressed. We present a solution to these problems by modifying the structure of the network through the addition of linear connections which carry shared weights. We call the new network structure the *linearly augmented feed-forward network*, and it is shown that the universal approximation theorems are still valid. Simulation experiments show the validity of the new method, and demonstrate that the new network is less sensitive to local minima and learns faster than the original network.

9.1   Introduction

One of the major problems with feed-forward network learning remains the accuracy and speed of the learning algorithms. Since the learning problem is a complex and highly nonlinear one [12,4], iterative learning procedures must be used to solve the optimisation problem [2,14]. A continuing desire to improve the behavior of the learning algorithm has lead to many excellent optimisation algorithms which are especially tailored for feed-forward network learning.

However, an important problem is the particular form of the error function that represents the learning problem. It has long been noted [10,16] that the derivatives of the error function are usually ill-conditioned. This ill-conditioning is reflected in error landscapes which contain many saddle points and flat areas.

Although this problem can be solved by using stochastic learning methods (e.g., [9,1,13]), these methods require many learning iterations in order to find an optimum, and are therefore not suited for problems where fast learning is a requirement. We therefore remain focused on gradient-based learning methods. Algorithms exist which attempt to find well-behaving minima [7], yet an important factor of the learning problem remains the structure of the feed-forward network.

In this chapter an explanation of the ill-conditioned learning problem is provided as well as a solution to alleviate this problem. Section 9.2 formally introduces the learning problem, and describes the problem of singularities in the learn matrices. In section 9.3 the cause of the singularities are analyzed and an adapted learning rule is introduced which alleviates this problem. Section 9.4 discusses a few applications.

9.2 The Learning Process

With a neural network $\mathcal{N} : \Re^N \times \Re^n \to \Re^M$ we create an approximation to a set of p learning samples $\{(\boldsymbol{x}_1, \boldsymbol{y}_1), (\boldsymbol{x}_2, \boldsymbol{y}_2), \ldots, (\boldsymbol{x}_p, \boldsymbol{y}_p)\}$, with $\boldsymbol{x}_i \in \Re^N$ and $\boldsymbol{y}_i \in \Re^M$, for which holds that $\forall 1 \leq i \leq p : \mathcal{F}(\boldsymbol{x}_i) = \boldsymbol{y}_i$. The function $\mathcal{F} : \Re^N \to \Re^M$ is called the **model function**.

Let n be the number of free parameters W of the network. In this particular case we are interested in approximating the learning samples rather than the underlying function $\mathcal{F}$, or assume that the p learning samples are representative for $\mathcal{F}$.

The oth output of the function that is represented by the neural network can be written as

$$\mathcal{N}(\boldsymbol{x}, W)_o = \sum_h w_{ho} s \left(\sum_i w_{ih} x_i + \theta_h \right) + \theta_o \tag{9.1}$$

where $s(x)$ the transfer function, o indicates an output unit, h a hidden unit, i an input unit. The symbol w_{ho} indicates an element of W corresponding with the connection from hidden unit h to output unit o; w_{ih} is similarly used for a connection from input unit i to hidden unit h. Finally, θ is a bias weight and therefore an element of W. An exemplar feed-forward network with one input and output unit and two hidden units is depicted in figure 9.1.

The learning task consists of minimizing an **approximation error**, which is usually defined as

$$E_\mathcal{N}(W) = \sum_{i=1}^{p} \| \mathcal{N}(\boldsymbol{x}_i, W) - \boldsymbol{y}_i \| \tag{9.2}$$

where for $\| \cdot \|$ we prefer to use the L_2 norm. We will leave the subscript $\mathcal{N}$ out when no confusion arises. $E(W)$ is (highly) nonlinear in W, such that iterative search techniques are required to find the W for which $E(W)$ is sufficiently small.

Finally we define the **residual pattern error**

$$e_{M(i-1)+j} = \| \mathcal{N}(\boldsymbol{x}_i, W)_j - \boldsymbol{y}_{ij} \|, \tag{9.3}$$

i.e., the error in the j'th output value for the i'th learning sample.

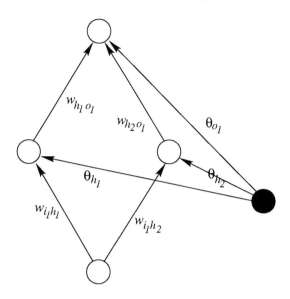

Fig. 9.1. An exemplar feed-forward neural network. The circles represent neurons; the black filled circle is a bias unit, and always carries an activation value of '1'.

9.2.1 Learning Methodology

Gradient based learning methods are characterized by considering low-order terms from the Taylor expansion to the approximation error,

$$E(W + W_0) = E(W_0) + \nabla E|_{W_0} W + W^T \nabla^2 E|_{W_0} W + \ldots \quad (9.4)$$

In most cases more than the second order term is neglected. We define

$$\tilde{E}_1(W + W_0) = E(W_0) + \nabla E|_{W_0} W \quad (9.5)$$

and

$$\tilde{E}_2(W + W_0) = \tilde{E}_1(W) + W^T \nabla^2 E|_{W_0} W \quad (9.6)$$

being the first-order and second-order approximation to E, respectively.

By locally considering the approximation error as a first- or second-order function of W, we can use several existing approximation methodologies to minimize E. When, according to the local second-order approximation information, E is minimized, the local information is updated and a second minimization step is carried out. This process is repeated until a minimum is found.

Well-known minimization methods are steepest descent (a variant of which is known as error backpropagation), conjugate gradient optimisation, Levenberg-Marquardt optimisation, variable metric methods, and (quasi-) Newton methods. Each of these methods has its advantages and disadvantages which are discussed elsewhere [14].

The optimisation methods work in principle as follows. By considering the second-order approximation to E, an optimum can be analytically found if ∇E

and $\nabla^2 E$ are known. After the optimum has been located, ∇E and $\nabla^2 E$ are recomputed using the local information, and the minimum is relocated. This process is repeated until a minimum is found.

Naturally, the success of this approach stands or falls with the form of the error function. If the error function is not too complex but smooth, and can be reasonably approximated by a quadratic function, the discussed optimisation methods are a reliable and fast way of finding minima. In feed-forward network training, however, the error functions appear to have a large number of flat areas where minimization is a difficult task due to limited floating point accuracy.

9.2.2 Condition of the Learning Problem

The flat areas of the error function can be formalized as follows. We define the $(Mp) \times n$ Jacobian matrix as

$$J \equiv \begin{pmatrix} \nabla e_1^T \\ \nabla e_2^T \\ \vdots \\ \nabla e_{Mp}^T \end{pmatrix} \quad \text{where} \quad \nabla e_k \equiv \begin{pmatrix} \frac{\partial e_k}{\partial w_1} \\ \frac{\partial e_k}{\partial w_2} \\ \vdots \\ \frac{\partial e_k}{\partial w_n} \end{pmatrix} \tag{9.7}$$

such that we can write J as

$$J \equiv \begin{pmatrix} \frac{\partial e_1}{\partial w_1} & \frac{\partial e_1}{\partial w_2} & \cdots & \frac{\partial e_1}{\partial w_n} \\ \frac{\partial e_2}{\partial w_1} & \frac{\partial e_2}{\partial w_2} & \cdots & \frac{\partial e_2}{\partial w_n} \\ \vdots & & \ddots & \vdots \\ \frac{\partial e_{Mp}}{\partial w_1} & \frac{\partial e_{Mp}}{\partial w_2} & \cdots & \frac{\partial e_{Mp}}{\partial w_n} \end{pmatrix}. \tag{9.8}$$

In the learning process we thus encounter that $\nabla E = 2J^T e$.

We furthermore define the Hessian $H = \nabla^2 E$, which is the matrix of second order derivatives of E. We are interested in the eigenvalues and eigenvectors of H. Since H is a positive semidefinite symmetric matrix close to a local minimum, its eigenvalues are all positive real numbers. When an eigenvalue is very small, the effect of moving in the direction of the corresponding eigenvector on the approximation error is very small. This means that, in that direction, the error function is (nearly) singular. The singularity of the error function can be expressed in the condition of H, which is defined as the quotient of its largest and its smallest eigenvalues.

As mentioned above, a bad condition of H may occur at minima or flat points in the error function $E(W)$. It appears that feed-forward learning tasks are generally characterized by having a singular or near-singular Hessian matrix. Although the said learning methods are mathematically not influenced by a badly conditioned Hessian, it does lead to inaccuracies due to the limited floating point accuracy of the digital computer.

9.2.3 What Causes the Singularities

Reference [10] lists a few cases in which the Hessian may become singular or near-singular. The listed reasons are associated with the ill character of the sigmoid transfer function. Typical for this function is the fact that $\lim_{x \to \infty} s(x) = c_+$ and $\lim_{x \to -\infty} s(x) = c_-$. Also, bad conditioning can be the result of uncentered data, and can also be alleviated [11].

Assuming that some neuron is in this 'saturated' state for all learning patterns, its input weights will have a delta equal to 0 (according to the backpropagation rule) such that these weights will never change. For each of the incoming weights of this neuron, this leads to a 0-row in H, and therefore singularity.

However, there is another important reason for singularity, which may especially occur after network initialization. When a multi-layer feed-forward network has a small (e.g., less than 0.1) weight leaving from a hidden unit, the influence of the *weights that feed into this hidden unit* is significantly reduced. Therefore the $\partial e/\partial w_k$ will be close to 0, leading to near-null rows in J and a near-singular H.

We observe that this kind of singularity is very common and touches a characteristic problem in feed-forward network learning: The gradients in the lower-layer weights are influenced by the higher-layer weights. A related problem is the influence that the change of the weights between the hidden and output units have on the change of the input weights; when they rapidly and frequently change, as will be the case during the initial stages of learning, the lower weights will have nonsensical repeated perturbations.

9.2.4 Definition of Minimum

A minimum is said to be reached when the derivative of the error function is zero, i.e., $\partial E/\partial w_{1 \leq k \leq n} = 0$. Since the gradient of the error function equals the column-sum of the Jacobian, a minimum has been reached when

$$\forall 1 \leq k \leq n : \sum_{i=1}^{Mp} \frac{\partial e_i}{\partial w_k} = 0, \qquad (9.9)$$

i.e., when each of the columns adds up to 0. Eq. (9.9) defines the minimum for a batch-learning system: the gradient, when summed over all learning samples, should be 0. This also means, however, that it may occur that elements of a column-sum cancel each other out, even when not all elements of the Jacobian are 0. Differently put, it may occur that the gradient for some patterns are non-zero, whereas the gradients sum up to zero.

The optimal case is reached when *all* elements of the Jacobian are 0. This means, of course, that the residual error of each pattern is 0.

9.3 Local Minima are Caused by BackPropagation

In this section we propose a new neural network structure which alleviates the above problems. In the standard backpropagation learning procedure, the gradient of the error function with respect to the weights is determined by computing the following steps for each learning pattern:

1. For each output unit o, compute the delta $\delta_o = y_o - a_o$ where a_o is the activation value for that unit, when a learning sample is presented.
2. Compute the weight derivative for the weights w_{ho} from the hidden to output units:
$$\Delta w_{ho} = \delta_o a_h$$
where a_h is the activation value for the hidden unit.
3. Compute the delta for the hidden unit:
$$\delta_h = \sum_o \delta_o w_{ho} s'(a_h).$$
4. Compute the weight derivative for the weights w_{ih} from the input to hidden units:
$$\Delta w_{ih} = \delta_h a_i = \sum_o \delta_o w_{ho} s'(a_h) a_i. \qquad (9.10)$$

The gradient is then computed as the summation of the Δw's.

From (9.10) we can see that there are four cases when the gradient for a weight from an input to a hidden unit is negligible, such that the corresponding row and column in the Hessian are near-zero:

– When δ_o is small. This is correct, since that case means that the output of the network is close to its desired output.
– When w_{ho} is small. This is an undesired situation: A small weight from hidden to output unit paralyzes weights from input to hidden units. This is especially important since the weight might have to change its value later.
– When $s'(a_h)$ is small; this occurs when the weight w_{ih} is large. Again, this saturation type of paralysis is undesired.
– Finally, when a_i is small. This is desired: When the input value is insignificant, it should have no influence on the output.

9.3.1 A New Neural Network Structure

In order to alleviate these problems, we propose a change to the learning system of (9.10) as follows:
$$\Delta w_{ih} = \sum_o \delta_o (w_{ho} s'[a_h] + c) a_i = \delta_h a_i + c \sum_o \delta_o a_i.$$

By adding a constant c to the middle term, we can solve both paralysis problems. In effect, an extra connection from each input unit to each output unit is created, with a weight value coupled to the weight from the input to hidden unit.

9. Solving the Ill-Conditioning in Neural Network Learning

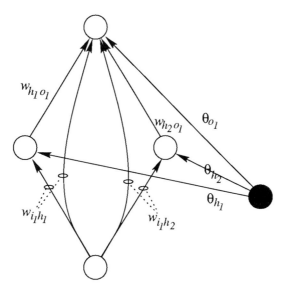

Fig. 9.2. An exemplar adapted feed-forward neural network.

Although c can be made a learnable parameter, in the sequel we will assume $c = 1$. The o'th output of the neural network is now computed as

$$\mathcal{M}(\boldsymbol{x}, W)_o = \sum_h \left(w_{ho} s \left[\sum_i w_{ih} x_i + \theta_h \right] + \sum_i w_{ih} x_i \right) + \theta_o. \qquad (9.11)$$

We call the new network the **linearly augmented feed-forward network**. The structure of this network is depicted in figure 9.2. Note the equivalence of $\mathcal{N}$ and $\mathcal{M}$, viz.,

$$\mathcal{M}(\boldsymbol{x}, W)_o \equiv \mathcal{N}(\boldsymbol{x}, W)_o + \sum_h \sum_i w_{ih} x_i. \qquad (9.12)$$

9.3.2 Influence on the Approximation Error E

Although the optimal W will be different for the networks $\mathcal{N}$ and $\mathcal{M}$, we can still compare the forms of the error functions $E_{\mathcal{N}}$ vs. $E_{\mathcal{M}}$. Using (9.2) and (9.11) we can compute the error in the approximation of a single learning sample $(\boldsymbol{x}, y)$ with N inputs, κ hidden units, and a single output:

$$E_{\mathcal{M}}(W)^2 = (\mathcal{M}[\boldsymbol{x}, W] - y)^2$$
$$= \left(\mathcal{N}(\boldsymbol{x}, W) - y + \sum_i \sum_h w_{ih} x_i \right)^2$$
$$= E_{\mathcal{N}}(W)^2 + 2 \sum_i \sum_h w_{ih} x_i (\mathcal{N}[\boldsymbol{x}, W] - y) + \left(\sum_i \sum_h w_{ih} x_i \right)^2. \quad (9.13)$$

The error for the network $\mathcal{M}$ differs from the error for $\mathcal{N}$ in two terms. When we consider $E_\mathcal{M}$ at those values of W where $E_\mathcal{N}$ is minimal, we can see that the difference between $E_\mathcal{N}$ and $E_\mathcal{M}$ consists of a normalization term $\|\sum_h w_h^T x\|$; w_h is the vector of weights connecting the inputs to hidden unit h. This non-negative term will only be zero when the vectors $w_1^T x$, $w_2^T x$, ..., $w_\kappa^T x$ cancel each other out for each input vector x which is in the training set. In words: $E_\mathcal{M}(W)^2$ introduces a penalty for hidden units doing the same thing, thus making the network use its resources more efficiently.

9.3.3 $\mathcal{M}$ and the Universal Approximation Theorems

It has been shown in various publications [3, 6, 8] that the ordinary feed-forward neural network $\mathcal{N}$ can represent any Borel-measurable function with a single layer of hidden units which have sigmoidal or Gaussian activation functions in the hidden layer.

Theorem 1. *The network $\mathcal{M}$ can represent any Borel-measurable function with a single layer of hidden units which have sigmoidal or Gaussian activation functions in the hidden layer.*

Proof. We show that any network $\mathcal{N}$ can be written as a network $\mathcal{M}$; therefore, the class of networks $\mathcal{M}$ are universal approximators.

By using (9.1) and (9.11),

$$\mathcal{N}(x, W)_o = \sum_{h=1}^{\kappa} w_{ho} s\left(\sum_i w_{ih} x_i + \theta_h\right) + \theta_o$$

$$= \sum_{h=1}^{\kappa} \left(w_{ho} s\left[\sum_i w_{ih} x_i + \theta_h\right] + \sum_i w_{ih} x_i\right) + \theta_o - \sum_{h=1}^{\kappa} \sum_i w_{ih} x_i$$

$$= \mathcal{M}(x, W)_o + \sum_{l=\kappa+1}^{2\kappa} \left(0 \left[\sum_i -w_{i,l-\kappa} x_i + 0\right] + \sum_i -w_{i,l-\kappa} x_i\right) + 0$$

$$= \mathcal{M}(x, W)_o + \mathcal{M}(x, V)_o$$

where V is a weight matrix such that the elements of V corresponding to the weights from hidden to output units are 0, and the other weights equal the negation of its W counterparts. Furthermore, bias weights are set to 0.

The sum $\mathcal{M}(x, W)_o + \mathcal{M}(x, V)_o$ represents two $\mathcal{M}$-networks, which can also be written as a single $\mathcal{M}(x, W')$-network with the double amount of hidden units, where $W' = [W V]$.

Using the theorems from [3, 6, 8] the proof is complete.

Note that it is also possible to write each $\mathcal{M}$-network as an $\mathcal{N}$-network, by doubling the number of hidden units and using infinitesimal weights from the input units to these hidden units, and their multiplicative inverse for the weights from these hidden to the output units.

Figure 9.3 shows the equivalence of an $\mathcal{N}$ and $\mathcal{M}$ network for the two-hidden unit case.

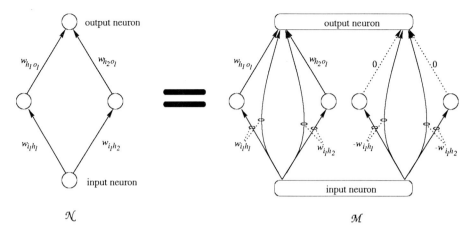

Fig. 9.3. Equivalence of an $\mathcal{N}$-network (left) and an $\mathcal{M}$-network (right). Note that bias units are left out for clarity.

9.3.4 Example

As an example, we train a network with a single hidden unit, and no bias connections, to represent the learning samples $(1,1)$ and $(2,2)$. The hidden unit has a sigmoid activation function. The function that is computed by the network thus is $\mathcal{N}(x,W) = w_2 s(w_1 x)$ for the original neural network, and $\mathcal{M}(x,W) = w_2 s(w_1 x) + w_1 x$ for the adapted neural network. We use the sigmoid function $s(x) = 1/(1+e^{-x})$ as activation function, which has the following well-known properties: $\lim_{x \to \infty} s(x) = 1$, $\lim_{x \to -\infty} s(x) = 0$, and $\lim_{x \to \pm\infty} s'(x) = 0$.

Figure 9.4 shows the error function and its derivatives for this neural network. In the top row of this figure we see the original neural network. Notice in the top middle figure, depicting $\partial E/\partial w_1$, that $\partial E/\partial w_1 \approx 0$ for small values of w_2. In other words, when w_2 is small, w_1 will hardly change its value. Similarly, when w_1 is large, then $\partial E/\partial w_1$ will be small due to the fact that the derivative of the transfer function is nearly 0. In the bottom row the modified neural network is depicted. Left, again, the error function. The middle figure clearly shows that the derivative has no areas anymore which are zero or very small. The right figure still shows that, if w_1 has a large negative value, $\partial E/\partial w_2$ is negligible: after all, the activation value of the hidden unit is near-zero.

9.4 Applications

We have applied the new learning scheme to several approximation problems. In all problems, each network has been run 3,000 times with different initial random values for the weights. In order to train the network, we used a Polak-Ribière conjugate gradient optimization technique with Powell restarts [14].

The applications with real data (problems 3 and 4) use two independent sets of data: a learn set and a cross-validation set. In all cases, the network was

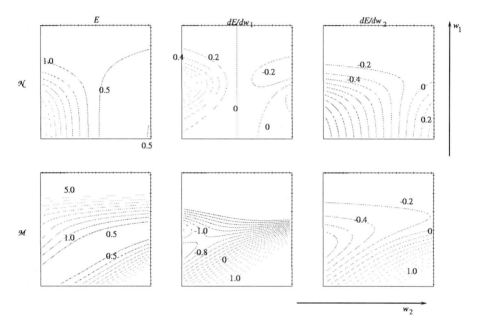

Fig. 9.4. Error function and derivatives using the original and adapted learning rule. The top row shows the error function for the original learning rule (left), as well as its derivative $\partial E/\partial w_1$ (centre) and $\partial E/\partial w_2$ (right). The bottom row shows the same graphs for the adapted learning rule. The contour lines have a distance of 0.5 (left graphs) and 0.2 (middle and right graphs).

trained until the error over the cross-validation set was minimal (i.e., up to but not beyond the point where the network started to over-fit). The approximation errors that are reported are computed using the samples in the cross-validation set.

Problem 1: (synthetic data) the XOR problem. The well-known XOR problem consists of representing four learning samples $[(0,0),0]$, $[(0,1),1]$, $[(1,0),1]$, and $[(1,1),0]$. The network has two inputs, two hidden units, and one output.

It has been noted [5] that the XOR problem is very atypical in neural network learning, because it is penalized for generalization. Nevertheless, the XOR problem is generally considered as a small standard learning benchmark problem.

Whereas the network $\mathcal{N}$ reaches local minima in 22.4% of the runs, the linearly augmented network $\mathcal{M}$ always reached a global minimum. For $\mathcal{N}$, the average number of steps to obtain an approximation error of 0.0 (within the 32-bit floating point accuracy of the computer) equals 189.1; for $\mathcal{M}$, 98.3 steps were required.

When using the ordinary error backpropagation learning rule (i.e., no conjugate gradient learning), the XOR learning problem has been reported to lead to 8.7% local minima [14].

Problem 2: (synthetic data) approximating $\tan(x)$. As a second test, the networks have been used for the approximation of the function $\mathcal{F}(x) = \tan(x)$. The function has been uniformly sampled in the domain $[0, \pi]$ using 20 learning samples in total. For the approximation we used a network with one input, five hidden units in a single hidden layer, and one output.

With network $\mathcal{N}$, 14.6% of the runs lead to a local minimum. The linearly augmented neural network is not perfect; it is stuck in a local minimum in 6.2% of the runs. In all other cases, a global minimum of very close to 0.0 was found.

Problem 3: (real data) approximating robot arm inverse kinematics. Thirdly we approximated data describing a hand-eye coordination learning problem with a Manutec R2 robot arm.

The data is organized as follows. There are five input variables, describing the current position of the robot arm in joint angles θ_2, θ_3, as well as the visual position of an object with respect to the hand-held camera in pixel coordinates x, y, z. The output part of the data consists of the required joint rotations $\Delta\theta_1$, $\Delta\theta_2$, and $\Delta\theta_3$ necessary to reach the object. See [15] for further description of this hand-eye coordination problem. In this particular problem, 1103 learning samples are used; 1103 samples are used for cross-validation. The optimal of six hidden units in a single layer [15] is used. Only the cross-validating data is used for evaluating the network.

With the normal network $\mathcal{N}$, in 2.3% of the cases the network got stuck in a minimum with an unacceptably high error for both the learn and test sets. This never occurred in the 3,000 trials in which the linearly augmented network was used. The cross-validated approximation error after 10,000 learning steps equals $4.20 \cdot 10^{-4}$ for network $\mathcal{N}$, and $4.04 \cdot 10^{-4}$ for network $\mathcal{M}$ (both values are average per learning sample). The new method shows a slight improvement here.

Problem 4: (real data) gear box deformation data. The final test consists of the following problem, which is encountered in the control of a newly developed lightweight robot arm. A gear box connects a DC motor with a robot arm segment. In order to position the robot arm segment at a desired joint angle θ_a, the DC motor has to exert a force τ. However, in the normal setup a joint angle decoder is only available at the DC motor side, which measures the angle θ_m. By mounting an extra decoder at the arm segment in a laboratory setup we can measure θ_a. The actual joint angle θ_a differs slightly from θ_m because of the gear-box elasticity. We attempt to learn θ_a from $(\theta_m, \dot\theta_m, \tau)$.

In order to learn these data, we use a network with 3 inputs, 6 hidden units in a single layer, and one output. The data consists of 4,000 learning samples as well as 4,000 samples used for the cross-validation. In each run, up to 10,000 learning iterations are performed but not beyond the point of overfitting.

Both network $\mathcal{N}$ as $\mathcal{M}$ do not get stuck in a minimum with an unacceptably high error for both the learn and test sets. A surprise, however, is encountered in the cross-validated approximation error that is computed after 10,000 learning steps. For network $\mathcal{N}$, this error equals $2.85 \cdot 10^{-4}$ per learning sample; for $\mathcal{M}$, however, this error is as low as $1.87 \cdot 10^{-6}$ per sample. Further data analysis has shown that the data has strong linear components, which explains the fact that the approximation error is two orders of magnitude lower.

Results are summarized in table 9.1.

Table 9.1. Results of the ordinary feed-forward network $\mathcal{N}$ and the linearly augmented feed-forward network $\mathcal{M}$ on the four problems.

		$\mathcal{N}$	$\mathcal{M}$
XOR	% local minima	22.4	0.0
	avg. steps	189.1	98.3
tan	% local minima	14.6	6.2
robot	% local minima	2.3	0.0
3D data	avg. error	$4.20 \cdot 10^{-4}$	$4.04 \cdot 10^{-4}$
gear box	% local minima	0.0	0.0
data	avg. error	$2.85 \cdot 10^{-4}$	$1.87 \cdot 10^{-6}$

9.5 Conclusion

It has been shown that the ordinary backpropagation learning rule leads to bad conditioning in the matrix of second-order derivatives of the error function which is encountered in feed-forward neural network learning. This again leads to local minima and saddle points in the error landscape. In order to alleviate this problem, we have introduced an adaptation to the backpropagation rule, which can be implemented as an adapted feed-forward neural network structure. We call this network the the *linearly augmented feed-forward neural network*. The adaptation leads to a learning rule which obtains stable values for the weights which connect the input units with the hidden units, even while the weights from hidden to output units change.

A mathematical analysis shows the validity of the method; in particular, the universal approximation theorems are shown to remain valid with the new neural network structure. The application of the new method to two sets of synthetic data and two sets of real data shows that the new method is much less sensitive to local minima, and reaches an optimum in fewer iterations.

Acknowledgments

The authors acknowledge the help of Alin Albu-Schäffer in supplying the gear box data.

References

1. E. Aarts and J. Korst. *Simulated Annealing and Boltzmann Machines*. John Wiley & Sons, 1989.
2. R. Battiti. First- and second-order methods for learning: Between steepest descent and Newton's method. *Neural Computation*, 4:141–166, 1992.
3. G. Cybenko. Approximation by superpositions of a sigmoidal function. *Mathematics of Control, Signals, and Systems*, 2(4):303–314, 1989.
4. B. DasGupta, H. T. Siegelmann, and E. D. Sontag. On the complexity of training neural networks with continuous activation functions. *IEEE Transactions on Neural Networks*, 6(6):1490–1504, November 1995.
5. S. E. Fahlman. An empirical study of learning speed in back-propagation networks. Technical Report CMU–CS–88–0–162, Carnegie Mellon University, September 1988.
6. K.-I. Funahashi. On the approximate realization of continuous mappings by neural networks. *Neural Networks*, 2(3):193–192, 1989.
7. S. Hochreiter and J. Schmidhuber. Flat minima. *Neural Computation*, 9(1):1–42, 1997.
8. K. Hornik, M. Stinchcombe, and H. White. Multilayer feedforward networks are universal approximators. *Neural Networks*, 2(5):359–366, 1989.
9. H. Robbins and S. Monro. A stochastic approximation method. *Annals of Mathematical Statistics*, 22(1):400–407, 1951.
10. S. Saarinen, R. Bramley, and G. Cybenko. Ill-conditioning in neural network training problems. *Siam Journal of Scientific Computing*, 14(3):693–714, May 1993.
11. N. N. Schraudolph. On centering neural network weight updates. Technical Report IDSIA-19-97, IDSIA, April 1997.
12. E. D. Sontag and H. J. Sussmann. Backpropagation can give rise to spurious local minima even for networks without hidden layers. *Complex Systems*, 3(1):91–106, February 1989.

13. K. P. Unnikrishnan and K. P. Venugopal. Alopex: A correlation based learning algorithm for feedforward and recurrent neural networks. *Neural Computation*, 6:469–490, 1994.
14. P. van der Smagt. Minimisation methods for training feed-forward networks. *Neural Networks*, 7(1):1–11, 1994.
15. P. van der Smagt. *Visual Robot Arm Guidance using Neural Networks*. PhD thesis, Dept of Computer Systems, University of Amsterdam, March 1995.
16. Q. J. Zhang, Y. J. Zhang, and W. Ye. Local-sparse connection multilayer networks. In *Proc. IEEE Conf. Neural Networks*, pages 1254–1257. IEEE, 1995.

10
Centering Neural Network Gradient Factors

Nicol N. Schraudolph

IDSIA, Corso Elvezia 36
6900 Lugano, Switzerland
nic@idsia.ch
http://www.idsia.ch/

Abstract. It has long been known that neural networks can learn faster when their input and hidden unit activities are centered about zero; recently we have extended this approach to also encompass the centering of error signals [15]. Here we generalize this notion to *all* factors involved in the network's gradient, leading us to propose centering the slope of hidden unit activation functions as well. Slope centering removes the linear component of backpropagated error; this improves credit assignment in networks with shortcut connections. Benchmark results show that this can speed up learning significantly without adversely affecting the trained network's generalization ability.

10.1 Introduction

Centering is a general methodology for accelerating learning in adaptive systems of the type exemplified by neural networks — that is, systems that are typically nonlinear, continuous, and redundant; that learn incrementally from examples, generally by some form of gradient descent. Its basic tenet is:

> All pattern-dependent factors entering the update equation for a neural network weight should be centered, i.e., have their average over patterns subtracted out.

Prior work. It is well-known that the inputs to an LMS adaptive filter should be centered to permit rapid yet stable adaptation [22], and it has been argued [12] that the same applies to input and hidden unit activity in a multi-layer network. Although Sejnowski [16] proposed a variant of Hebbian learning in which both the pre- and postsynaptic factors of the weight update are centered, the idea was not taken up when backpropagation became popular. The benefits of centering error signals in multi-layer networks were thus reported only recently [15]; here we finally suggest centering as a general methodology, and present backpropagation equations in which *all* factors are centered.

Independence of architecture. Although centering is introduced here in the context of feedforward networks with sigmoid activation functions, the approach itself has a far wider reach. The implementation details may vary, but in essence centering is not tied to any particular architecture: its principles are equally applicable to feedforward and recurrent networks, with sigmoid, radial, or other basis functions, with or without topological structure, time delays, multiplicative gates, *etc.* — in short, the host of architectural elements used in neural network design.

Independence of learning algorithm. Similarly, centering is not wedded to any particular learning algorithm either. It may be applied to deterministic (batch) or stochastic (online) gradient descent; more importantly, it may be freely combined with more sophisticated optimization techniques such as expectation maximization, conjugate gradient and quasi-Newton methods. It also leaves available the many useful tricks often employed with stochastic gradient descent, such as momentum, learning rate adaptation, gradient normalization, and so forth. Due to this flexibility, centering has the potential to further accelerate even the fastest of these methods.

Overview. Section 10.2 introduces the centering approach in terms of the modifications it mandates for ordinary backpropagation learning. We then discuss implementation details in Section 10.3 before presenting benchmark results in Section 10.4. Section 10.5 concludes our paper with a brief analysis of how centering facilitates faster learning.

10.2 Centered Backpropagation

The backpropagation learning algorithm is characterized by three equations, describing the forward propagation of activity, the backpropagation of error, and the modification of weights, respectively. Here are the implications of centering for each of these three equations:

10.2.1 Activity Propagation

Conventional. Consider a neural network with activation of node j given by

$$x_j = f_j(y_j), \quad y_j = \sum_{i \in A_j} w_{ij}\, \check{x}_i, \tag{10.1}$$

where f_j is a nonlinear (typically sigmoid) activation function, w_{ij} are the synaptic weights, and A_j denotes the set of *anterior* nodes feeding their activity $\check{x}_i$ into node j. Conventionally, the anterior nodes' output is fed forward directly, *i.e.*, $(\forall i)\ \check{x}_i \equiv x_i$. We imply that nodes are activated via Equation 10.1 in appropriately ordered (feedforward) sequence, and that some have their values clamped so as to represent external inputs to the network. In particular, we posit a bias input $x_0 \equiv 1$ and require that all nodes are connected to it: $(\forall j > 0)\ 0 \in A_j$.

Centered. As suggested by LeCun et al. [12], the activity of the network's input and hidden units should be centered to permit faster learning (see Chapter 1). We do so by setting

$$(\forall i > 0) \quad \check{x}_i = x_i - \langle x_i \rangle, \qquad (10.2)$$

where $\langle \cdot \rangle$ denotes averaging over training samples (see Section 10.3 for ways to implement this operator). Note that the bias input must *not* be centered — since $x_0 = \langle x_0 \rangle = 1$, it would otherwise become inoperative.

10.2.2 Weight Modification

Conventional. The weights w_{ij} of the network given in Equation 10.1 are typically optimized by gradient descent in some objective function E. With a local step size η_{ij} for each weight, this results in the weight update equation

$$\Delta w_{ij} = \eta_{ij}\, \delta_j\, \check{x}_i, \quad \text{where} \quad \delta_j = -\partial E/\partial y_j. \qquad (10.3)$$

Centered. We have recently proposed [15] that the error signals δ_j should be centered as well to achieve even faster convergence. This is done by updating all non-bias weights via

$$(\forall i > 0) \quad \Delta w_{ij} = \eta_{ij}\, \check{\delta}_j\, \check{x}_i, \quad \text{where} \quad \check{\delta}_j = \delta_j - \langle \delta_j \rangle. \qquad (10.4)$$

As before, this centered update must *not* be used for bias weights, for otherwise they would remain forever stuck (except for stochastic fluctuations) at their present values:

$$* \qquad \langle \Delta w_{0j} \rangle \propto \langle \check{\delta}_j \rangle = \langle \delta_j \rangle - \langle \delta_j \rangle = 0. \qquad (10.5)$$

Instead, bias weights are updated conventionally (Equation 10.3). Since this means that the average error $\langle \delta_j \rangle$ is given exclusively to the bias weight w_{0j}, we have previously called this technique *d.c. error shunting* [15].

10.2.3 Error Backpropagation

Conventional. For output units, the error δ_j can be computed directly from the objective function; for hidden units, it must be derived through error backpropagation:

$$\delta_j = f'_j(y_j)\, \gamma_j, \quad \gamma_j = \sum_{k \in P_j} w_{jk}\, \delta_k, \qquad (10.6)$$

where P_j denotes the set of *posterior* nodes fed from node j, and $f'_j(y_j)$ is the node's current *slope* — the derivative of its nonlinearity f_j at the present level of activation.

Centered. By inserting Equation 10.6 into Equation 10.4, we can express the weight update for hidden units as a triple product of their activity, backpropagated error, and slope:

$$\Delta w_{ij} \propto f'_j(y_j)\,\check{\gamma}_j\,\tilde{x}_i\,, \tag{10.7}$$

where $\check{\gamma}_j$ denotes backpropagated *centered* errors. It is not necessary to center the $\check{\gamma}_j$ explicitly since

$$\langle\check{\gamma}_j\rangle = \left\langle \sum_{k\in P_j} w_{jk}\,\check{\delta}_k \right\rangle = \sum_{k\in P_j} w_{jk}\,\langle\check{\delta}_k\rangle = 0\,. \tag{10.8}$$

By centering activity and error signals we have so far addressed two of the three factors in Equation 10.7, leaving the remaining third factor — the node's slope — to be dealt with. This is done by modifying the nonlinear part of error backpropagation (Equation 10.6) to

$$\check{\delta}_j = \check{f}'_j(y_j)\,\check{\gamma}_j\,, \quad \text{where} \quad \check{f}'_j(y_j) = f'_j(y_j) - \langle f'_j(y_j)\rangle\,. \tag{10.9}$$

Decimation of linear errors. Note that for a *linear* node n we would have $f'_n(y_n) \equiv 1$, and Equation 10.9 would always yield $\check{\delta}_n \equiv 0$. In other words, slope centering (for any node) blocks backpropagation of the *linear component* of error signals — that component which a linear node in the same position would receive. Viewed in terms of error decimation, we have thus taken the logical next step past error centering, which removed the d.c. (constant) component of error signals.

Shortcuts. It was important then to have a parameter — the bias weight — to receive and correct the d.c. error component about to be eliminated. Likewise, we now require additional weights to implement the linear mapping from anterior to posterior nodes that the unit in question is itself no longer capable of. Formally, we demand that for each node j for which Equation 10.9 is used, we have

$$(\forall i \in A_j)\ \ P_j \subseteq P_i\,. \tag{10.10}$$

We refer to connections that bypass a node (or layer) in this fashion as *shortcuts*. It has been noted before that neural network learning sometimes improves with the addition of shortcut weights. In our own experiments (see Section 10.4), however, we find that it is slope centering that makes shortcut weights genuinely useful.

A complementary approach? In Chapter 9, van der Smagt and Hirzinger also advocate shortcuts as a means for accelerating neural network learning. Note, however, that their use of shortcuts is quite different from ours: in order to improve the conditioning of a neural network, they add shortcut connections whose weights are coupled to (shared with) *existing* weights. They thus suitably modify the network's topology without adding new weight parameters, or deviating from a strict gradient-based optimization framework. By contrast, we deliberately decimate the linear component of the gradient for hidden units in order to focus them on their nonlinear task. We then use shortcuts with *additional* weight parameters to take care of the linear mapping that the hidden units now ignore.

While both these approaches use shortcuts to achieve their ends, from another perspective they appear almost complementary: whereas we eliminate the linear component from our gradient, van der Smagt and Hirzinger in fact *add* just such a component to theirs. It may even be possible to profitably combine the two approaches in a single — admittedly rather complicated — neural network architecture.

10.3 Implementation Techniques

We can distinguish a variety of approaches to centering a variable in a neural network in terms of how the averaging operator $\langle \cdot \rangle$ is implemented. Specifically, averaging may be performed either exactly or approximately, and applied either *a priori*, or adaptively during learning in either batch (deterministic) or online (stochastic) settings:

centering method:	approximate	exact
a priori	by design	extrinsic
adaptive { online	running average	—
adaptive { batch	previous batch	two-pass, single-pass

10.3.1 A Priori Methods

By design. Some of the benefits of centering may be reaped without *any* modification of the learning algorithm, simply by setting up the system appropriately. For instance, the hyperbolic tangent (tanh) function with its symmetric range from -1 to 1 will typically produce better-centered output than the commonly used logistic sigmoid $f(y) = 1/(1 + e^{-y})$ ranging from 0 to 1, and is therefore the preferred activation function for hidden units [12]. Similarly, the input representation can (and should) be chosen such that inputs will be roughly centered. When using shortcuts, one may even choose *a priori* to subtract a constant (say, half their maximum) from hidden unit slopes to improve their centering.

We refer to these approximate methods as centering *by design*. Though inexact, they provide convenient and easily implemented tricks to speed up neural network learning. Regardless of whether further acceleration techniques will be required or not, it is generally a good idea to keep centering in mind as a design principle when setting up learning tasks for neural networks.

Extrinsic. Quantities that are extrinsic to the network — *i.e.*, not affected by its weight changes — may often be centered exactly prior to learning. In particular, for any given training set the network's inputs can be centered in this fashion. Even in online settings where the training set is not known in advance, it is sometimes possible to perform such *extrinsic* centering based upon prior knowledge of the training data: instead of a time series $x(t)$ one might for instance present the temporal difference signal $x(t) - x(t-1)$ as input to the network, which will be centered if $x(t)$ is stationary.

10.3.2 Adaptive Methods

Online. When learning online, the immediate environment of a single weight within a multi-layer network is highly non-stationary, due to the simultaneous adaptation of other weights, if not due to the learning task itself. A uniquely defined average of some signal $x(t)$ to be centered is therefore not available online, and we must make do with *running averages* — smoothed versions of the signal itself. A popular smoother is the *exponential trace*

$$\bar{x}(t+1) = \alpha \bar{x}(t) + (1-\alpha) x(t), \qquad (10.11)$$

which has the advantage of being *history-free* and *causal*, i.e., requiring neither past nor future values of x for the present update. The free parameter α (with $0 \leq \alpha \leq 1$) determines the time scale of averaging. Its choice is not trivial: if it is too small, $\bar{x}$ will be too noisy; if it is too large, the average will lag too far behind the (drifting) signal.

Note that the computational cost of centering by this method is proportional to the number of nodes in the network. In densely connected networks, this is dwarfed by the number of weights, so that the propagation of activities and error signals through these weights dominates the computation. The cost of online centering will therefore make itself felt in small or sparsely connected networks only.

Two-pass batch. A simple way to implement exact centering in a batch learning context is to perform *two* passes through the training set for each weight update: the first to calculate the required averages, the second to compute the resulting weight changes. This obviously may increase the computational cost of network training by almost a factor of two. For relatively small networks and training sets, the activity and error for each node and pattern can be stored during the first pass, so that the second pass only consists of the weight update (Equation 10.4). Where this is not possible, a feedforward-only first pass (Equation 10.1) is sufficient to compute average activities and slopes; error centering may then be implemented via one of the other methods described here.

Previous batch. To avoid the computational overhead of a two-pass method, one can use the averages collected over the *previous* batch in the computation of weight changes for the current batch. This approximation assumes that the averages involved do not change too much from batch to batch; this may result in stability problems in conjunction with very high learning rates. Computationally this method is quite attractive in that it is cheaper still than the online technique described above. When mini-batches are used for training, both approaches can be combined profitably by centering with an exponential trace over mini-batch averages.

Single-pass batch. It is possible to perform exact centering in just a *single* pass through a batch of training patterns. This is done by expanding the triple product of the fully centered batch weight update (*cf.* Equation 10.7). Using f'_j as a shorthand for $f'_j(y_j)$, we have

$$\begin{aligned}
\Delta w_{ij} &\propto \langle (x_i - \langle x_i \rangle)(\gamma_j - \langle \gamma_j \rangle)(f'_j - \langle f'_j \rangle) \rangle \\
&= \langle x_i \gamma_j f'_j \rangle - \langle \langle x_i \rangle \gamma_j f'_j \rangle - \langle x_i \langle \gamma_j \rangle f'_j \rangle - \langle x_i \gamma_j \langle f'_j \rangle \rangle \\
&\quad + \langle \langle x_i \rangle \langle \gamma_j \rangle f'_j \rangle + \langle \langle x_i \rangle \gamma_j \langle f'_j \rangle \rangle + \langle x_i \langle \gamma_j \rangle \langle f'_j \rangle \rangle - \langle \langle x_i \rangle \langle \gamma_j \rangle \langle f'_j \rangle \rangle \\
&= \langle x_i \gamma_j f'_j \rangle - \langle x_i \rangle \langle \gamma_j f'_j \rangle - \langle \gamma_j \rangle \langle x_i f'_j \rangle - \langle x_i \gamma_j \rangle \langle f'_j \rangle + 2 \langle x_i \rangle \langle \gamma_j \rangle \langle f'_j \rangle \quad (10.12)
\end{aligned}$$

In addition to the ordinary (uncentered) batch weight update term $\langle x_i \gamma_j f'_j \rangle$ and the individual averages $\langle x_i \rangle$, $\langle \gamma_j \rangle$, and $\langle f'_j \rangle$, the single-pass centered update (10.12) thus also requires collection of the sub-products $\langle x_i \gamma_j \rangle$, $\langle x_i f'_j \rangle$, and $\langle \gamma_j f'_j \rangle$. Due to the extra computation involved, the single-pass batch update is not necessarily more efficient than a two-pass method. It is simplified considerably, however, when not all factors are involved — for instance, when activities have already been centered *a priori* so that $\langle x_i \rangle \approx 0$.

Note that the expansion technique shown here may be used to derive an exact single-pass batch method for *any* weight update that involves the addition (or subtraction) of some quantity that must be computed from the entire batch of training patterns. This includes algorithms such as BCM learning [4, 10] and binary information gain optimization [14].

10.4 Empirical Results

While activity centering has long been part of backpropagation lore, and empirical results for error centering have been reported previously [15], slope centering is being proposed for the first time here. It is thus too early to assess its general applicability or utility; here we present a number of experiments designed to show the typical effect that centering has on speed and reliability of convergence as well as generalization performance in feedforward neural networks trained by accelerated backpropagation methods.

The next section describes the general setup and acceleration techniques used in all our experiments. Subsequent sections then present our respective results for two well-known benchmarks: the toy problem of symmetry detection in binary patterns, and a difficult vowel recognition task.

10.4.1 Setup of Experiments

Benchmark design. For each benchmark task we performed a number of *experiments* to compare performance with *vs.* without various forms of centering. Each experiment consisted of 100 *runs* starting from different initial weights but identical in all other respects. For each run, networks were initialized with random weights from a zero-mean Gaussian distribution with standard deviation 0.3. All experiments were given the same sequence of random numbers for their 100 weight initializations; the seed for this sequence was picked only after the design of the benchmark had been finalized.

Training modality. In order to make the results as direct an assessment of centering as possible, training was done in batch mode so as to avoid the additional free parameters (*e.g.*, smoothing time constants) required by online methods. Where not done *a priori*, centering was then implemented with the exact two-pass batch method. In addition, we always updated the hidden-to-output weights of the network *before* backpropagating error through them. This is known to sometimes improve convergence behavior [17], and we have found it to increase stability at the large step sizes we desire.

Competitive controls. The ordinary backpropagation (plain gradient descent) algorithm has many known defects, and a large number of acceleration techniques has been proposed for it. We informally tested a number of such techniques, then picked the combination that achieved the fastest reliable convergence. This combination — *vario-η* and *bold driver* — was then used for all experiments reported here. Thus any performance advantage for centering reported thereafter has been realized *on top of* a state-of-the-art accelerated gradient method as control.

Vario-η [23, page 48]. This interesting technique — also described in Chapter 17 — sets the local learning rate for each weight inversely proportional to the standard deviation of its stochastic gradient. The weight change thus becomes

$$\Delta w_{ij} = \frac{-\eta\, g_{ij}}{\varrho + \sigma(g_{ij})}, \quad \text{where } g_{ij} \equiv \frac{\partial E}{\partial w_{ij}} \quad \text{and } \sigma(u) \equiv \sqrt{\langle u^2 \rangle - \langle u \rangle^2}, \tag{10.13}$$

with the small positive constant ϱ preventing division by near-zero values. Vario-η can be used in both batch and online modes, and is quite effective in that it not only performs gradient normalization, but also adapts step sizes to the level of noise in the local gradient signal.

We used vario-η for all experiments reported here, with $\varrho = 0.1$. In a batch implementation this leaves only one free parameter to be determined: the global learning rate η.

Bold driver [11, 21, 2, 3]. This algorithm for adapting the global learning rate η is simple and effective, but only works for batch learning. Starting from some initial value, η is increased by a certain factor after each batch in which the error did not increase by more than a very small constant ε (required for numerical stability). Whenever the error rises by more than ε, however, the last weight change is undone, and η decreased sharply.

All experiments reported here were performed using bold driver with a learning rate increment of 2%, a decrement of 50%, and $\varepsilon = 10^{-10}$. These values were found to provide fast, reliable convergence across all experiments. Due to the amount of recomputation they require, we do count the "failed" epochs (whose weight changes are subsequently undone) in our performance figures.

10.4.2 Symmetry Detection Problem

In our first benchmark, a fully connected feedforward network with 8 inputs, 8 hidden units and a single output is to learn the symmetry detection task: given an 8-bit binary pattern at the input, it is to signal at the output whether the pattern is symmetric about its middle axis (target = 1) or not (target = 0). This is admittedly a toy problem, although not a trivial one.

Since the target is binary, we used a logistic output unit and cross-entropy loss function. For each run the network was trained on all 256 possible patterns until the root-mean-square error of its output over the batch fell below 0.01. We recorded the number of epochs required to reach this criterion, but did not test for generalization ability on this task.

Error and activity centering. In our first set of experiments we examined the separate and combined effect of centering the network's activity and/or error signals. For convenience, activity centering was performed *a priori* by using -1 and 1 as input levels, and the hyperbolic tangent (tanh) as activation function for hidden units. The off-center control experiments were done with 0 and 1 as input levels and the logistic activation function $f(y) = 1/(1 + e^{-y})$. Note that all differences between the tanh and logistic nonlinearities are eliminated by the vario-η algorithm, *except* for the eccentricity of their respective outputs.

Results. Table 10.1 shows that centering either activity or error signals produced an approximate 7-fold increase in convergence speed. In no instance was a run that used one (or both) of these centering methods slower than the corresponding control without centering. The similar magnitude of the speed-up suggests that it may be due to the improved conditioning of the Hessian achieved by centering either errors or activities (see Section 10.5). Note, however, that activity centering beat error centering almost 2/3 of the time in the direct comparison.

On the other hand, error centering appeared to improve the *reliability* of convergence: it cut the convergence time's coefficient of variation (the ratio between its standard deviation and mean, henceforth: c.v.) in half while activity centering

error signals:	conventional				centered
activities:	mean ± st.d. quartiles	direct comparison: # of faster runs			mean ± st.d. quartiles
off-center (0/1)	669 ± 308 453/580/852	0	0 — 100 0 35	7	97.5 ± 21.8 82/95.5/109
centered (-1/1)	93.1 ± 46.7 67.5/79.5/94	100	63 × 100 14 — 84	93	65.4 ± 15.9 57/62/70

Table 10.1. The effect of centering activities and/or error signals on the symmetry detection task without shortcuts. Reported are the empirical mean, standard deviation, and $25^{th}/50^{th}/75^{th}$ percentile (rounded to three significant digits) of the number of epochs required to converge to criterion. Also shown is the result of directly comparing runs with identical random seeds. The number of runs in each comparison may sum to less than 100 due to ties.

left it unchanged. We speculate that this may be the beneficial effect of centering on the *backpropagated* error, which does not occur for activity centering.

Finally, a further speedup of 50% (while maintaining the lower c.v.) occurred when both activity and error signals were centered. This may be attributed to the fact that our centering of hidden unit activity *by design* (*cf.* Section 10.3) was only approximate. To assess the significance of these effects, note that since the data was collected over 100 runs, the standard error of the reported mean time to convergence is $1/\sqrt{100} = 1/10$ its reported standard deviation.

Shortcuts and slope centering. In the second set of experiments we left both activity and error signals centered, and examined the separate and combined effect of adding shortcuts and/or slope centering. Note that since the complement of a symmetric bit pattern is also symmetric, the symmetry detection task has *no* linear component at all — we would therefore expect shortcuts to be of minimal benefit here.

Results. Table 10.2 shows that indeed adding shortcuts alone was not beneficial — in fact it slowed down convergence in over 80% of the cases, and significantly increased the c.v. Subsequent addition of slope centering, however, brought about an almost 3-fold increase in learning speed, and restored the original c.v. of about 1/4. When used together, slope centering and shortcuts never increased convergence time, and on average cut it in half. By contrast, slope centering *without* shortcuts failed to converge at all about 1/3 of the time. This may come as a surprise, considering that the given task had no linear component. However, consider the following:

slopes: topology:		conventional					centered
		mean ± st.d. quartiles	direct comparison: # of faster runs				mean ± st.d. quartiles
short- cuts?	no	65.4 ± 15.9 57/62/70	81	0	61	*	51.6 ± 16.2 43/64.5/∞
			—	×	—		
	yes	90.4 ± 31.1 69.5/80/102	17	39	99	95	33.1 ± 8.6 28/31/35
			0 – 100				

* Mean and standard deviation exclude 34 runs which did not converge.

Table 10.2. The effect of centering slopes and/or adding shortcuts on the symmetry detection task with centered activity and error signals. Results are shown in the same manner as in Table 10.1.

Need for shortcuts. Due to the monotonicity of their nonlinear transfer function, hidden units always carry some linear moment, in the sense of a positive correlation between their net input and output. In the absence of shortcuts, the hidden units must arrange themselves so that their linear moments together match the overall linear component of the task (here: zero). This adaptation process is normally driven by the linear component of the error — which slope centering removes.

The remaining nonlinear error signals can still jostle the hidden units into an overall solution, but such an indirect process is bound to be unreliable: as it literally removes slope from the error surface, slope centering creates numerous local minima. Shortcut weights turn these local minima into global ones by modeling the missing (linear) component of the gradient, thereby freeing the hidden units from any responsibility to do so.

In summary, while a network without shortcuts trained with slope centering may converge to a solution, the addition of shortcut weights is necessary to ensure that slope centering will not be detrimental to the learning process. Conversely, slope centering can prevent shortcuts from acting as redundant "detractors" that impede learning instead of assisting it. These two techniques should therefore always be used in conjunction.

10.4.3 Vowel Recognition Problem

Our positive experiences with centering on the symmetry detection task immediately raise two further questions: 1) will these results transfer to more challenging, realistic problems, and 2) is the gain in learning speed — as often happens — bought at the expense of generalization ability? In order to address these questions, we conducted further experiments with the speaker-independent vowel recognition data due to Deterding [5], a popular benchmark for which good generalization performance is rather difficult to achieve.

The task. The network's task is to recognize the eleven steady-state vowels of British English in a speaker-independent fashion, given 10 spectral features (specifically: LPC-derived log area ratios) of the speech signal. The data consists of 990 patterns to be classified: 6 instances for each of the 11 vowels spoken by each of 15 speakers. We follow the convention of splitting it into a training set containing the data from the first 8 (4 male, 4 female) speakers, and a test set containing those of the remaining 7 (4 male, 3 female). Note that there is no separate validation set available.

Prior work. Robinson [13] pioneered the use of Deterding's data as a benchmark by comparing the performance of a number of neural network architectures on it. Interestingly, none of his methods could outperform the primitive single nearest neighbor approach (which misclassifies 44% of test patterns), thus posing a challenge to the pattern recognition community. Trained on the task as formulated above, conventional backpropagation networks in fact appear to reach their limits at error rates of around 42% [6, 9], while an adaptive nearest neighbor technique can achieve 38% [7]. In Chapter 7, Flake reports comparably favorable results for RBF networks as well as his own hybrid architectures. Even better performance can be obtained by using speaker sex/identity information [19, 20], or by training a separate model for each vowel [8]. By combining these two approaches, a test set error of 23% has been reached [18], the lowest we are aware of to date.

Training and testing. We trained fully connected feedforward networks with 10 inputs, 22 hidden units, and 11 logistic output units by minimization of cross-entropy loss. The target was 1 for the output corresponding to the correct vowel, 0 for all others. Activity centering was done *a priori* by explicitly centering the inputs (separately for training and test set), and by using the tanh nonlinearity for hidden units. The uncentered control experiments used the original input data, and logistic activation functions.

The relatively small size of our networks enabled us to run all experiments out to 2 000 epochs of training. After each epoch, the network's generalization ability was measured in terms of its misclassification rate on the test set. For the purpose of testing, a maximum likelihood approach was adopted: the network's highest output for a given test pattern was taken to indicate its classification of that pattern.

First results. Figure 10.1 shows how the average test set error (over 100 runs) evolved during training in each of the 8 experiments we performed for this benchmark. For all curves, error bars were at most the size of the marks shown along the curve, and have therefore been omitted for clarity. Following our experience on the symmetry detection task, shortcuts and slope centering were always used in conjunction whereas activity and error centering were examined independently. The following effects can be discerned:

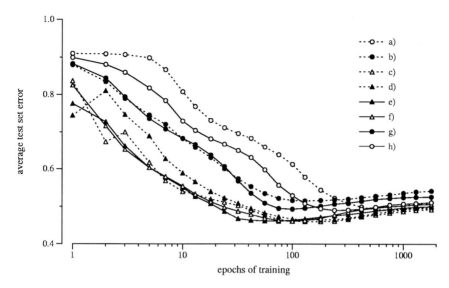

Fig. 10.1. Evolution of the average test set error while learning the vowel recognition task with activity centering (triangular marks), error centering (filled marks), and/or slope centering with shortcut weights (solid lines), *vs.* their uncentered controls. Experiments are denoted a)–h) as in Table 10.3.

1. All experiments with activity centering (triangular marks) clearly outperformed all experiments without it (circular marks) in both average convergence speed and minimum average test set error.
2. All experiments with shortcuts and slope centering (solid lines) outperformed the corresponding experiment without them (dashed lines).
3. With one notable exception (experiment d), error centering (filled marks) sped up convergence significantly. Its effect was greatest in the experiments without activity centering.
4. The best experiment in terms of both convergence speed and minimum average test set error was e), the fully centered one; the worst was a), the fully conventional one.

The qualitative picture that emerges is that centering appears to significantly speed up convergence without adversely affecting the trained network's generalization ability. We will now attempt to quantify this finding.

Quantifying the effect. Since the curves in Figure 10.1 are in fact superpositions of 100 nonlinear curves each, they are ill-suited to quantitative analysis: value and location of the minimum average test set error do not tell us anything about the distribution of such minima across individual runs — not even their average value or location. In order to obtain such quantitative results, we need to identify an appropriate minimum in test set error for each run. This will allow us to directly compare runs with identical initial weights across experiments, as well as to characterize the distribution of minima within each experiment by aggregate statistics (*e.g.*, mean, standard deviation, quartiles) for both the minimum test set error, and the time taken to reach it.

A fair and consistent strategy to identify minima suitable for the quantitative comparisons we have in mind is not trivial to design. Individual runs may have multiple minima in test set error, or none at all. If we were to just use the global minimum over the duration of the run (2 000 epochs), we would not be able to distinguish a fast method which makes some insignificant improvement to a long-found minimum late in the run from a slow method which takes that long to reach its first minimum. Given that we are concerned with both the quality of generalization performance and the speed with which it is achieved, a greedy strategy for picking appropriate minima is indicated.

Identification of minima. We follow the evolution of test set error over the course of each run, noting new minima as we encounter them. If the best value found so far is not improved upon within a certain period of time, we pick it as *the* minimum of that run for the purpose of quantitative analysis. The appropriate length of waiting period before giving up on further improvement is a difficult issue — see Chapter 2 for a discussion. For a fair comparison between faster and slower optimization methods, it should be proportional to the time it took to reach the minimum in question: a slow run then has correspondingly more time to improve its solution than a fast one.

Unfortunately this approach fails if a minimum of test set error occurs during the initial transient, within the first few epochs of training: the waiting period would then be too short, causing us to give up prematurely. On the other hand, we cannot wait longer than the overall duration of the run. We therefore stop looking for further improvement in a run after $\min(2\,000,\ 2\epsilon+100)$ epochs, where ϵ records when the network first achieved the lowest test set error seen so far in that run. Only 9 out of the 800 runs reported here expired at the upper limit of 2 000 epochs, so we are confident that its imposition did not significantly skew our results.

Test set used as validation set. Note that since we use the test set to determine at which point to compare performance, we have effectively appropriated it as a validation set. The minimum test set errors reported below are therefore *not* unbiased estimators for the network's ability to generalize to novel speakers, and should not be compared to proper measurements of this ability (for which the test set must not affect the training procedure in any way). Nonetheless, let us not forget that the lowest test set error does measure the network's generalization ability in a consistent fashion after all: even though these scores are all biased to favor a particular set of novel speakers (the test set), by no means does this render their comparison *against each other* insignificant.

Overview of results. Table 10.3 summarizes quantitative results obtained in this fashion for the vowel recognition problem. To assess their significance, recall

experiment	features: centering activ.	features: error	features: slope	shortcuts	performance measure: minimum test set error mean ± st.d. quartiles	dir. comparison: # of better runs					epochs required mean ± st.d. quartiles	dir. comparison: # of faster runs				
a)					48.0 ± 3.6 45.7/47.3/50.0	67	17	13	19	37	554 ± 321 365/486/691	3	10	4	1	14
b)		✓			49.1 ± 2.9 47.0/49.6/50.9	31 10					125 ± 82 67.5/104/163	97 51				
c)	✓				43.9 ± 2.5 42.3/43.9/46.0		82 51				156 ± 110 75/137/215		90 47			
d)	✓	✓			44.3 ± 2.3 42.9/44.2/45.9	89	46 49	84 65			158 ± 141 72/124/186	48	52 21	96 85		
e)	✓	✓	✓	✓	44.2 ± 2.5 42.3/44.4/46.3	70	49 47		80 68		72.4 ± 55.5 37.5/51.5/81	76	78 75		99 92	
f)	✓		✓	✓	44.2 ± 2.8 42.3/44.4/46.1		51 68				113 ± 64 69.5/97.5/148		24 88			
g)		✓	✓	✓	46.8 ± 3.7 44.0/47.0/48.9	27 43					126 ± 139 64/94/138	22 84				
h)			✓	✓	46.5 ± 3.1 44.5/46.8/48.5	56	31	29	30	61	270 ± 164 162/235/316	15	12	15	8	86

Table 10.3. Minimum test set error (misclassification rate in %), and the number of epochs required to reach it, for the vowel recognition task. Except for the different layout, results are reported in the same manner as in Tables 10.1 and 10.2. Due to space limitations, only selected pairs of experiments are compared directly.

that the standard error in the mean of a performance measure reported here is $1/\sqrt{100} = 1/10$ of its reported standard deviation. Figure 10.2 depicts the same data (except for the direct comparisons) graphically in form of cumulative histograms for the minimum test set error and the number of epochs required to reach it.

The results generally confirm the trends observed in Figure 10.1. Runs in the fully centered experiment e) clearly converged most rapidly — and to test set errors that were among the best. Compared to the conventional setup a), full centering converged almost 8 times faster on average while generalizing better 80% of the time.

Generalization performance. Average misclassification rates on the test set ranged from 44% to 49%, which we consider a fair result given our comparatively small networks. They cluster into three groups: networks with activity centering

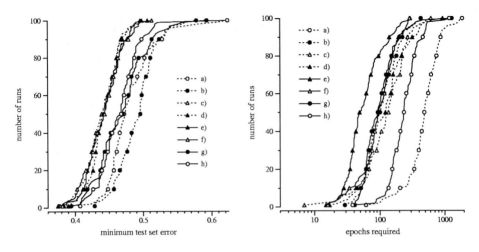

Fig. 10.2. Cumulative histograms for the minimum test set error (left), and the number of epochs required to reach it (right), for the vowel recognition task. Curves are labelled as in Figure 10.1, and marked every 10th percentile.

achieved around 44%, the two others with shortcuts and slope centering came in under 47%, while the remaining two only reached 48–49%. The cumulative histogram (Figure 10.2, left) shows that all activity-centered networks had an almost identical distribution of minimum test set errors.

Note that centering the inputs changes the task, and that the addition of shortcuts changes the network topology. It is possible that this — rather than centering *per se* — accounts for their beneficial effect on generalization. Error centering was the one feature in our experiments that changed the dynamics of learning exclusively. Its addition appeared to slightly *worsen* generalization, particularly in the absence of other forms of centering. This could be caused by a reduction (due to centering) of the effective number of parameters in what is already a rather small model. Such an effect should not overly concern us: one could easily recoup the lost degrees of freedom by slightly increasing the number of hidden units for centered networks.

Convergence speed. All three forms of centering examined here clearly sped up convergence, both individually and in combination. A slight anomaly appeared in that the addition of error centering in going from experiment c) to d) had no significant effect on the average number of epochs required. A look at the cumulative histogram (Figure 10.2, right) reveals that while experiment d) is ahead between the 20th and 80th percentile, c) had fewer unusually slow runs than d), and a few exceptionally fast ones.

With the other forms of centering in place, the addition of error centering was unequivocally beneficial: average convergence time decreased from 113 epochs in

f) to 72.4 epochs in e). The histogram shows that the fully centered e) is far ahead of the competition through almost the entire percentile range.

Finally, it is interesting to note that the addition of shortcuts and slope centering, both on their own and to a network with activity and error centering, roughly doubled the convergence speed — the same magnitude of effect as observed on the symmetry detection task.

10.5 Discussion

The preceding section has shown that centering can indeed have beneficial effects on the learning speed and generalization ability of a neural network. Why is this so? In what follows, we offer an explanation from three (partially overlapping) perspectives, considering in turn the effect of centering on the condition number of the Hessian, the level of noise in the local gradient, and the credit assignment between different parts of the network.

Conditioning the Hessian. It is well known that the minimal convergence time for first-order gradient descent on quadratic error surfaces is inversely related to the condition number of the Hessian matrix, *i.e.*, the ratio between its largest and its smallest eigenvalue. A common strategy for accelerating gradient descent is therefore to seek to improve the condition number of the Hessian.

For a single linear node $y = w^T x$ with squared loss function, the Hessian is simply the covariance matrix of the inputs: $H = \langle x x^T \rangle$. Its largest eigenvalue is typically caused by the d.c. component of x [12]. Centering the inputs removes that eigenvalue, thus conditioning the Hessian and permitting larger step sizes (*cf.* Chapter 1). For batch learning, error centering has exactly the same effect on the local weight update:

$$\Delta w \propto \langle (\delta - \langle \delta \rangle) x \rangle = \langle \delta x \rangle - \langle \delta \rangle \langle x \rangle = \langle \delta (x - \langle x \rangle) \rangle \qquad (10.14)$$

Error centering does go further than activity centering, however, in that it also affects the error backpropagated to anterior nodes. Moreover, Equation 10.14 does not hold for online learning, where the gradient is noisy.

Noise reduction. It can be shown that centering improves the signal-to-noise ratio of the local gradient. Omitting the slope factor for the sake of simplicity, consider the noisy weight update

$$\Delta w_{ij} \propto (\delta_j + \phi)(x_i + \xi) = \delta_j x_i + \xi \delta_j + \phi x_i + \phi \xi \qquad (10.15)$$

where ϕ and ξ are the noise terms, presumed to be zero-mean, and independent of activity, error, and each other. In the expansion on the right-hand side, the first term is the desired (noise-free) weight update while the others represent noise that contaminates it. While the last (pure noise) term cannot be helped, we can reduce the variance of the two mixed terms by centering δ_j and x_i so as to minimize $\langle \delta_j^2 \rangle$ and $\langle x_i^2 \rangle$, respectively.

One might of course contend that in doing so, we are also shrinking the signal $\delta_j x_i$, so that in terms of the signal-to-noise ratio we are no better — in fact, worse — off than before. This cuts right to the heart of the matter, for centering rests upon the notion that the error signal relevant to a non-bias, non-shortcut weight *is* the fully centered weight update, and that any d.c. components in $\delta_j x_i$ should therefore also be regarded as a form of noise. This presumption can of course be maintained only because we do have bias and shortcut weights to address the error components that centering removes.

Improved credit assignment. From the perspective of a network that has these additional parameters, then, centering is a way to improve the assignment of responsibility for the network's errors: constant errors are shunted to the bias weights, linear errors to the shortcut weights, and the remainder of the network is bothered only with those parts of the error signal that actually require a nonlinearity. Centering thus views hidden units as a scarce resource that should only be called upon where necessary. Given the computational complications that arise in the training of nonlinear nodes, we submit that this is an appropriate and productive viewpoint.

Future work. While the results reported here are quite promising, more experiments are required to assess the general applicability and effectiveness of centering. For feedforward networks, we would like to explore the use of centering with multiple hidden layers, stochastic (online) gradient descent, and for function approximation (rather than classification) problems. The centering approach *per se*, however, is rather more general than that, and so further ahead we anticipate its application to a range of more sophisticated network architectures, learning algorithms, and problem domains.

Acknowledgments

I would like to thank the editors of this book as well as my colleagues Jürgen Schmidhuber, Marco Wiering, and Rafał Sałustowicz for their helpful comments. This work was supported by the Swiss National Science Foundation under grant numbers 2100–045700.95/1 and 2000–052678.97/1.

References

1. J. Anderson and E. Rosenfeld, editors. *Neurocomputing: Foundations of Research.* MIT Press, Cambridge, 1988.
2. R. Battiti. Accelerated back-propagation learning: Two optimization methods. *Complex Systems*, 3:331–342, 1989.
3. R. Battiti. First- and second-order methods for learning: Between steepest descent and Newton's method. *Neural Computation*, 4(2):141–166, 1992.
4. E. Bienenstock, L. Cooper, and P. Munro. Theory for the development of neuron selectivity: Orientation specificity and binocular interaction in visual cortex. *Journal of Neuroscience*, 2, 1982. Reprinted in [1].

5. D. H. Deterding. *Speaker Normalisation for Automatic Speech Recognition*. PhD thesis, University of Cambridge, 1989.
6. M. Finke and K.-R. Müller. Estimating a-posteriori probabilities using stochastic network models. In M. C. Mozer, P. Smolensky, D. S. Touretzky, J. L. Elman, and A. S. Weigend, editors, *Proceedings of the 1993 Connectionist Models Summer School*, Boulder, CO, 1994. Lawrence Erlbaum Associates, Hillsdale, NJ.
7. T. J. Hastie and R. J. Tibshirani. Discriminant adaptive nearest neighbor classification. *IEEE Transactions on Pattern Analysis and Machine Intelligence*, 18(6):607–616, 1996.
8. M. Herrmann. On the merits of topography in neural maps. In T. Kohonen, editor, *Proceedings of the Workshop on Self-Organizing Maps*, pages 112–117. Helsinki University of Technology, 1997.
9. S. Hochreiter and J. Schmidhuber. Feature extraction through LOCOCODE. To appear in *Neural Computation*, 1998.
10. N. Intrator. Feature extraction using an unsupervised neural network. *Neural Computation*, 4(1):98–107, 1992.
11. A. Lapedes and R. Farber. A self-optimizing, nonsymmetrical neural net for content addressable memory and pattern recognition. *Physica*, D 22:247–259, 1986.
12. Y. LeCun, I. Kanter, and S. A. Solla. Eigenvalues of covariance matrices: Application to neural-network learning. *Physical Review Letters*, 66(18):2396–2399, 1991.
13. A. J. Robinson. *Dynamic Error Propagation Networks*. PhD thesis, University of Cambridge, 1989.
14. N. N. Schraudolph and T. J. Sejnowski. Unsupervised discrimination of clustered data via optimization of binary information gain. In S. J. Hanson, J. D. Cowan, and C. L. Giles, editors, *Advances in Neural Information Processing Systems*, volume 5, pages 499–506. Morgan Kaufmann, San Mateo, CA, 1993.
15. N. N. Schraudolph and T. J. Sejnowski. Tempering backpropagation networks: Not all weights are created equal. In D. S. Touretzky, M. C. Mozer, and M. E. Hasselmo, editors, *Advances in Neural Information Processing Systems*, volume 8, pages 563–569. The MIT Press, Cambridge, MA, 1996.
16. T. J. Sejnowski. Storing covariance with nonlinearly interacting neurons. *Journal of Mathematical Biology*, 4:303–321, 1977.
17. S. Shah, F. Palmieri, and M. Datum. Optimal filtering algorithms for fast learning in feedforward neural networks. *Neural Networks*, 5:779–787, 1992.
18. J. B. Tenenbaum and W. T. Freeman. Separating style and content. In M. C. Mozer, M. I. Jordan, and T. Petsche, editors, *Advances in Neural Information Processing Systems*, volume 9, pages 662–668. The MIT Press, Cambridge, MA, 1997.
19. P. D. Turney. Exploiting context when learning to classify. In *Proceedings of the European Conference on Machine Learning*, pages 402–407, 1993.
20. P. D. Turney. Robust classification with context-sensitive features. In *Proceedings of the Sixth International Conference on Industrial and Engineering Applications of Artificial Intelligence and Expert Systems*, pages 268–276, 1993.
21. T. P. Vogl, J. K. Mangis, A. K. Rigler, W. T. Zink, and D. L. Alkon. Accelerating the convergence of the back-propagation method. *Biological Cybernetics*, 59:257–263, 1988.
22. B. Widrow, J. M. McCool, M. G. Larimore, and C. R. Johnson, Jr. Stationary and nonstationary learning characteristics of the LMS adaptive filter. *Proceedings of the IEEE*, 64(8):1151–1162, 1976.

23. H. G. Zimmermann. Neuronale Netze als Entscheidungskalkül. In H. Rehkugler and H. G. Zimmermann, editors, *Neuronale Netze in der Ökonomie: Grundlagen und finanzwirtschaftliche Anwendungen*, pages 1–87. Vahlen Verlag, Munich, 1994.

11
Avoiding Roundoff Error in Backpropagating Derivatives

Tony Plate

School of Mathematical and Computing Sciences,
Victoria University, Wellington, New Zealand.
tap@mcs.vuw.ac.nz   http://www.mcs.vuw.ac.nz/~tap/

Abstract. One significant source of roundoff error in backpropagation networks is the calculation of derivatives of unit outputs with respect to their total inputs. The roundoff error can lead result in high relative error in derivatives, and in particular, derivatives being calculated to be zero when in fact they are small but non-zero. This roundoff error is easily avoided with a simple programming trick which has a small memory overhead (one or two extra floating point numbers per unit) and an insignificant computational overhead.

11.1 Introduction

Backpropagating derivatives is an essential part of training multilayer networks. Accuracy of these derivatives is important to many training methods, especially ones which use second-order information, such as conjugate gradients. The standard formula for backpropagating error derivatives (eg., as given in Ripley [3], Bishop [1], and Rumelhart, Hinton, and Williams [4]) use floating point arithmetic in such a way that can result in significant roundoff error. In particular, small derivatives are rounded off to zero. These errors can cause second-order methods to become confused (because of inaccurate gradients) and can also cause the weight search to stop or be very slow because some derivatives are calculated to be zero when in fact they are small but non-zero. This chapter explains how this particular source of roundoff error can be avoided simply and cheaply. The method applies to both logistic and tanh units, and to the sum-squared and cross-entropy error functions.

In this chapter, the symbol "$=$" is used to denote mathematical equality, and the symbol "$\leftarrow$" is used to denote an assignment to some floating-point variable. Floating point values are denoted by an asterisk, eg., x_i^* is the floating point version of x_i.

11.2 Roundoff error in sigmoid units

Consider a non-input unit whose output y_i is computed as the logistic function of its total input x_i:

$$y_i = \frac{1}{1+\exp(-x_i)} \quad \text{and} \quad y_i^* \leftarrow \frac{1}{1+\exp(-x_i^*)}.$$

In the backpropagation phase of training, we need to calculate the partial derivative of the error with respect to the total input. This is done using the chain rule:

$$\frac{\partial E}{\partial x_i} = \frac{\partial E}{\partial y_i} \frac{\partial y_i}{\partial x_i}.$$

The standard way of calculating $\frac{\partial y_i}{\partial x_i}$ is to use the formula relating it to y_i, which for the logistic function is

$$\left(\frac{\partial y_i}{\partial x_i}\right)^* \leftarrow y_i^*(1-y_i^*).$$

This computation is a potential source of roundoff error: if the actual value of y_i is so close to one that y_i^* is exactly one, then $(\frac{\partial y_i}{\partial x_i})^*$ will equal zero.

Figures 11.1 show the values of this expression calculated in single and double precision floating point arithmetic. In single precision, when x_i is greater than about 17.33, $y_i^*(1-y_i^*)$ evaluates to zero. For x_i values slightly lower than 17.33 there is significant quantization. In double precision $y_i^*(1-y_i^*)$ evaluates to zero when x_i is greater than about 36.74.

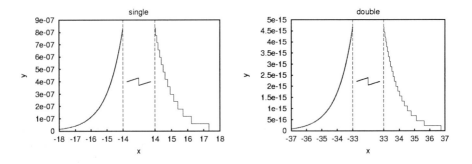

Fig. 11.1. Quantization in calculated values of unit derivatives ($\frac{\partial y}{\partial x}$) for logistic units, computed using the formula $y^*(1-y^*)$ where $y^* \leftarrow 1/(1+\exp(-x^*))$. Roundoff error causes quantization on the right hand side of each plot.

Note that in Figure 11.1 the roundoff error due to the calculation of $1-y^*$ only occurs for positive x. For negative x, y^* approaches zero, and is accurately represented, and the relative roundoff error in $1-y^*$ is insignificant. This provides a clue as to how to avoid the roundoff error for positive x: don't compute $1-y^*$ when y^* is close to one. Indeed, these roundoff errors can be avoided entirely by storing an extra value with each unit, which is $z_i = 1 - y_i$, calculated accurately in floating point arithmetic as follows:

$$y_i^* \leftarrow \frac{1}{1+\exp(-x_i^*)} \quad \text{and} \quad z_i^* \leftarrow \frac{\exp(-x_i^*)}{1+\exp(-x_i^*)}.$$

Together, these two floating points numbers accurately represent the unit output at its extremes: y_i^* is an accurate representation when the output is close to zero, and z_i^* is an accurate representation of one minus the output when the output is close to one. With them, $\frac{\partial y_i}{\partial x_i}$ is simply and accurately calculated as follows:

$$\left(\frac{\partial y_i}{\partial x_i}\right)^* \leftarrow y_i^* z_i^*.$$

Implementing these calculations requires the storage of one extra floating point number per unit (ie., z_i^*), and the computation of one extra division per unit. This extra resource usage is insignificant because the total resource requirements for a forward- and back-propagation pass are proportional to the number of weights in the network, which is usually of the order of the square of the number of units.

The value y_i is also used in calculating partial derivatives for weights on connections emerging from unit i. However, the errors in the representations of values of y_i close to one do not cause high relative errors in these weight derivatives (except in special circumstances where derivatives from different examples cancel each other, but there is no simple remedy in these rare situations). Hence, it is generally safe to use just y_i^* for these calculations.

11.2.1 Sum-squared error computations

Roundoff errors can also occur in the calculation of errors and their derivatives. This is unlikely to be important if targets are rounded-off versions of true targets, eg., targets like 0.3129871 or 0.9817523, because such targets have as much roundoff error as the values of unit outputs.

However, if sum-squared error is used, and targets are all 0 or 1 and are accurate (i.e., 1 is not a rounded-off version of 0.99999999 or 1.00000001), and unit i computes a logistic function of its total input, then the following formulas can be used (where t_i is the target for unit i, $E = \frac{1}{2}(t_i - y_i)^2$, and $z_i = 1 - y_i$, calculated as before):

$$E^* \leftarrow \begin{cases} \frac{1}{2}(t_i^* - y_i^*)^2 & \text{if } y_i^* \leq 0.5 \\ \frac{1}{2}((t_i^* - 1) + z_i^*)^2 & \text{if } y_i^* > 0.5 \end{cases}$$

$$\left(\frac{\partial E}{\partial y_i}\right)^* \leftarrow \begin{cases} y_i^* - t_i^* & \text{if } y_i^* \leq 0.5 \\ (1 - t_i^*) - z_i^* & \text{if } y_i^* > 0.5 \end{cases}$$

These formulas could be split into a greater number of simpler cases if t_i were always 0 or 1, but as they are they are correct for general values of t_i and accurate when t_i is 0 or 1.

11.2.2 Single logistic-output cross-entropy computations

If the network has a single output unit (unit i) which represents probability in a two-class classification problem, then the cross-entropy error function [1] for one example is as follows:

$$E = t_i \log y_i + (1 - t_i)\log(1 - y_i).$$

This error function is most appropriately used with a logistic function on the output unit. In this case, errors and partial derivatives can be calculated from the following formulas, and the calculations will be accurate for $t_i = 0$ or 1 (where $z_i = 1 - y_i$, calculated as before). To get accurate results, it is necessary to use the function[1] $\mathrm{log1p}(x)$, which computes $\log(1 + x)$ accurately for tiny x.

$$E^* \leftarrow \begin{cases} t_i^* \log y_i^* + (1-t_i^*)\mathrm{log1p}(-y_i^*) & \text{if } y_i^* \leq 0.5 \\ t_i^* \mathrm{log1p}(-z_i^*) + (1-t_i^*)\log z_i^* & \text{if } y_i^* > 0.5 \end{cases}$$

$$\left(\frac{\partial E}{\partial x_i}\right)^* \leftarrow \begin{cases} y_i^* - t_i^* & \text{if } y_i^* \leq 0.5 \\ 1 - t_i^* - z_i^* & \text{if } y_i^* > 0.5 \end{cases}$$

11.2.3 Other approaches to avoiding zero-derivatives with the logistic function

Fahlman [2] suggested adding 0.1 to $\frac{\partial y_i}{\partial x_i}$ in order to decrease learning time by eliminating flat spots in the training surface (ie., spots with small derivatives), and also to avoid zero-derivatives due to roundoff. In Fahlman's experiments this technique improved the learning time. However, it also makes the derivatives incorrect with respect to the error function. This may not matter in networks where the actual output numbers do not mean very much other than "high" or "low". However, in networks where achieving accurate estimates of the targets is important, eg., where the targets are probabilities, or targets are continuous values in a modeling task, this technique is undesirable as it causes the gradient search to not minimize the error, but to minimize some other quantity instead.

Another technique sometimes used with neural networks is to transform targets from the range $[0, 1]$ to $[0.1, 0.9]$. Again, a secondary motivation is to avoid zero-derivatives, and again, this technique should not be used where actual output values have any more significance than "high" or "low".

---

[1] $\mathrm{log1p}(x)$ is available in Unix math libraries.

11.3 Softmax and cross-entropy computations

In networks which must classify instances into one of k mutually exclusive classes, cross-entropy is often used as the error measure, together with the softmax output function. The softmax output function allows the outputs of the k output units to be interpreted as probabilities: it forces each output to be between zero and one, and forces their total to be one. Assuming that outputs units are numbered 1 to k, the equations for softmax and cross-entropy (for one example) are as follows:

$$y_i = \frac{\exp(x_i)}{\sum_{j=1}^{k} \exp(x_j)} \quad \text{and} \quad E = \sum_{j=1}^{k} t_j \log y_j.$$

The derivative of the error with respect to x has a very simple form:

$$\frac{\partial E}{\partial x_i} = y_i - t_i.$$

This equation is subject to high relative error due to roundoff when t_i is exactly one and y_i is close to one. This can happen often in training neural nets, and can lead to derivatives being calculated as zero when in fact they are just small.

This roundoff error, and also possible overflow in the computation of y_i, can be avoided by using the following computations, which use extra floating point variables to store the values of $z_i = 1 - y_i$ and $x'_i = x_i - \max_j x_j$:

$$m^* \leftarrow \max_j x_j^*$$

$$x_i'^* \leftarrow x_i^* - m^*$$

$$Q^* \leftarrow \sum_j \exp(x_j'^*)$$

$$y_i^* \leftarrow \frac{1}{Q^*} \exp(x_i'^*)$$

$$z_i^* \leftarrow \begin{cases} 1 - y_i^* & \text{if } y_i^* \leq 0.5 \\ \frac{1}{Q^*} \sum_{j \neq i} \exp(x_j'^*) & \text{if } y_i^* > 0.5 \end{cases}$$

$$\left(\frac{\partial E}{\partial x_i}\right)^* \leftarrow \begin{cases} y_i^* - t_i^* & \text{if } y_i^* \leq 0.5 \\ (1 - t_i^*) - z_i^* & \text{if } y_i^* > 0.5 \end{cases}$$

$$E^* \leftarrow \sum_j \begin{cases} t_j^* \log(y_j^*) & \text{if } y_i^* \leq 0.5 \\ t_j^* \log 1 p(-z_j^*) & \text{if } y_i^* > 0.5 \end{cases}$$

The space overhead is low: at most two extra floating point variables per output unit, depending on the implementation. The time overhead is also low. The only lengthy additional computation is the second alternative for z_i^*, which can be performed for at most one i because it is impossible for more than one y_i to be greater than 0.5.

11.4 Roundoff error in tanh units

The same ideas apply to the calculation of derivatives in tanh units, which are described by the following equations:

$$y_i = \frac{1-\exp(-2x_i)}{1+\exp(-2x_i)} \quad \text{and} \quad \frac{\partial y_i}{\partial x_i} = (1-y_i)(1+y_i).$$

For tanh units, the output y_i can take on values between -1 and 1. Hence, to represent outputs accurately at the extremes we need two extra floating point numbers to store the values $z_i = 1 - y_i$ and $u_i = 1 + y_i$. Then the following expressions evaluated in floating point arithmetic will avoid unnecessary roundoff error:

$$v_i^* \leftarrow \exp(-2x_i^*) \qquad\qquad y_i^* \leftarrow u_i^* - 1$$
$$u_i^* \leftarrow \frac{2}{1+v_i^*} \qquad\qquad \left(\frac{\partial y_i}{\partial x_i}\right)^* \leftarrow z_i^* u_i^*$$
$$z_i^* \leftarrow v_i^* u_i^*$$

The tanh function is not often used on output units, but if desired, formulas for accurately calculating errors and derivatives when targets are always 1 or -1 are easily derived.

11.5 Why bother?

Since the derivatives which are affected by roundoff error are very small, the proposal to calculate them accurately might provoke the response "why bother?" For on-line learning methods (stochastic gradient), there is probably no point, as computing small gradients to high accuracy is unlikely to make any difference. However, for nets trained in batch mode with second-order methods such as conjugate gradients, small derivatives can be quite important, for the following reasons:

1. Quantization in error or derivatives can confuse line search routines, so avoiding quantization is a good thing.
2. Many small derivatives can add up.
3. Computing small but non-zero derivatives allows training methods to continue as opposed to stopping because of zero derivatives. Some training methods can make significant progress with small derivatives, so it is possible that the weights will move out of the flat area of the error function.

Indeed, there is little reason not to compute these values accurately, as the extra storage and computations are insignificant.

References

1. C. Bishop. *Neural Networks for Pattern Recognition*. Oxford University Press, 1995.

2. S.E. Fahlman. Fast-learning variations on back-propagation: An empirical study. In D. Touretzky, G. Hinton, and T. Sejnowski, editors, *Proceedings of the 1988 Connectionist Models Summer School*, pages 38–51, San Mateo, 1989. Morgan Kaufmann.
3. B.D. Ripley. *Pattern Recognition and Neural Networks*. Cambridge University Press, 1995.
4. D. E. Rumelhart, G. E. Hinton, and Williams R. J. Learning internal representations by error propagation. In *Parallel distributed processing: Explorations in the microstructure of cognition*, volume 1, pages 318–362. MIT Press, Cambridge, MA, 1986.

Representing and Incorporating Prior Knowledge in Neural Network Training

Preface

The present section focuses on tricks for four important aspects in learning: (1) incorporation of prior knowledge, (2) choice of representation for the learning task, (3) unequal class prior distributions, and finally (4) large network training.

Patrice Simard, et al. review the famous **tangent distance** and **tangent propagation algorithms**. Included are tricks for speeding and increasing the stability that were not published in previous work. One important trick for obtaining good performance in their methods is **smoothing** (see p. 253) of the input data, which is illustrated for 2D handwritten character recognition images. To obtain the tangent we must compute the derivative, however, this obviously cannot be calculated for discrete images. To compute tangent vectors, it must be possible to interpolate between the pixel values of the images (or the features of the images). Since many interpolations are possible, a "smoothing" regularizer is used because it has the extra benefit of imposing some control on the locality of the transformation invariance. However, care must be taken not to over smooth (or useful features are washed away). Another trick is to use the so-called **elastic tangent distance** (see p. 253) which eliminates the problem of singular systems which arises when there is zero distance between two patterns, or when their tangent vectors are parallel. Finally, through a very refined **hierarchy of resolution and accuracy** (see p. 255) the tangent distance algorithm can be sped up by two or tree orders of magnitude over a straight forward implementation.

Tangentprop takes the invariance idea even further by making it possible to incorporate local or global invariances and prior knowledge directly into the loss function of backpropagation and to **backpropagate tangents** efficiently (see p. 262). The chapter concludes with a digression on the mathematical background (Lie groups) and the simplifications that follow from the theory of this very successful (record breaking OCR) algorithm.

In the next chapter Larry Yaeger, et al. present an entire collection of tricks for an application that also is of strong commercial interest, on-line handwritten character recognition. These tricks were used by the authors to develop the recognizer in the Apple Computer's Newton MessagePad® and eMate® products. Their recognizer consists of three main components: a segmenter, a classifier, and a context driven search component. Of particular interest is the intricate **representation of stroke information** (see p. 278) that serves as input to the classifier. Prior knowledge has led the authors to include specific stroke features, grayscale character images, and coarse character attributes (stroke count and character aspect ratio), all combined into an ensemble of networks. The ensemble decisions are then combined in algorithmic search through a hypothesis space of dictionaries and combinations of dictionaries comprising a broad coverage, weakly applied language model.

An extensive set of tricks for training the network are also discussed, including:

Normalizing output error: Assist output activities of secondary choices to have non-zero values (***NormOutErr***). This enhances the robustness for the subsequent integrated search procedure since the search can now also take the n best choices into account (p. 280).

Negative training: Reduce the effect of invalid character segmentation (p. 282).

Stroke warping: Generate randomly warped versions of patterns to increase the generalization ability (p. 283).

Frequency balancing: Reduce the problem of unbalanced class priors using the simple trick of repeating low frequency classes more often in order to force the network to allocate more resources to these cases (p. 284).

Error emphasis: Account for different and uncommon writing *styles* by presenting difficult or unusual patterns more often (p. 285).

Quantized weights: Enable the neural network classifier to run with only one-byte weights and train with a temporary additional two bytes (p. 287).

In chapter 14, Steve Lawrence, et al. discuss several different tricks for alleviating the problem of unbalanced class prior probabilities. Some theoretical explanations are also provided. In the first trick, **prior scaling** (p. 300), weight updates are scaled so that the total expected update for each class is equal. The second trick, **probabilistic sampling** (p. 302), slightly modifies the frequency balancing in chapter 13. Here, a class is first chosen and then from within this class a sample is drawn. The next trick is referred to as **post scaling** (p. 302). The network is trained as usual but the network outputs are rescaled after training. This method can also be used to optimize other criteria that are different from the loss function the network has been trained with. Finally the authors propose to **equalize class memberships** (p. 303) by either subsampling the class with higher frequency or by duplicating patterns from the class with lower frequency (however they report that this trick works least efficiently). The effectiveness of each of these tricks is examined and compared for an ECG classification problem.

Training problems with thousands of classes and millions of examples, as are common for speech and handwritten character recognition problems, pose a major challenge. While many of the training techniques discussed so far work well for moderate size nets, they can fail miserably for these extremely large problems. In chapter 15, Jürgen Fritsch and Michael Finke design a representation and architecture for such large scale learning problems and, like the previous two chapters, they also tackle the problem of unbalanced class priors (since not all of the 24k subphonemes are equally probable). They exemplify their approach by building a large vocabulary speech recognizer. In the first step they break down the task into a hierarchy of smaller decision problems of controllable size (**divide and conquer**, p. 317) and estimate the conditional probabilities for each node of the decision tree with a neural network. The network training uses **mini batches** and **individual adaptive learning rates** that are increased if progress is made in weight space and decreased if the fluctuations in weight space are too high (p. 336). These estimated probabilities – modeled by every single neural network node – are combined to give an overall estimate of the class decision probabilities.

The authors either determine the **decision tree** structure manually or estimate it by their **ACID clustering** algorithm (p. 328). Interestingly, the manual structure design was outperformed by the proposed agglomerative clustering scheme. No doubt that prior knowledge helps to achieve better classification results. However, this astonishing result indicates that human prior knowledge, although helpful in general, is suboptimal for structuring such a large task, particularly since automatic clustering allows for fine-grain subdivision of the classification task and aims for uniformity of priors. This desirable goal is hardly achievable by manual construction of the classification hierarchy. Furthermore, the human prior knowledge also does not provide the best basis from which a machine learning algorithm can learn optimally, a fact that is important to keep in mind for other applications as well.

Jenny & Klaus

12
Transformation Invariance in Pattern Recognition – Tangent Distance and Tangent Propagation

Patrice Y. Simard[1], Yann A. LeCun[1], John S. Denker[1], and Bernard Victorri[2]

[1] Image Processing Services Research Lab, AT&T Labs - Research, 100 Schulz Drive, Red Bank, NJ 07701-7033, USA
patrice@research.att.com, http://www.research.att.com/info/patrice
[2] CNRS, ELSAP, ENS, 1 rue Maurice Arnoux, F-92120 MONTROUGE, France.

Abstract. In pattern recognition, statistical modeling, or regression, the amount of data is a critical factor affecting the performance. If the amount of data and computational resources are unlimited, even trivial algorithms will converge to the optimal solution. However, in the practical case, given limited data and other resources, satisfactory performance requires sophisticated methods to regularize the problem by introducing *a priori* knowledge. Invariance of the output with respect to certain transformations of the input is a typical example of such *a priori* knowledge. In this chapter, we introduce the concept of tangent vectors, which compactly represent the essence of these transformation invariances, and two classes of algorithms, "tangent distance" and "tangent propagation", which make use of these invariances to improve performance.

12.1 Introduction

Pattern Recognition is one of the main tasks of biological information processing systems, and a major challenge of computer science. The problem of pattern recognition is to classify objects into categories, given that objects in a particular category may have widely-varying features, while objects in different categories may have quite similar features. A typical example is handwritten digit recognition. Characters, typically represented as fixed-size images (say 16 by 16 pixels), must be classified into one of 10 categories using a *classification function*. Building such a classification function is a major technological challenge, as irrelevant variabilities among objects of the same class must be eliminated, while meaningful differences between objects of different classes must be identified. These classification functions for most real pattern recognition tasks are too complicated to be synthesized "by hand" using only what humans know about the task. Instead, we use sophisticated techniques that combine humans' *a priori* knowledge with information automatically extracted from a set of labeled examples (the training set). These techniques can be divided into two camps, according to the number of parameters they require: the "memory based" algorithms, which in effect store a sizeable subset of the entire training set, and the "learned-function" techniques, which learn by adjusting a comparatively small number of parameters. This distinction is arbitrary because the patterns stored by a memory-based algorithm can be considered the parameters of a very complex learned function. The distinction is however useful in this work, because memory based algorithms often rely on a metric which can be modified to incorporate transformation invariances, while learned-function algorithms consist of selecting a classification function, the derivatives of which can be constrained to reflect the same transformation invariances. The two methods for incorporating invariances are different enough to justify two independent sections.

12.1.1 Memory based algorithms

To compute the classification function, many practical pattern recognition systems, and several biological models, simply store all the examples, together with their labels, in a memory. Each incoming pattern can then be compared to all the stored prototypes, and the labels associated with the prototypes that best match the input determine the output. The above method is the simplest example of the *memory-based* models. Memory-based models require three things: a *distance measure* to compare inputs to prototypes, an *output function* to produce an output by combining the labels of the prototypes, and a *storage scheme* to build the set of prototypes.

All three aspects have been abundantly treated in the literature. Output functions range from simply voting the labels associated with the k closest prototypes (K-Nearest Neighbors), to computing a score for each class as a linear combination of the distances to all the prototypes, using fixed [21] or learned [5] coefficients. Storage schemes vary from storing the entire training set, to picking appropriate subsets of it (see [8], chapter 6, for a survey) to learned-functions such as learning vector quantization (LVQ) [17] and gradient descent. Distance measures can be as simple as the Euclidean distance, assuming the patterns and prototypes are represented as vectors, or more complex as in the generalized quadratic metric [10] or in elastic matching methods [15].

A simple but inefficient pattern recognition method is to use a simple distance measure, such as Euclidean distance between vectors representing the raw input, combined with a very large set of prototypes. This method is inefficient because almost all possible instances of a category must be present in the prototype set. In the case of handwritten digit recognition, this means that digits of each class in all possible positions, sizes, angles, writing styles, line thicknesses, skews, etc... must be stored. In real situations, this approach leads to impractically large prototype sets or to mediocre recognition accuracy as illustrated in Figure 12.1. An unlabeled image of a thick, slanted "9" must be classified by finding the closest prototype image out of two images representing respectively a thin, upright "9" and a thick, slanted "4". According to the Euclidean distance (sum of the squares of the pixel to pixel differences), the "4" is closer. The result is an incorrect classification. The classical way of dealing with this problem is to

12. Tangent Distance and Tangent Propagation

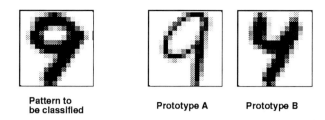

Pattern to be classified   Prototype A   Prototype B

Fig. 12.1. According to the Euclidean distance the pattern to be classified is more similar to prototype B. A better distance measure would find that prototype A is closer because it differs mainly by a rotation and a thickness transformation, two transformations which should leave the classification invariant.

use a so-called *feature extractor* whose purpose is to compute a representation of the patterns that is minimally affected by transformations of the patterns that do not modify their category. For character recognition, the representation should be invariant with respect to position, size changes, slight rotations, distortions, or changes in line thickness. The design and implementation of feature extractors is the major bottleneck of building a pattern recognition system. For example, the problem illustrated in Figure 12.1 can be solved by deslanting and thinning the images.

An alternative to this is to use an invariant *distance measure* constructed in such a way that the distance between a prototype and a pattern will not be affected by irrelevant transformations of the pattern or of the prototype. With an invariant distance measure, each prototype can match many possible instances of pattern, thereby greatly reducing the number of prototypes required.

The natural way of doing this is to use "deformable" prototypes. During the matching process, each prototype is deformed so as to best fit the incoming pattern. The quality of the fit, possibly combined with a measure of the amount of deformation, is then used as the distance measure [15]. With the example of Figure 12.1, the "9" prototype would be rotated and thickened so as to best match the incoming "9". This approach has two shortcomings. First, a set of allowed deformations must be designed based on *a priori* knowledge. Fortunately, this is feasible for many tasks, including character recognition. Second, the search for the best-matching deformation is often enormously expensive, and/or unreliable. Consider the case of patterns that can be represented by vectors. For example, the pixel values of a 16 by 16 pixel character image can be viewed as the components of a 256-dimensional vector. One pattern, or one prototype, is a point in this 256-dimensional space. Assuming that the set of allowable transformations is continuous, the set of all the patterns that can be obtained by transforming one prototype using one or a combination of allowable transformations is a surface in the 256-D pixel space. More precisely, when a pattern P is transformed (e.g. rotated) according to a transformation $s(P, \alpha)$ which depends on one parameter α (e.g. the angle of the rotation), the set of all the transformed patterns

$$S_P = \{x \mid \exists \boldsymbol{\alpha} \text{ for which } x = s(P, \alpha)\} \tag{12.1}$$

is a one-dimensional curve in the vector space of the inputs. In the remainder of this chapter, we will always assume that we have chosen s be differentiable with respect to both P and α and such that $s(P,0) = P$.

When the set of transformations is parameterized by n parameters α_i (rotation, translation, scaling, etc.), the intrinsic dimension of the manifold S_P is at most n. For example, if the allowable transformations of character images are horizontal and vertical shifts, rotations, and scaling, the surface will be a 4-dimensional manifold.

In general, the manifold will not be linear. Even a simple image translation corresponds to a highly non-linear transformation in the high-dimensional pixel space. For example, if the image of an "8" is translated upward, some pixels oscillate from white to black and back several times. Matching a deformable prototype to an incoming pattern now amounts to finding the point on the surface that is at a minimum distance from the point representing the incoming pattern. This non-linearity makes the matching much more expensive and unreliable. Simple minimization methods such as gradient descent (or conjugate gradient) can be used to find the minimum-distance point, however, these methods only converge to a *local* minimum. In addition, running such an iterative procedure for each prototype is usually prohibitively expensive.

If the set of transformations happens to be linear in pixel space, then the manifold is a linear subspace (a hyperplane). The matching procedure is then reduced to finding the shortest distance between a point (vector) and a hyperplane, which is an easy-to-solve quadratic minimization problem. This special case has been studied in the statistical literature and is sometimes referred to as Procrustes analysis [24]. It has been applied to signature verification [12] and on-line character recognition [26].

This chapter considers the more general case of non-linear transformations such as geometric transformations of gray-level images. Remember that even a simple image translation corresponds to a highly non-linear transformation in the high-dimensional pixel space. The main idea of the chapter is to approximate the surface of possible transforms of a pattern by its tangent plane at the pattern, thereby reducing the matching to finding the shortest distance between two planes. This distance is called the *tangent distance*. The result of the approximation is shown in Figure 12.2, in the case of rotation for handwritten digits. At the top of the figure, is the theoretical curve in pixel space which represents equation (12.1), together with its linear approximation. Points of the transformation curve are depicted below for various amounts of rotation (each angle corresponds to a value of α). The bottom of Figure 12.2 depicts the linear approximation of the curve $s(P,\alpha)$ given by the Taylor expansion of s around $\alpha = 0$:

$$s(P,\alpha) = s(P,0) + \alpha \frac{\partial s(P,\alpha)}{\partial \alpha} + O(\alpha^2) \approx P + \alpha T. \qquad (12.2)$$

This linear approximation is completely characterized by the point P and the tangent vector $T = \frac{\partial s(P,\alpha)}{\partial \alpha}$. Tangent vectors, also called the Lie derivatives of the transformation s, will be the subject of section 12.4. As can be seen from

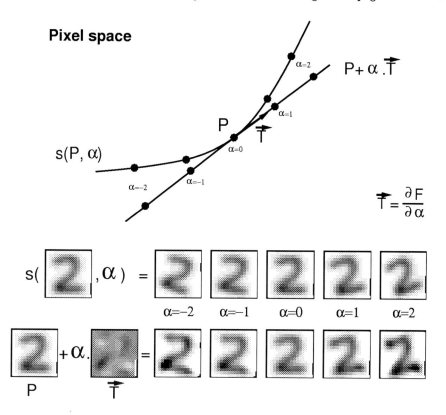

Fig. 12.2. Top: Representation of the effect of the rotation in pixel space. Middle: Small rotations of an original digitized image of the digit "2", for different angle values of α. Bottom: Images obtained by moving along the tangent to the transformation curve for the same original digitized image P by adding various amounts (α) of the tangent vector T.

Figure 12.2, for reasonably small angles ($\|\alpha\| < 1$), the approximation is very good.

Figure 12.3 illustrates the difference between the Euclidean distance, the full invariant distance (minimum distance between manifolds) and the tangent distance. In the figure, both the prototype and the pattern are deformable (two-sided distance), but for simplicity or efficiency reasons, it is also possible to deform only the prototype or only the unknown pattern (one-sided distance).

Although in the following we will concentrate on using tangent distance to recognize images, the method can be applied to many different types of signals: temporal signals, speech, sensor data...

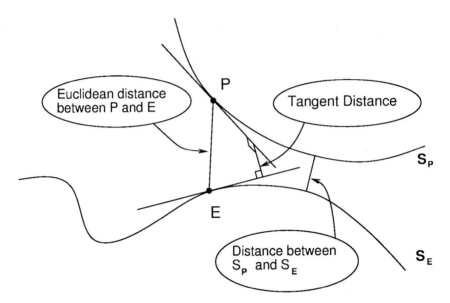

Fig. 12.3. Illustration of the Euclidean distance and the tangent distance between P and E. The curves S_p and S_e represent the sets of points obtained by applying the chosen transformations (for example translations and rotations) to P and E. The lines going through P and E represent the tangent to these curves. Assuming that working space has more dimensions than the number of chosen transformations (on the diagram, assume one transformation in a 3-D space) the tangent spaces do not intersect and the tangent distance is uniquely defined.

12.1.2 Learned-function algorithms

Rather than trying to keep a representation of the training set, it is also possible to choose a classification function by learning a set of parameters. This is the approach taken in neural networks, curve fitting, regression, et cetera.

We assume all data is drawn independently from a given statistical distribution $\mathcal{P}$, and our learning machine is characterized by the set of functions it can implement, $G_w(x)$, indexed by the vector of parameters w. We write $F(x)$ to represent the "correct" or "desired" labeling of the point x. The task is to find a value for w such that G_w best approximates F. We can use a finite set of training data to help find this vector. We assume the correct labeling $F(x)$ is known for all points in the training set. For example, G_w may be the function computed by a neural net having weights w, or G_w may be a polynomial having coefficients w. Without additional information, finding a value for w is an ill-posed problem unless the number of parameters is small and/or the size of the training set is large. This is because the training set does not provide enough information to distinguish the best solution among all the candidate ws. This problem is illustrated in Figure 12.4 (left). The desired function F (solid line) is to be approximated by a functions G_w (dotted line) from four examples $\{(x_i, F(x_i))\}_{i=1,2,3,4}$. As exemplified in the picture, the fitted function G_w

largely disagrees with the desired function F between the examples, but it is not possible to infer this from the training set alone. Many values of w can generate many different functions G_w, some of which may be terrible approximations of F, even though they are in complete agreement with the training set. Because of this, it is customary to add "regularizers", or additional constraints, to restrict the search of an acceptable w. For example, we may require the function G_w to be "smooth", by adding the constraint that $\|w\|^2$ should be minimized. It is important that the regularizer reflects a property of F, hence regularizers depend on *a priori* knowledge about the function to be modeled.

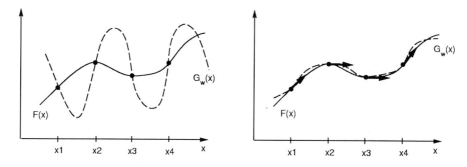

Fig. 12.4. Learning a given function (solid line) from a limited set of examples (x_1 to x_4). The fitted curves are shown by dotted line. Left: The only constraint is that the fitted curve goes through the examples. Right: The fitted curves not only go through each example but also its derivatives evaluated at the examples agree with the derivatives of the given function.

Selecting a good family $\mathcal{G} = \{G_w, w \in \Re^q\}$ of functions is a difficult task, sometimes known as "model selection" [16, 14]. If $\mathcal{G}$ contains a large family of functions, it is more likely that it will contain a good approximation of F (the function we are trying to approximate), but it is also more likely that the selected candidate (using the training set) will generalize poorly because many functions in $\mathcal{G}$ will agree with the training data and take outrageous values between the training samples. If, on the other hand, $\mathcal{G}$ contains a small family of functions, it is more likely that a function G_w which fits the data will be a good approximation of F. The capacity of the family of functions $\mathcal{G}$ is often referred to as the VC dimension [28, 27]. If a large amount of data is available, $\mathcal{G}$ should contain a large family of functions (high VC dimension), so that more functions can be approximated, and in particular, F. If, on the other hand, the data is scarce, $\mathcal{G}$ should be restricted to a small family of functions (low VC dimension), to control the values between the (more distant) samples[1]. The VC dimension

---

[1] Note that this point of view also applies to memory based systems. In the case where *all* the training data can be kept in memory, however, the VC dimension is infinite, and the formalism is meaningless. The VC dimension is a learning paradigm and is not useful unless learning is involved.

can also be controlled by putting a knob on how much effect is given to some regularizers. For instance it is possible to control the capacity of a neural network by adding "weight decay" as a regularizer. Weight decay is a heuristic that favors smooth classification functions, by making a tradeoff by decreasing $\|w\|^2$ at the cost, usually, of slightly increased error on the training set. Since the optimal classification function is not necessarily smooth, for instance at a decision boundary, the weight decay regularizer can have adverse effects.

As mentioned earlier, the regularizer should reflect interesting properties (*a priori* knowledge) of the function to be learned. If the functions F and G_w are assumed to be differentiable, which is generally the case, the search for G_w can be greatly improved by requiring that G_w's derivatives evaluated at the points $\{x_i\}$ are more or less equal (this is the regularizer knob) to the derivatives of F at the same points (Figure 12.4 right). This result can be extended to multidimensional inputs. In this case, we can impose the equality of the derivatives of F and G_w in *certain directions*, not necessarily in all directions of the input space. Such constraints find immediate use in traditional pattern recognition problems. It is often the case that *a priori* knowledge is available on how the desired function varies with respect to some transformations of the input. It is straightforward to derive the corresponding constraint on the directional derivatives of the fitted function G_w in the directions of the transformations (previously named tangent vectors). Typical examples can be found in pattern recognition where the desired classification function is known to be invariant with respect to some transformation of the input such as translation, rotation, scaling, etc., in other words, the directional derivatives of the classification function in the directions of these transformations is zero.

This is illustrated in Figure 12.4. The right part of the figure shows how the additional constraints on G_w help generalization by constraining the values of G_w outside the training set. For every transformation which has a known effect on the classification function, a regularizer can be added in the form of a constraint on the directional derivative of G_w in the direction of the tangent vector (such as the one depicted in Figure 12.2), computed from the curve of transformation.

The next section will analyze in detail how to use a distance based on tangent vector in memory based algorithms. The subsequent section will discuss the use of tangent vectors in neural network, with the tangent propagation algorithm. The last section will compare different algorithms to compute tangent vectors.

12.2 Tangent Distance

The Euclidean distance between two patterns P and E is in general not appropriate because it is sensitive to irrelevant transformations of P and of E. In contrast, the transformed distance $\mathcal{D}(E, P)$ is defined to be the minimal distance between the two manifolds S_P and S_E, and is therefore invariant with respect to the transformation used to generate S_P and S_E (see Figure 12.3). Unfortunately, these manifolds have no analytic expression in general, and finding the distance between them is a difficult optimization problem with multiple local minima. Besides, true invariance is not necessarily desirable since a rotation of a "6" into a "9" does not preserve the correct classification.

Our approach consists of computing the minimum distance between the linear surfaces that best approximate the non-linear manifolds S_P and S_E. This solves three problems at once: 1) linear manifolds have simple analytical expressions which can be easily computed and stored, 2) finding the minimum distance between linear manifolds is a simple least squares problem which can be solved efficiently and, 3) this distance is locally invariant but not globally invariant. Thus the distance between a "6" and a slightly rotated "6" is small but the distance between a "6" and a "9" is large. The different distances between P and E are represented schematically in Figure 12.3.

The figure represents two patterns P and E in 3-dimensional space. The manifolds generated by s are represented by one-dimensional curves going through E and P respectively. The linear approximations to the manifolds are represented by lines tangent to the curves at E and P. These lines do not intersect in 3 dimensions and the shortest distance between them (uniquely defined) is $D(E, P)$. The distance between the two non-linear transformation curves $\mathcal{D}(E, P)$ is also shown on the figure.

An efficient implementation of the tangent distance $D(E, P)$ will be given in the next section, using image recognition as an illustration. We then compare our methods with the best known competing methods. Finally we will discuss possible variations on the tangent distance and how it can be generalized to problems other than pattern recognition.

12.2.1 Implementation

In this section we describe formally the computation of the tangent distance. Let the function s transform an image P to $s(P, \alpha)$ according to the parameter α. We require s to be differentiable with respect to α and P, and require $s(P, 0) = P$. If P is a 2 dimensional image for instance, $s(P, \alpha)$ could be a rotation of P by the angle α. If we are interested in all transformations of images which conserve distances (isometry), $s(P, \alpha)$ would be a rotation by α_θ followed by a translation by α_x, α_y of the image P. In this case $\alpha = (\alpha_\theta, \alpha_x, \alpha_y)$ is a vector of parameters of dimension 3. In general, $\alpha = (\alpha_1, \ldots, \alpha_m)$ is of dimension m.

Since s is differentiable, the set $S_P = \{x \mid \exists \alpha \text{ for which } x = s(P, \alpha)\}$ is a differentiable manifold which can be approximated to the first order by a hyperplane T_P. This hyperplane is tangent to S_P at P and is generated by the columns of matrix

$$L_P = \left.\frac{\partial s(P, \alpha)}{\partial \alpha}\right|_{\alpha=0} = \left[\frac{\partial s(P, \alpha)}{\partial \alpha_1}, \ldots, \frac{\partial s(P, \alpha)}{\partial \alpha_m}\right]_{\alpha=0} \quad (12.3)$$

which are vectors tangent to the manifold. If E and P are two patterns to be compared, the respective tangent planes T_E and T_P can be used to define a new distance D between these two patterns. The tangent distance $D(E, P)$ between

E and P is defined by

$$D(E,P) = \min_{x \in T_E, y \in T_P} \|x - y\|^2. \tag{12.4}$$

The equation of the tangent planes T_E and T_P is given by:

$$E'(\alpha_E) = E + L_E \alpha_E \tag{12.5}$$
$$P'(\alpha_P) = P + L_P \alpha_P \tag{12.6}$$

where L_E and L_P are the matrices containing the tangent vectors (see equation (12.3)) and the vectors α_E and α_P are the coordinates of E' and P' (using bases L_E and L_P) in the corresponding tangent planes. Note that E', E, L_E and α_E denote vectors and matrices in linear equations (12.5). For example, if the pixel space was of dimension 5, and there were two tangent vectors, we could rewrite equation (12.5) as

$$\begin{bmatrix} E'_1 \\ E'_2 \\ E'_3 \\ E'_4 \\ E'_5 \end{bmatrix} = \begin{bmatrix} E_1 \\ E_2 \\ E_3 \\ E_4 \\ E_5 \end{bmatrix} + \begin{bmatrix} L_{11} & L_{12} \\ L_{21} & L_{22} \\ L_{31} & L_{32} \\ L_{41} & L_{42} \\ L_{51} & L_{52} \end{bmatrix} \begin{bmatrix} \alpha_1 \\ \alpha_2 \end{bmatrix}. \tag{12.7}$$

The quantities L_E and L_P are attributes of the patterns so in many cases they can be precomputed and stored.

Computing the tangent distance

$$D(E,P) = \min_{\alpha_E, \alpha_P} \|E'(\alpha_E) - P'(\alpha_P)\|^2 \tag{12.8}$$

amounts to solving a linear least squares problem. The optimality condition is that the partial derivatives of $D(E,P)$ with respect to α_P and α_E should be zero:

$$\frac{\partial D(E,P)}{\partial \alpha_E} = 2(E'(\alpha_E) - P'(\alpha_P))^\top L_E = 0 \tag{12.9}$$

$$\frac{\partial D(E,P)}{\partial \alpha_P} = 2(P'(\alpha_P) - E'(\alpha_E))^\top L_P = 0. \tag{12.10}$$

Substituting E' and P' by their expressions yields to the following linear system of equations, which we must solve for α_P and α_E:

$$L_P^\top (E - P - L_P \alpha_P + L_E \alpha_E) = 0 \tag{12.11}$$
$$L_E^\top (E - P - L_P \alpha_P + L_E \alpha_E) = 0. \tag{12.12}$$

The solution of this system is

$$(L_{PE} L_{EE}^{-1} L_E^\top - L_P^\top)(E - P) = (L_{PE} L_{EE}^{-1} L_{EP} - L_{PP})\alpha_P \tag{12.13}$$
$$(L_{EP} L_{PP}^{-1} L_P^\top - L_E^\top)(E - P) = (L_{EE} - L_{EP} L_{PP}^{-1} L_{PE})\alpha_E \tag{12.14}$$

where $L_{EE} = L_E^\top L_E$, $L_{PE} = L_P^\top L_E$, $L_{EP} = L_E^\top L_P$ and $L_{PP} = L_P^\top L_P$. LU decompositions of L_{EE} and L_{PP} can be precomputed. The most expensive part in solving this system is evaluating L_{EP} (L_{PE} can be obtained by transposing L_{EP}). It requires $m_E \times m_P$ dot products, where m_E is the number of tangent vectors for E and m_P is the number of tangent vectors for P. Once L_{EP} has been computed, α_P and α_E can be computed by solving two (small) linear systems of respectively m_E and m_P equations. The tangent distance is obtained by computing $\|E'(\alpha_E) - P'(\alpha_P)\|$ using the value of α_P and α_E in equations (12.5) and (12.6). If n is the dimension of the input space (i.e. the length of vector E and P), the algorithm described above requires roughly $n(m_E+1)(m_P+1)+3(m_E^3+m_P^3)$ multiply-adds. Approximations to the tangent distance can however be computed more efficiently.

12.2.2 Some illustrative results

Local Invariance: The "local[2] invariance" of tangent distance can be illustrated by transforming a reference image by various amounts and measuring its distance to a set of prototypes.

The bottom of Figure 12.5 shows 10 typical handwritten digit images. One of them – the digit "3" – is chosen to be the reference. The reference is translated horizontally by the amount indicated in the abscissa. There are ten curves for Euclidean distance and ten more curves for tangent distance, measuring the distance between the translated reference and one of the 10 digits.

Since the reference was chosen from the 10 digits, it is not surprising that the curve corresponding to the digit "3" goes to 0 when the reference is not translated (0 pixel translation). It is clear from the figure that if the reference (the image "3") is translated by more than 2 pixels, the Euclidean distance will confuse it with other digits, namely "8" or "5". In contrast, there is no possible confusion when tangent distance is used. As a matter of fact, in this example, the tangent distance correctly identifies the reference up to a translation of 5 pixels! Similar curves were obtained with all the other transformations (rotation, scaling, etc...).

The "local" invariance of tangent distance with respect to small transformations generally implies more accurate classification for much larger transformations. This is the single most important feature of tangent distance.

The locality of the invariance has another important benefit: Local invariance can be enforced with *very few* tangent vectors. The reason is that for infinitesimal (local) transformations, there is a direct correspondence[3] between the tangent vectors of the tangent plane and the various compositions of transformations.

---

[2] Local invariance refers to invariance with respect to small transformations (i.e. a rotation of a very small angle). In contrast, global invariance refers to invariance with respect to arbitrarily large transformations (i.e. a rotation of 180 degrees). Global invariance is not desirable in digit recognition, since we need to distinguish "6" from a "9".

[3] an isomorphism actually, see "Lie algebra" in [6].

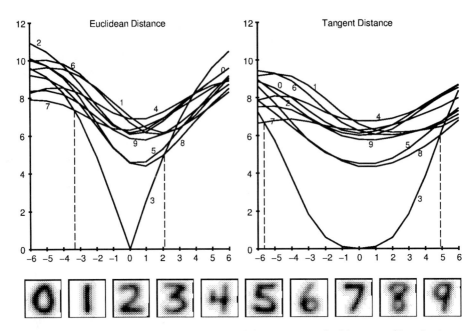

Fig. 12.5. Euclidean and tangent distances between 10 typical images of handwritten digits and a translated image of the digit "3". The abscissa represents the amount of horizontal translation (measured in pixels).

For example, the three tangent vectors for X-translation, Y-translation and rotations around the origin, generate a tangent plane corresponding to all the possible compositions of horizontal translations, vertical translations and rotations. The resulting tangent distance is then locally invariant to *all* the translations and *all* the rotations (around any center). Figure 12.6 further illustrates this phenomenon by displaying points in the tangent plane generated from only 5 tangent vectors. Each of these images looks like it has been obtained by applying various combinations of scaling, rotation, horizontal and vertical skewing, and thickening. Yet, the tangent distance between any of these points and the original image is 0.

Handwritten Digit Recognition: Experiments were conducted to evaluate the performance of tangent distance for handwritten digit recognition. An interesting characteristic of digit images is that we can readily identify a set of local transformations which do not affect the identity of the character, while covering a large portion of the set of possible *instances* of the character. Seven such image transformations were identified: X and Y translations, rotation, scaling, two hyperbolic transformations (which can generate shearing and squeezing), and line thickening or thinning. The first six transformations were chosen to span the set of all possible linear coordinate transforms in the image plane. (Nevertheless, they correspond to highly non-linear transforms in pixel space.) Additional

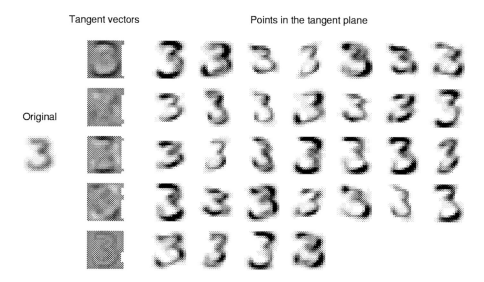

Fig. 12.6. Left: Original image. Middle: 5 tangent vectors corresponding respectively to the 5 transformations: scaling, rotation, expansion of the X axis while compressing the Y axis, expansion of the first diagonal while compressing the second diagonal and thickening. Right: 32 points in the tangent space generated by adding or subtracting each of the 5 tangent vectors.

transformations have been tried with less success. Three databases were used to test our algorithm:

US Postal Service database: The database consisted of 16 × 16 pixel size-normalized images of handwritten digits, coming from US mail envelopes. The training and testing set had respectively 9709 and 2007 examples.

NIST1 database: The second experiment was a competition organized by the National Institute of Standards and Technology (NIST) in Spring 1992. The object of the competition was to classify a test set of 59,000 handwritten digits, given a training set of 223,000 patterns.

NIST2 database: The third experiment was performed on a database made out of the training and testing database provided by NIST (see above). NIST had divided the data into two sets which unfortunately had different distributions. The training set (223,000 patterns) was easier than the testing set (59,000 patterns). In our experiments we combined these two sets 50/50 to make a training set of 60,000 patterns and testing/validation sets of 10,000 patterns each, all having the same characteristics.

For each of these three databases we tried to evaluate human performance to benchmark the difficulty of the database. For USPS, two members of our group went through the test set and both obtained a 2.5% raw error performance. The human performance on NIST1 was provided by the National Institute of Standard and Technology. The human performance on NIST2 was measured on a small subsample of the database and must therefore be taken with caution. Several of the leading algorithms where tested on each of these databases.

The first experiment used the K-Nearest Neighbor algorithm, using the ordinary Euclidean distance. The prototype set consisted of all available training examples. A 1-Nearest Neighbor rule gave optimal performance in USPS while a 3-Nearest Neighbors rule performed better in NIST2.

The second experiment was similar to the first, but the distance function was changed to tangent distance with 7 transformations. For the USPS and NIST2 databases, the prototype set was constructed as before, but for NIST1 it was constructed by cycling through the training set. Any patterns which were misclassified were added to the prototype set. After a few cycles, no more prototypes are added (the training error was 0). This resulted in 10,000 prototypes. A 3-Nearest Neighbors rule gave optimal performance on this set.

Other algorithms such as neural nets [18, 20], optimal margin classifier [7], local learning [3] and boosting [9] were also used on these databases. A case study can be found in [20].

The results are summarized in Table 12.2.2. As illustrated in the table, the

	Human	K-NN	T.D.	Lenet1	Lenet4	OMC	LL	Boost
USPS	2.5	5.7	2.5	4.2		4.3	3.3	2.6
NIST1	1.6		3.2		3.7			4.1
NIST2	0.2	2.4	1.1	1.7	1.1	1.1	1.1	0.7

Table 12.1. Results: Performances in % of errors for (in order) human, K-nearest neighbor, tangent distance, Lenet1 (simple neural network), Lenet4 (large neural network), optimal margin classifier (OMC), local learning (LL) and boosting (Boost).

tangent distance algorithm equals or outperforms all other algorithms we tested, in all cases except one: Boosted LeNet 4 was the winner on the NIST2 database. This is not surprising. The K-nearest neighbor algorithm (with no preprocessing) is very unsophisticated in comparison to local learning, optimal margin classifier, and boosting. The advantage of tangent distance is the *a priori* knowledge of transformation invariance embedded into the distance. When the training data is sufficiently large, as is the case in NIST2, some of this knowledge can be picked up from the data by the more sophisticated algorithms. In other words, the value of *a priori* knowledge decreases as the size of the training set increases.

12.2.3 How to make tangent distance work

This section is dedicated to the technological "know how" which is necessary to make tangent distance work with various applications. "Tricks" of this sort are usually not published for various reasons (they are not always theoretically sound, page area is too valuable, the tricks are specific to one particular application, commercial competitive considerations discourage telling everyone how to reproduce the result, etc.), but they are often a determining factor in making the technology a success. Several of these techniques will be discussed here.

Smoothing the input space: This is the single most important factor in obtaining good performance with tangent distance. By definition, the tangent vectors are the Lie derivatives of the transformation function $s(P, \alpha)$ with respect to α. They can be written as:

$$L_P = \frac{\partial s(P, \alpha)}{\partial \alpha}\bigg| = \lim_{\epsilon \to 0} \frac{s(P, \epsilon) - s(P, 0)}{\epsilon}. \tag{12.15}$$

It is therefore very important that s be differentiable (and well behaved) with respect to α. In particular, it is clear from equation (12.15) that $s(P, \epsilon)$ must be computed for ϵ arbitrarily small. Fortunately, even when P can only take discrete values, it is easy to make s differentiable. The trick is to use a smoothing interpolating function C_σ as a preprocessing for P, such that $s(C_\sigma(P), \alpha)$ is differentiable (with respect to $C_\sigma(P)$ and α, not with respect to P). For instance, if the input space for P is binary images, $C_\sigma(P)$ can be a convolution of P with a Gaussian function of standard deviation σ. If $s(C_\sigma(P), \alpha)$ is a translation of α pixels, the derivative of $s(C_\sigma(P), \alpha)$ can easily be computed since $s(C_\sigma(P), \epsilon)$ can be obtained by translating Gaussian functions. This preprocessing will be discussed in more details in section 12.4.

The smoothing factor σ controls the locality of the invariance. The smoother the transformation curve defined by s is, the longer the linear approximation will be valid. In general the best smoothing is the maximum smoothing which does not blur the features. For example, in handwritten character recognition with 16x16 pixel images, a Gaussian function with a standard deviation of 1 pixel yielded the best results. Increased smoothing led to confusion (such as a "5" mistaken for "6" because the lower loop had been closed by the smoothing) and decreased smoothing didn't make full use of the invariance properties.

If the available computation time allows it, the best strategy is to extract features first, smooth shamelessly, and then compute the tangent distance on the smoothed features.

Controlled deformation: The linear system given in equation (12.8) is singular if some of the tangent vectors for E or P are parallel. Although the probability of this happening is zero when the data is taken from a real-valued continuous distribution (as is the case in handwritten character recognition), it is possible that a pattern may be duplicated in both the training and the test set, resulting in a division by zero error. The fix is quite simple and elegant. Equation (12.8) can be replaced by equation:

$$D(E, P) = \min_{\alpha_E, \alpha_P} \|E + L_E \alpha_E - P - L_P \alpha_P\|^2 + k\|L_E \alpha_E\|^2 + k\|L_P \alpha_P\|^2. \tag{12.16}$$

The physical interpretation of this equation, depicted in Figure 12.7, is that the point $E'(\alpha_E)$ on the tangent plane T_E is attached to E with a spring with

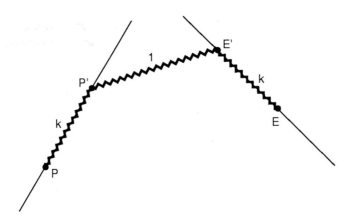

Fig. 12.7. The tangent distance between E and P is the elastic energy stored in each of the three springs connecting P, P', E' and E. P' and E' can move without friction along the tangent planes. The spring constants are indicated on the figure.

spring constant k and to $P'(\alpha_p)$ (on the tangent plane T_P) with spring constant 1, and $P'(\alpha_p)$ is also attached to P with spring constant k. (All three springs have zero natural length.) The new tangent distance is the total potential elastic energy stored of all three springs at equilibrium. As for the standard tangent distance, the solution can easily be obtained by differentiating equation (12.16) with respect to α_E and α_P. The differentiation yields:

$$L_P^T(E - P - L_P(1+k)\alpha_P + L_E\alpha_E) = 0 \qquad (12.17)$$
$$L_E^T(E - P - L_P\alpha_P + L_E(1+k)\alpha_E) = 0. \qquad (12.18)$$

The solution of this system is

$$(L_{PE}L_{EE}^{-1}L_E^T - (1+k)L_P^T)(E - P) = (L_{PE}L_{EE}^{-1}L_{EP} - (1+k)^2 L_{PP})\alpha_P \quad (12.19)$$
$$(L_{EP}L_{PP}^{-1}L_P^T - (1+k)L_E^T)(E - P) = ((1+k)^2 L_{EE} - L_{EP}L_{PP}^{-1}L_{PE})\alpha_E \quad (12.20)$$

where $L_{EE} = L_E^T L_E$, $L_{PE} = L_P^T L_E$, $L_{EP} = L_E^T L_P$ and $L_{PP} = L_P^T L_P$. The system has the same complexity as the vanilla tangent distance except that, it always has a solution for $k \geq 0$, and is more numerically stable. Note that for $k = 0$, it is equivalent to the standard tangent distance, while for $k = \infty$, we have the Euclidean distance. This approach is also very useful when the number of tangent vectors is greater or equal than the number of dimensions of the space. The standard tangent distance would most likely be zero (when the tangent spaces intersect), but the "spring" tangent distance still expresses valuable information about the invariances.

If the number of dimension of the input space is large compared to the number of tangent vectors, keeping k as small as possible is better because it doesn't interfere with the "sliding" along the tangent plane (E' and P' are less constrained).

Contrary to intuition, there is no danger of sliding too far in high dimensional space because tangent vectors are always roughly orthogonal and they could only slide far if they were parallel.

Hierarchy of distances: If several invariances are used, classification using tangent distance alone would be quite expensive. Fortunately, if a typical memory based algorithm is used, for example, K-nearest neighbors, it is quite unnecessary to compute the full tangent distance between the unclassified pattern and all the labeled samples. In particular, if a crude estimate of the tangent distance indicates with a sufficient confidence that a sample is very far from the pattern to be classified, no more computation is needed to know that this sample is not one of the K-nearest neighbors. Based on this observation one can build a hierarchy of distances which can greatly reduce the computation of each classification. Let's assume, for instance, that we have m approximations D_i of the tangent distance, ordered such that D_1 is the crudest approximation of the tangent distance and D_m is exactly tangent distance (for instance D_1 to D_5 could be the Euclidean distance with increasing resolution, and D_6 to D_{10} each add a tangent vector at full resolution).

The basic idea is to keep a pool of all the prototypes which could potentially be the K-nearest neighbors of the unclassified pattern. Initially the pool contains all the samples. Each of the distances D_i corresponds to a stage of the classification process. The classification algorithm has 3 steps at each stage, and proceeds from stage 1 to stage m or until the classification is complete: Step 1: the distance D_i between all the samples in the pool and the unclassified pattern is computed. Step 2: A classification and a confidence score is computed with these distances. If the confidence is good enough, let's say better than C_i (for instance, if all the samples left in the pool are in the same class) the classification is complete, otherwise proceed to step 3. Step 3: The K_i closest samples, according to distance D_i are kept in the pool, while the remaining samples are discarded.

Finding the K_i closest samples can be done in $O(p)$ (where p is the number of samples in the pool) since these elements need not to be sorted [22, 2]. The reduced pool is then passed to stage $i + 1$.

The two constants C_i and K_i must be determined in advance using a validation set. This can easily be done graphically by plotting the error as a function of K_i and C_i at each stage (starting with all K_i equal to the number of labeled samples and $C_i = 1$ for all stages). At each stage there is a minimum K_i and minimum C_i which give optimal performance on the validation set. By taking larger values, we can decrease the probability of making errors on the test sets. The slightly worse performance of using a hierarchy of distances is often well worth the speed-up. The computational cost of a pattern classification is then equal to:

$$\text{computational cost} \approx \sum_i \begin{array}{c}\text{number of}\\\text{prototypes}\\\text{at stage } i\end{array} \times \begin{array}{c}\text{distance}\\\text{complexity}\\\text{at stage } i\end{array} \times \begin{array}{c}\text{probability}\\\text{to reach}\\\text{stage } i\end{array} \quad (12.21)$$

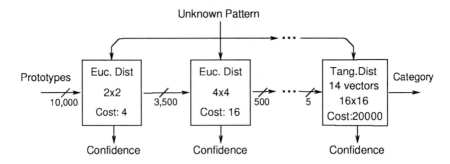

Fig. 12.8. Pattern recognition using a hierarchy of distances. The filter proceeds from left (starting with the whole database) to right (where only a few prototypes remain). At each stage distances between prototypes and the unknown pattern are computed and sorted; then the best candidate prototypes are selected for the next stage. As the complexity of the distance increases, the number of prototypes decreases, making computation feasible. At each stage a classification is attempted and a confidence score is computed. If the confidence score is high enough, the remaining stages are skipped.

All this is better illustrated with an example as in Figure 12.8. This system was used for the USPS experiment described in a previous section. In classification of handwritten digits (16x16 pixel images), D_1, D_2, and D_3, were the Euclidean distances at resolution 2×2, 4×4 and 8×8 respectively. D_4 was the one sided tangent distance with X-translation, on the sample side only, at resolution 8×8. D_5 was the double sided tangent distance with X-translation at resolution 16×16. Each of the subsequent distances added one tangent vector on each side (Y-translation, scaling, rotation, hyperbolic deformation1, hyperbolic deformation2 and thickness) until the full tangent distance was computed (D_{11}).

Table 12.2 shows the expected number of multiply-adds at each of the stages. It should be noted that the full tangent distance need only be computed for 1 in 20 unknown patterns (probability 0.05), and only with 5 samples out of the original 10,000. The net speed up was in the order of 500, compared with computing the full tangent distance between every unknown pattern and every sample (this is 6 times faster than computing the Euclidean distance at full resolution).

Multiple iterations: Tangent distance can be viewed as one iteration of a Newton-type algorithm which finds the points of minimum distance on the true transformation manifolds. The vectors α_E and α_P are the coordinates of the two closest points in the respective tangent spaces, but they can also be interpreted as the value for the real (non-linear) transformations. In other words, we can use α_E and α_P to compute the points $s(E, \alpha_E)$ and $s(P, \alpha_P)$, the real non-linear transformation of E and P. From these new points, we can recompute the tangent vectors, and the tangent distance and reiterate the process. If the appropriate conditions are met, this process can converge to a local minimum in the distance between the two transformation manifolds of P and E.

i	# of T.V.	Reso	# of proto (K_i)	# of prod	Probab	# of mul/add
1	0	4	9709	1	1.00	40,000
2	0	16	3500	1	1.00	56,000
3	0	64	500	1	1.00	32,000
4	1	64	125	2	0.90	14,000
4	2	256	50	5	0.60	40,000
6	4	256	45	7	0.40	32,000
7	6	256	25	9	0.20	11,000
8	8	256	15	11	0.10	4,000
9	10	256	10	13	0.10	3,000
10	12	256	5	15	0.05	1,000
11	14	256	5	17	0.05	1,000

Table 12.2. Summary computation for the classification of 1 pattern: The first column is the distance index, the second column indicates the number of tangent vectors (0 for the Euclidean distance), and the third column indicates the resolution in pixels, the fourth is K_i or the number of prototypes on which the distance D_i must be computed, the fifth column indicates the number of additional dot products which must be computed to evaluate distance D_i, the sixth column indicates the probability of not skipping that stage after the confidence score has been used, and the last column indicates the total average number of multiply-adds which must be performed (product of column 3 to 6) at each stage.

12.3 Tangent Propagation

The previous section dealt with memory-based techniques. We now apply tangent-distance principles to learned-function techniques.

The key idea is to incorporate the invariance directly into the learned classification function. In this section, we present an algorithm, called "tangent propagation", in which gradient descent is used to propagate information about the invariances of the training data. The process is a generalization of the widely-used "back propagation" method, which propagates information about the training data itself.

We again assume all data is drawn independently from a given statistical distribution $\mathcal{P}$, and our learning machine is characterized by the set of functions in can implement, $G_w(x)$, indexed by the vector of parameters w. Ideally, we would like to find w which minimizes the energy function

$$\mathcal{E} = \int \|G_w(x) - F(x)\|^2 d\mathcal{P}(x) \tag{12.22}$$

where $F(x)$ represents the "correct" or "desired" labeling of the point x. In the real world we must estimate this integral using only a finite set of training points B drawn the distribution $\mathcal{P}$. That is, we try to minimize

$$E_p = \sum_{i=1}^{p} \|G_w(x_i) - F(x_i)\| \tag{12.23}$$

where the sum runs over the training set B. An estimate of w can be computed by following a gradient descent using the weight-update rule:

$$\Delta w = -\eta \frac{\partial E_p}{\partial w}. \tag{12.24}$$

Let's consider an input transformation $s(x, \alpha)$ controlled by a parameter α. As always, we require that s is differentiable and that $s(x, 0) = x$. Now, in addition to the known labels of the training data, we assume that $\frac{\partial F(s(x_i, \alpha))}{\partial \alpha}$ is known at $\alpha = 0$ for each point x in the training set. To incorporate the invariance property into $G_w(x)$, we add that the following constraint on the derivative:

$$E_r = \sum_{i=1}^{p} \left| \frac{\partial G_w(s(x_i, \alpha))}{\partial \alpha} - \frac{\partial F(s(x_i, \alpha))}{\partial \alpha} \right|^2_{\alpha=0} \tag{12.25}$$

should be small at $\alpha = 0$. In many pattern classification problems, we are interested in the local classification invariance property for $F(x)$ with respect to the transformation s (the classification does not change when the input is slightly transformed), so we can simplify equation (12.25) to:

$$E_r = \sum_{i=1}^{p} \left| \frac{\partial G_w(s(x_i, \alpha))}{\partial \alpha} \right|^2_{\alpha=0} \quad (12.26)$$

since $\frac{\partial F(s(x_i, \alpha))}{\partial \alpha} = 0$. To minimize this term we can modify the gradient descent rule to use the energy function

$$E = \eta E_p + \mu E_r \quad (12.27)$$

with the weight update rule:

$$\Delta w = -\frac{\partial E}{\partial w}. \quad (12.28)$$

The learning rates (or regularization parameters) η and μ are tremendously important, because they determine the tradeoff between learning the invariances (based on the chosen directional derivatives) versus learning the label itself (i.e. the zeroth derivative) at each point in the training set.

The local variation of the classification function, which appears in equation (12.26) can be written as:

$$\left. \frac{\partial G_w(s(x,\alpha))}{\partial \alpha} \right|_{\alpha=0} = \left. \frac{\partial G_w(s(x,\alpha))}{\partial s(x,\alpha)} \frac{\partial s(x,\alpha)}{\partial \alpha} \right|_{\alpha=0} = \nabla_x G_w(x). \left. \frac{\partial s(x,\alpha)}{\partial \alpha} \right|_{\alpha=0} \quad (12.29)$$

since $s(x, \alpha) = x$ if $\alpha = 0$ and where $\nabla_x G_w(x)$ is the Jacobian of $G_w(x)$ for pattern x, and $\partial s(\alpha, x)/\partial \alpha$ is the *tangent vector* associated with transformation s as described in the previous section. Multiplying the tangent vector by the Jacobian involves one forward propagation through a "linearized" version of the network. If α is multi-dimensional, the forward propagation must be repeated for each tangent vector.

The theory of Lie algebras [11] ensures that compositions of local (small) transformations correspond to linear combinations of the corresponding tangent vectors (this result will be discussed further in section 12.4). Consequently, if $E_r(x) = 0$ is verified, the network derivative in the direction of a linear combination of the tangent vectors is equal to the same linear combination of the desired derivatives. In other words, if the network is successfully trained to be locally invariant with respect to, say, horizontal translations and vertical translations, it will be invariant with respect to compositions thereof.

It is possible to devise an efficient algorithm, "tangent prop", for performing the weight update (equation (12.28)). It is analogous to ordinary backpropagation, but in addition to propagating neuron activations, it also propagates the tangent vectors. The equations can be easily derived from Figure 12.9.

12.3.1 Local rule

The forward propagation equation is:

$$a_i^l = \sum_j w_{ij}^l x_j^{l-1} \qquad x_i^l = \sigma(a_i^l) \quad (12.30)$$

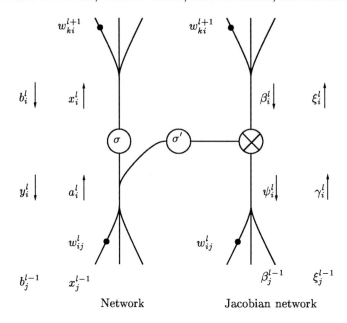

Fig. 12.9. Forward propagated variables (a, x, γ, ξ), and backward propagated variables (b, y, β, ψ) in the regular network (roman symbols) and the Jacobian (linearized) network (greek symbols). Converging forks (in the direction in which the signal is traveling) are sums, diverging forks just duplicate the values.

where σ is a non linear differentiable function (typically a sigmoid). The forward propagation starts at the first layer ($l = 1$), with x^0 being the input layer, and ends at the output layer ($l = L$). Similarly, The tangent forward propagation (tangent prop) is defined by:

$$\gamma_i^l = \sum_j w_{ij}^l \xi_j^{l-1} \qquad \xi_i^l = \sigma'(a_i^l)\gamma_i^l. \qquad (12.31)$$

The tangent forward propagation starts at the first layer ($l = 1$), with ξ^0 being the tangent vector $\frac{\partial s(x,\alpha)}{\partial \alpha}$, and ends at the output layer ($l = L$). The tangent gradient backpropagation can be computed using the chain rule:

$$\frac{\partial E}{\partial \xi_i^l} = \sum_k \frac{\partial E}{\partial \gamma_k^{l+1}} \frac{\partial \gamma_k^{l+1}}{\partial \xi_i^l} \qquad \frac{\partial E}{\partial \gamma_i^l} = \frac{\partial E}{\partial \xi_i^l} \frac{\partial \xi_i^l}{\partial \gamma_i^l} \qquad (12.32)$$

$$\beta_i^l = \sum_k \psi_k^{l+1} w_{ki}^{l+1} \qquad \psi_i^l = \beta_i^l \sigma'(a_i^l). \qquad (12.33)$$

The tangent backward propagation starts at the output layer ($l = L$), with ξ^L being the network variation $\frac{\partial G_w(s(x,\alpha))}{\partial \alpha}$, and ends at the input layer. Similarly, the gradient backpropagation equation is:

$$\frac{\partial E}{\partial x_i^l} = \sum_k \frac{\partial E}{\partial a_k^{l+1}} \frac{\partial a_k^{l+1}}{\partial x_i^l} \qquad \frac{\partial E}{\partial a_i^l} = \frac{\partial E}{\partial x_i^l}\frac{\partial x_i^l}{\partial a_i^l} + \frac{\partial E}{\partial \xi_i^l}\frac{\partial \xi_i^l}{\partial a_i^l} \qquad (12.34)$$

$$b_i^l = \sum_k y_k^{l+1} w_{ki}^{l+1} \qquad y_i^l = b_i^l \sigma'(a_i^l) + \beta_i \sigma''(a_i^l)\gamma_i^l. \qquad (12.35)$$

The standard backward propagation starts at the output layer ($l = L$), with $x^L = G_w(x^0)$ being the network output, and ends at the input layer. Finally, the weight update is:

$$\Delta w_{ij}^l = -\frac{\partial E}{\partial a_i^l}\frac{\partial a_i^l}{\partial w_{ij}^l} - \frac{\partial E}{\partial \gamma_i^l}\frac{\partial \gamma_i^l}{\partial w_{ij}^l} \qquad (12.36)$$

$$\Delta w_{ij}^l = -y_i^l x_j^{l-1} - \psi_i^l \xi_j^{l-1}. \qquad (12.37)$$

The computation requires one forward propagation and one backward propagation per pattern and per tangent vector during training. After the network is trained, it is approximately locally invariant with respect to the chosen transformation. After training, the evaluation of the learned function is in all ways identical to a network which is not trained for invariance (except that the weights have different values).

12.3.2 Results

Two experiments illustrate the advantages of tangent prop. The first experiment is a classification task, using a small (linearly separable) set of 480 binary images of handwritten digits. The training sets consist of 10, 20, 40, 80, 160 or 320 patterns, and the test set contains 160 patterns. The patterns are smoothed using a Gaussian kernel with standard deviation of one half pixel. For each of the training set patterns, the tangent vectors for horizontal and vertical translation are computed. The network has two hidden layers with locally connected shared weights, and one output layer with 10 units (5194 connections, 1060 free parameters) [19]. The generalization performance as a function of the training set size for traditional backprop and tangent prop are compared in Figure 12.10. We have conducted additional experiments in which we implemented not only translations but also rotations, expansions and hyperbolic deformations. This set of 6 generators is a basis for all linear transformations of coordinates for two dimensional images. It is straightforward to implement other generators including gray-level-shifting, "smooth" segmentation, local continuous coordinate transformations and independent image segment transformations.

The next experiment is designed to show that in applications where data is highly correlated, tangent prop yields a large speed advantage. Since the distortion model implies adding lots of highly correlated data, the advantage of tangent prop over the distortion model becomes clear.

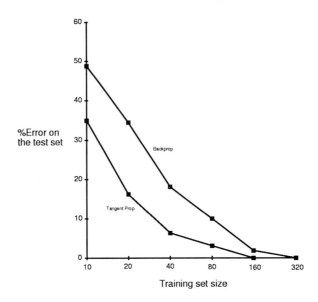

Fig. 12.10. Generalization performance curve as a function of the training set size for the tangent prop and the backprop algorithms

The task is to approximate a function that has plateaus at three locations. We want to enforce local invariance near each of the training points (Figure 12.11, bottom). The network has one input unit, 20 hidden units and one output unit. Two strategies are possible: either generate a small set of training points covering each of the plateaus (open squares on Figure 12.11 bottom), or generate one training point for each plateau (closed squares), and enforce local invariance around them (by setting the desired derivative to 0). The training set of the former method is used as a measure of performance for both methods. All parameters were adjusted for approximately optimal performance in all cases. The learning curves for both models are shown in Figure 12.11 (top). Each sweep through the training set for tangent prop is a little faster since it requires only 6 forward propagations, while it requires 9 in the distortion model. As can be seen, stable performance is achieved after 1300 sweeps for the tangent prop, versus 8000 for the distortion model. The overall speedup is therefore about 10.

Tangent prop in this example can take advantage of a very large regularization term. The distortion model is at a disadvantage because the only parameter that effectively controls the amount of regularization is the magnitude of the distortions, and this cannot be increased to large values because the right answer is only invariant under *small* distortions.

12.3.3 How to make tangent prop work

Large network capacity: Relatively few experiments have been done with tangent propagation. It is clear, however, that the invariance constraint can be

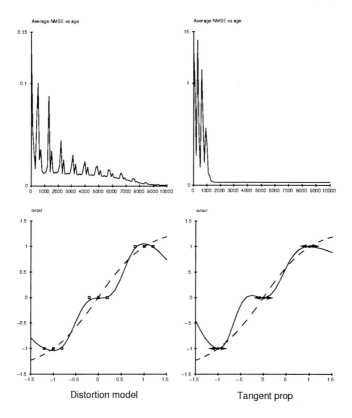

Fig. 12.11. Comparison of the distortion model (left column) and tangent prop (right column). The top row gives the learning curves (error versus number of sweeps through the training set). The bottom row gives the final input-output function of the network; the dashed line is the result for unadorned back prop.

extremely beneficial. If the network does not have enough capacity, it will not benefit from the extra knowledge introduced by the invariance.

Interleaving of the tangent vectors: Since the tangent vectors introduce even more correlation inside the training set, a substantial speed up can be obtained by alternating a regular forward and backward propagation with a tangent forward and backward propagation (even if there are several tangent vectors, only one is used at each pattern). For instance, if there were 3 tangent vectors, the training sequence could be:

$$x_1, t_1(x_1), x_2, t_2(x_2), x_3, t_3(x_3), x_4, t_1(x_4), x_5, t_2(x_5), \ldots \quad (12.38)$$

where x_i means a forward and backward propagation for pattern i and $t_j(x_i)$ means a tangent forward and backward propagation of tangent vector j of pattern i. With such interleaving, the learning converges faster than grouping all the tangent vectors together. Of course, this only makes sense with on-line updates as opposed to batch updates.

12.4 Tangent Vectors

In this section, we consider the general paradigm for transformation invariance and for the tangent vectors which have been used in the two previous sections. Before we introduce each transformation and their corresponding tangent vectors, a brief explanation is given of the theory behind the practice. There are two aspects to the problem. First it is possible to establish a formal connection between groups of transformations of the input space (such as translation, rotation, etc. of $\Re^2$) and their effect on a functional of that space (such as a mapping of $\Re^2$ to $\Re$, which may represent an image, in continuous form). The theory of Lie groups and Lie algebra [6] allows us to do this. The second problem has to do with coding. Computer images are finite vectors of discrete variables. How can a theory which was developed for differentiable functionals of $\Re^2$ to $\Re$ be applied to these vectors? We first give a brief explanation of the theorems of Lie groups and Lie algebras which are applicable to pattern recognition. Next, we explore solutions to the coding problem. Finally some examples of transformation and coding are given for particular applications.

12.4.1 Lie groups and Lie algebras

Consider an input space $\mathcal{I}$ (the plane $\Re^2$ for example) and a differentiable function f which maps points of $\mathcal{I}$ to $\Re$.

$$f : X \in I \longmapsto f(X) \in \Re. \qquad (12.39)$$

The function $f(X) = f(x, y)$ can be interpreted as the continuous (defined for all points of $\Re^2$) equivalent of the discrete computer image $P[i, j]$.

Next, consider a family of transformations t_α, parameterized by α, which maps bijectively a point of $\mathcal{I}$ to a point of $\mathcal{I}$

$$t_\alpha : X \in \mathcal{I} \longmapsto t_\alpha(X) \in \mathcal{I}. \qquad (12.40)$$

We assume that t_α is differentiable with respect to α and X, and that t_0 is the identity. For example t_α could be the group of affine transformations of $\Re^2$:

$$t_\alpha : \begin{pmatrix} x \\ y \end{pmatrix} \longmapsto \begin{pmatrix} x + \alpha_1 x + \alpha_2 y + \alpha_5 \\ \alpha_3 x + y + \alpha_4 y + \alpha_6 \end{pmatrix} \quad \text{with} \quad \begin{vmatrix} 1+\alpha_1 & \alpha_2 \\ \alpha_3 & 1+\alpha_4 \end{vmatrix} \neq 0. \qquad (12.41)$$

This is a Lie group[4] with 6 parameters. Another example is the group of direct isometry:

$$t_\alpha : \begin{pmatrix} x \\ y \end{pmatrix} \longmapsto \begin{pmatrix} x\cos\theta - y\sin\theta + a \\ x\sin\theta + y\cos\theta + b \end{pmatrix} \qquad (12.42)$$

---
[4] A Lie group is a group that is also a differentiable manifold such that the differentiable structure is compatible with the group structure.

which is a Lie group with 3 parameters.

We now consider the functional $s(f, \alpha)$, defined by

$$s(f, \alpha) = f \circ t_\alpha^{-1}. \tag{12.43}$$

This functional s, which takes another functional f as an argument, should remind the reader of Figure 12.2 where P, the discrete equivalent of f, is the argument of s.

The Lie algebra associated with the action of t_α on f is the space generated by the m local transformations L_{α_i} of f defined by:

$$L_{\alpha_i}(f) = \left.\frac{\partial s(f, \alpha)}{\partial \alpha_i}\right|_{\alpha=0}. \tag{12.44}$$

We can now write the local approximation of s as:

$$s(f, \alpha) = f + \alpha_1 L_{\alpha_1}(f) + \alpha_2 L_{\alpha_2}(f) + \ldots + \alpha_m L_{\alpha_m}(f) + o(\|\alpha\|^2)(f). \tag{12.45}$$

This equation is the continuous equivalent of equation (12.2) used in the introduction.

The following example illustrates how L_{α_i} can be computed from t_α. Let's consider the group of direct isometry defined in equation (12.42) (with parameter $\alpha = (\theta, a, b)$ as before, and $X = (x, y)$)

$$s(f, \alpha)(X) = f((x-a)\cos\theta + (y-b)\sin\theta, -(x-a)\sin\theta + (y-b)\cos\theta). \tag{12.46}$$

If we differentiate around $\alpha = (0, 0, 0)$ with respect to θ, we obtain

$$\frac{\partial s(f, \alpha)}{\partial \theta}(X) = y\frac{\partial f}{\partial x}(x, y) + (-x)\frac{\partial f}{\partial y}(x, y) \tag{12.47}$$

that is

$$L_\theta = y\frac{\partial}{\partial x} + (-x)\frac{\partial}{\partial y}. \tag{12.48}$$

The transformation $L_a = -\frac{\partial}{\partial x}$ and $L_b = -\frac{\partial}{\partial y}$ can be obtained in a similar fashion. All local transformations of the group can be written as

$$s(f, \alpha) = f + \theta(y\frac{\partial f}{\partial x} + (-x)\frac{\partial f}{\partial y}) - a\frac{\partial f}{\partial x} - b\frac{\partial f}{\partial y} + o(\|\alpha\|^2)(f) \tag{12.49}$$

which corresponds to a linear combination of the 3 basic operators L_θ, L_a and L_b[5]. The property which is most important to us is that the 3 operators generate the whole space of local transformations. The result of applying the operators to a function f, such as a 2D image for example, is the set of vectors which we have been calling "tangent vector" in the previous sections. Each point in the tangent space correspond to a unique transformation and conversely any transformation of the Lie group (in the example all rotations of any angle and center together with all translations) corresponds to a point in the tangent plane.

---

[5] These operators are said to generate a Lie algebra, because on top of the addition and multiplication by a scalar, there is a special multiplication called "Lie bracket" defined by $[L_1, L_2] = L_1 \circ L_2 - L_2 \circ L_1$. In the above example we have $[L_\theta, L_a] = L_b$, $[L_a, L_b] = 0$, and $[L_b, L_\theta] = L_a$.

12.4.2 Tangent vectors

The last problem which remains to be solved is the problem of coding. Computer images, for instance, are coded as a finite set of discrete (even binary) values. These are hardly the differentiable mappings of $\mathcal{I}$ to $\Re$ which we have been assuming in the previous subsection.

To solve this problem we introduce a smooth interpolating function C which maps the discrete vectors to a continuous mapping of $\mathcal{I}$ to $\Re$. For example, if P is a image of n pixels, it can be mapped to a continuously valued function f over $\Re^2$ by convolving it with a two dimensional Gaussian function g_σ of standard deviation σ. This is because g_σ is a differentiable mapping of $\Re^2$ to $\Re$, and P can be interpreted as a sum of impulse functions. In the two dimensional case we can write the new interpretation of P as:

$$P'(x,y) = \sum_{i,j} P[i][j]\delta(x-i)\delta(y-j) \qquad (12.50)$$

where $P[i][j]$ denotes the finite vector of discrete values, as stored in a computer. The result of the convolution is of course differentiable because it is a sum of Gaussian functions. The Gaussian mapping is given by:

$$C_\sigma : P \longmapsto f = P' * g_\sigma. \qquad (12.51)$$

In the two dimensional case, the function f can be written as:

$$f(x,y) = \sum_{i,j} P[i][j] g_\sigma(x-i, y-j). \qquad (12.52)$$

Other coding functions C can be used, such as cubic spline or even bilinear interpolation. Bilinear interpolation between the pixels yields a function f which is differentiable almost everywhere. The fact that the derivatives have two values at the integer locations (because the bilinear interpolation is different on both side of each pixels) is not a problem in practice – just choose one of the two values.

The Gaussian mapping is preferred for two reasons: First, the smoothing parameter σ can be used to control the locality of the invariance. This is because when f is smoother, the local approximation of equation (12.45) is valid for larger transformations. And second, when combined with the transformation operator L, the derivative can be applied on the closed form of the Gaussian function. For instance, if the X-translation operator $L = \frac{\partial}{\partial x}$ is applied to $f = P' * g_\sigma$, the actual computation becomes:

$$L_X(f) = \frac{\partial}{\partial x}(P' * g_\sigma) = P' * \frac{\partial g_\sigma}{\partial x}. \qquad (12.53)$$

because of the differentiation properties of convolution when the support is compact. This is easily done by convolving the original image with the X-derivative of

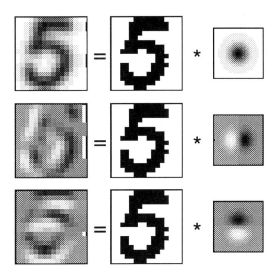

Fig. 12.12. Graphic illustration of the computation of f and two tangent vectors corresponding to $L_x = \partial/\partial x$ (X-translation) and $L_x = \partial/\partial y$ (Y-translation), from a binary image I. The Gaussian function $g(x,y) = \exp(-\frac{x^2+y^2}{2\sigma^2})$ has a standard deviation of $\sigma = 0.9$ in this example although its graphic representation (small images on the right) have been rescaled for clarity.

the Gaussian function g_σ. This operation is illustrated in Figure 12.12. Similarly, the tangent vector for scaling can be computed with

$$L_S(f) = \left(x\frac{\partial}{\partial x} + y\frac{\partial}{\partial y}\right)(I * g_\sigma) = x(I * \frac{\partial g_\sigma}{\partial x}) + y(I * \frac{\partial g_\sigma}{\partial y}). \quad (12.54)$$

This operation is illustrated in Figure 12.13.

12.4.3 Important transformations in image processing

This section summarizes how to compute the tangent vectors for image processing (in 2D). Each discrete image I_i is convolved with a Gaussian of standard deviation g_σ to obtain a representation of the continuous image f_i, according to equation:

$$f_i = I_i * g_\sigma. \quad (12.55)$$

The resulting image f_i will be used in all the computations requiring I_i (except for computing the tangent vector). For each image I_i, the tangent vectors are computed by applying the operators corresponding to the transformations of interest to the expression $I_i * g_\sigma$. The result, which can be precomputed, is an image which is the tangent vector. The following list contains some of the most useful tangent vectors:

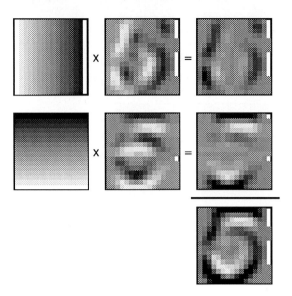

Fig. 12.13. Graphic illustration of the computation of the tangent vector $T_u = D_x S_x + D_y S_y$ (bottom image). In this example the displacement for each pixel is proportional to the distance of the pixel to the center of the image ($D_x(x,y) = x - x_0$ and $D_y(x,y) = y - y_0$). The two multiplications (horizontal lines) as well as the addition (vertical right column) are done pixel by pixel.

X-translation: This transformation is useful when the classification function is known to be invariant with respect to the input transformation:

$$t_\alpha : \begin{pmatrix} x \\ y \end{pmatrix} \longmapsto \begin{pmatrix} x + \alpha \\ y \end{pmatrix}. \quad (12.56)$$

The Lie operator is defined by:

$$L_X = \frac{\partial}{\partial x}. \quad (12.57)$$

Y-translation: This transformation is useful when the classification function is known to be invariant with respect to the input transformation:

$$t_\alpha : \begin{pmatrix} x \\ y \end{pmatrix} \longmapsto \begin{pmatrix} x \\ y + \alpha \end{pmatrix}. \quad (12.58)$$

The Lie operator is defined by:

$$L_Y = \frac{\partial}{\partial y}. \quad (12.59)$$

12. Tangent Distance and Tangent Propagation

Rotation: This transformation is useful when the classification function is known to be invariant with respect to the input transformation:

$$t_\alpha : \begin{pmatrix} x \\ y \end{pmatrix} \longmapsto \begin{pmatrix} x \cos \alpha - y \sin \alpha \\ x \sin \alpha + y \cos \alpha \end{pmatrix}. \tag{12.60}$$

The Lie operator is defined by:

$$L_R = y \frac{\partial}{\partial x} + (-x) \frac{\partial}{\partial y}. \tag{12.61}$$

Scaling: This transformation is useful when the classification function is known to be invariant with respect to the input transformation:

$$t_\alpha : \begin{pmatrix} x \\ y \end{pmatrix} \longmapsto \begin{pmatrix} x + \alpha x \\ y + \alpha y \end{pmatrix}. \tag{12.62}$$

The Lie operator is defined by:

$$L_S = x \frac{\partial}{\partial x} + y \frac{\partial}{\partial y}. \tag{12.63}$$

Parallel hyperbolic transformation: This transformation is useful when the classification function is known to be invariant with respect to the input transformation:

$$t_\alpha : \begin{pmatrix} x \\ y \end{pmatrix} \longmapsto \begin{pmatrix} x + \alpha x \\ y - \alpha y \end{pmatrix}. \tag{12.64}$$

The Lie operator is defined by:

$$L_S = x \frac{\partial}{\partial x} - y \frac{\partial}{\partial y}. \tag{12.65}$$

Diagonal hyperbolic transformation: This transformation is useful when the classification function is known to be invariant with respect to the input transformation:

$$t_\alpha : \begin{pmatrix} x \\ y \end{pmatrix} \longmapsto \begin{pmatrix} x + \alpha y \\ y + \alpha x \end{pmatrix}. \tag{12.66}$$

The Lie operator is defined by:

$$L_S = y \frac{\partial}{\partial x} + x \frac{\partial}{\partial y}. \tag{12.67}$$

The resulting tangent vector is is the norm of the gradient of the image, which is very easy to compute.

Thickening: This transformation is useful when the classification function is known to be invariant with respect to variation of thickness. This is known in morphology as dilation and its inverse, erosion. It is very useful in certain domains (such as handwritten character recognition because) thickening and thinning are natural variations which correspond to the pressure applied on a pen, or to different absorbtion properties of the ink on the paper. A dilation (resp. erosion) can be defined as the operation of replacing each value f(x,y) by the largest (resp. smallest) value of $f(x', y')$ found within a neighborhood of a certain shape, centered at (x, y). The region is called the structural element. We will assume that the structural element is a sphere of radius α. We define the thickening transformation as the function which takes the function f and generates the function f'_α defined by:

$$f'_\alpha(X) = \max_{\|r\| \leq \alpha} f(X + r) \quad \text{for } \alpha \geq 0 \tag{12.68}$$

$$f'_\alpha(X) = \min_{\|r\| \leq -\alpha} f(X + r) \quad \text{for } \alpha \leq 0. \tag{12.69}$$

The derivative of the thickening for $\alpha \geq 0$ can be written as:

$$\lim_{\alpha \to 0} \frac{f'(X) - f(X)}{\alpha} = \lim_{\alpha \to 0} \frac{\max_{\|r\| \leq \alpha} f(X + r) - f(X)}{\alpha}. \tag{12.70}$$

$f(X)$ can be put within the max expression because it does not depend on $\|r\|$. Since $\|\alpha\|$ tends toward 0, we can write:

$$f(X + r) - f(X) = r \cdot \nabla f(X) + O(\|r\|^2) \approx r \cdot \nabla f(X). \tag{12.71}$$

The maximum of

$$\max_{\|r\| \leq \alpha} f(X + r) - f(X) = \max_{\|r\| \leq \alpha} r \cdot \nabla f(X) \tag{12.72}$$

is attained when r and $\nabla f(X)$ are co-linear, that is when

$$r = \alpha \frac{\nabla f(X)}{\|\nabla f(X)\|} \tag{12.73}$$

assuming $\alpha \geq 0$. It can easily be shown that this equation holds when α is negative, because we then try to minimize equation (12.69). We therefore have:

$$\lim_{\alpha \to 0} \frac{f'_\alpha(X) - f(X)}{\alpha} = \|\nabla f(X)\| \tag{12.74}$$

which is the tangent vector of interest. Note that this is true for α positive or negative. The same tangent vector describes both thickening and thinning. Alternatively, we can use our computation of the displacement r and define the following transformation of the input:

$$t_\alpha(f) : \begin{pmatrix} x \\ y \end{pmatrix} \longmapsto \begin{pmatrix} x + \alpha r_x \\ y + \alpha r_y \end{pmatrix} \tag{12.75}$$

where

$$(r_x, r_y) = r = \alpha \frac{\nabla f(X)}{\|\nabla f(X)\|}. \tag{12.76}$$

This transformation of the input space is different for each pattern f (we do not have a Lie group of transformations, but the field structure generated by the (pseudo Lie) operator is still useful. The operator used to find the tangent vector is defined by:

$$L_T = \|\nabla\| \tag{12.77}$$

which means that the tangent vector image is obtained by computing the normalized gray level gradient of the image at each point (the gradient at each point is normalized).

The last 5 transformations are depicted in Figure 12.14 with the tangent vector. The last operator corresponds to a thickening or thinning of the image. This unusual transformation is extremely useful for handwritten character recognition.

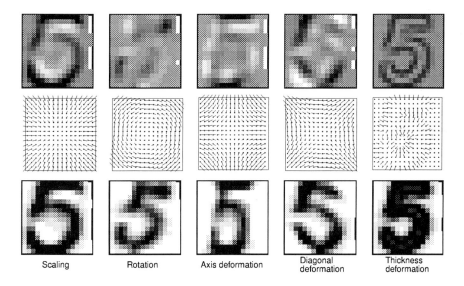

Fig. 12.14. Illustration of 5 tangent vectors (top), with corresponding displacements (middle) and transformation effects (bottom). The displacement D_x and D_y are represented in the form of vector field. It can be noted that the tangent vector for the thickness deformation (right column) correspond to the norm of the gradient of the gray level image.

12.5 Conclusion

The basic tangent distance algorithm is quite easy to understand and implement. Even though hardly any preprocessing or learning is required, the performance is surprisingly good and compares well to the best competing algorithms. We believe that the main reason for this success is its ability to incorporate *a priori* knowledge into the distance measure. The only algorithm which performed better than tangent distance on one of the three databases was boosting, which has similar *a priori* knowledge about transformations built into it.

Many improvements are of course possible. For instance, smart preprocessing can allow us to measure the tangent distance in a more appropriate "feature" space, instead of the original pixel space. In image classification, for example, the features could be horizontal and vertical edges. This would most likely further improve the performance[6] The only requirement is that the preprocessing must be differentiable, so that the tangent vectors can be computed (propagated) into the feature space.

It is also straightforward to modify more complex algorithms such as LVQ (learning vector quantization) to use a tangent distance. In this case even the tangent vectors can be trained. The derivation has been done for batch training [13] and for on-line training [23] of the tangent vectors. When such training is performed, the *a priori* knowledge comes from other constraints imposed on the tangent vectors (for instance how many tangent vectors are allowed, which classes of transformation do they represent, etc).

Finally, many optimizations which are commonly used in distance based algorithms can be used as successfully with tangent distance to speed up computation. The multi-resolution approach have already been tried successfully [25]. Other methods like "multi-edit-condensing" [1, 30] and K-d tree [4] are also possible.

The main advantage of tangent distance is that it is a modification of a standard distance measure to allow it to incorporate *a priori* knowledge that is specific to the problem at hand. Any algorithms based on a common distance measure (as it is often the case in classification, vector quantization, predictions, etc...) can potentially benefit from a more problem-specific distance. Many of these "distance based" algorithms do not require any learning, which means that they can be adapted instantly by just adding new patterns in the database. These additions are leveraged by the *a priori* knowledge put in the tangent distance.

The two drawbacks of tangent distance are its memory and computational requirements. The most computationally and memory efficient algorithms generally involve learning [20]. Fortunately, the concept of tangent vectors can also be used in learning. This is the basis for the tangent propagation algorithm. The concept is quite simple: instead of learning a classification function from examples of its values, one can also use information about its derivatives. This

---

[6] There may be an additional cost for computing the tangent vectors in the feature space if the feature space is very complex.

information is provided by the tangent vectors. Unfortunately, not many experiments have been done in this direction. The two main problems with tangent propagation are that the capacity of the learning machine has to be adjusted to incorporate the additional information pertinent to the tangent vectors, and that training time must be increased. After training, the classification time and complexity are unchanged, but the classifier's performance is improved.

To a first approximation, using tangent distance or tangent propagation is like having a much larger database. If the database was plenty large to begin with, tangent distance or tangent propagation would not improve the performance. To a better approximation, tangent vectors are like using a distortion model to magnify the size of the training set. In many cases, using tangent vectors will be preferable to collecting (and labeling!) vastly more training data, and preferable (especially for memory-based classifiers) to dealing with all the data generated by the distortion model. Tangent vectors provide a compact and powerful representation of *a priori* knowledge which can easily be integrated in the most popular algorithms.

Acknowledgement P.S. and Y.L. gratefully acknowledge NSF grant No. INT-9726745.

References

1. P. A.Devijver and J. Kittler. *Pattern Recognition, A Statistical Approache*. Prentice Hall, Englewood Cliffs, 1982.
2. A. V. Aho, J. E. Hopcroft, and J. D. Ullman. *Data Structure and Algorithms*. Addison-Wesley, 1983.
3. L. Bottou and V. N. Vapnik. Local learning algorithms. *Neural Computation*, 4(6):888–900, 1992.
4. A. J. Broder. Strategies for efficient incremental nearest neighbor search. *Pattern Recognition*, 23:171–178, 1990.
5. D. S. Broomhead and D. Lowe. Multivariable functional interpolation and adaptive networks. *Complex Systems*, 2:321–355, 1988.
6. Y. Choquet-Bruhat, C. DeWitt-Morette, and M. Dillard-Bleick. *Analysis, Manifolds and Physics*. North-Holland, Amsterdam, Oxford, New York, Tokyo, 1982.
7. C. Cortes and V. Vapnik. Support vector networks. *Machine Learning*, 20:273–297, 1995.
8. B. V. Dasarathy. *Nearest Neighbor (NN) Norms: NN Pattern classification Techniques*. IEEE Computer Society Press, Los Alamitos, California, 1991.
9. H. Drucker, R. Schapire, and P. Y. Simard. Boosting performance in neural networks. *International Journal of Pattern Recognition and Artificial Intelligence*, 7, No. 4:705–719, 1993.
10. K. Fukunaga and T. E. Flick. An optimal global nearest neighbor metric. *IEEE transactions on Pattern analysis and Machine Intelligence*, 6, No. 3:314–318, 1984.
11. R. Gilmore. *Lie Groups, Lie Algebras and some of their Applications*. Wiley, New York, 1974.
12. T. Hastie, E. Kishon, M. Clark, and J. Fan. A model for signature verification. Technical Report 11214-910715-07TM, AT&T Bell Laboratories, July 1991.

13. T. Hastie and P. Y. Simard. Metrics and models for handwritten character recognition. *Statistical Science*, 13, 1998.
14. T. J. Hastie and R. J. Tibshirani. *Generalized Linear Models*. Chapman and Hall, London, 1990.
15. G. E. Hinton, C. K. I. Williams, and M. D. Revow. Adaptive elastic models for hand-printed character recognition. In *Advances in Neural Information Processing Systems*, pages 512–519. Morgan Kaufmann Publishers, 1992.
16. A. E. Hoerl and R. W. Kennard. Ridge regression: Biased estimation for non-orthogonal problems. *Technometrics*, 12:55–67, 1970.
17. T. Kohonen. Self-organization and associative memory. In *Springer Series in Information Sciences*, volume 8. Springer-Verlag, 1984.
18. Y. Le Cun, B. Boser, J. S. Denker, D. Henderson, R. E. Howard, W. Hubbard, and L. D. Jackel. Handwritten digit recognition with a back-propagation network. In David Touretzky, editor, *Advances in Neural Information Processing Systems*, volume 2, (Denver, 1989), 1990. Morgan Kaufman.
19. Y. LeCun. Generalization and network design strategies. In R. Pfeifer, Z. Schreter, F. Fogelman, and L. Steels, editors, *Connectionism in Perspective*, Zurich, Switzerland, 1989. Elsevier. an extended version was published as a technical report of the University of Toronto.
20. Y. LeCun, L. D. Jackel, L. Bottou, C. Cortes, J. S. Denker, H. Drucker, I. Guyon, U. A. Muller, E. Sackinger, P. Simard, and V. Vapnik. Learning algorithms for classification: A comparison on handwritten digit recognition. In J. H. Oh, C. Kwon, and S. Cho, editors, *Neural Networks: The Statistical Mechanics Perspective*, pages 261–276. World Scientific, 1995.
21. E. Parzen. On estimation of a probability density function and mode. *Ann. Math. Stat.*, 33:1065–1076, 1962.
22. W. H. Press, B. P. Flannery, Teukolsky S. A., and Vetterling W. T. *Numerical Recipes*. Cambridge University Press, Cambridge, 1988.
23. H. Schwenk. The diabolo classifier. *Neural Computation*, in press, 1998.
24. R. Sibson. Studies in the robustness of multidimensional scaling: Procrustes statistices. *J. R. Statist. Soc.*, 40:234–238, 1978.
25. P. Y. Simard. Efficient computation of complex distance metrics using hierarchical filtering. In *Advances in Neural Information Processing Systems*. Morgan Kaufmann Publishers, 1994.
26. F. Sinden and G. Wilfong. On-line recognition of handwritten symbols. Technical Report 11228-910930-02IM, AT&T Bell Laboratories, June 1992.
27. V. N. Vapnik. *Estimation of dependences based on empirical data*. Springer Verlag, 1982.
28. V. N. Vapnik and A. Ya. Chervonenkis. On the uniform convergence of relative frequencies of events to their probabilities. *Th. Prob. and its Applications*, 17(2):264–280, 1971.
29. N. Vasconcelos and A. Lippman. Multiresolution tangent distance for affine-invariant classification. In *Advances in Neural Information Processing Systems*, volume 10, pages 843–849. Morgan Kaufmann Publishers, 1998.
30. J. Voisin and P. Devijver. An application of the multiedit-condensing technique to the reference selection problem in a print recognition system. *Pattern Recogntion*, 20 No 5:465–474, 1987.

13
Combining Neural Networks and Context-Driven Search for On-Line, Printed Handwriting Recognition in the Newton

Larry S. Yaeger[1], Brandyn J. Webb[2], and Richard F. Lyon[3]

[1] Apple Computer 5540 Bittersweet Rd. Beanblossom, IN 46160
[2] The Future 4578 Fieldgate Rd. Oceanside, CA 92056
[3] Foveonics, Inc. 10131-B Bubb Rd. Cupertino, CA 95014
larryy@pobox.com, http://www.beanblossom.in.us/larryy/

Abstract. While on-line handwriting recognition is an area of long-standing and ongoing research, the recent emergence of portable, pen-based computers has focused urgent attention on usable, practical solutions. We discuss a combination and improvement of classical methods to produce robust recognition of hand-printed English text, for a recognizer shipping in new models of Apple Computer's Newton MessagePad®and eMate®. Combining an artificial neural network (ANN), as a character classifier, with a context-driven search over segmentation and word recognition hypotheses provides an effective recognition system. Long-standing issues relative to training, generalization, segmentation, models of context, probabilistic formalisms, etc., need to be resolved, however, to get excellent performance. We present a number of recent innovations in the application of ANNs as character classifiers for word recognition, including integrated multiple representations, normalized output error, negative training, stroke warping, frequency balancing, error emphasis, and quantized weights. User-adaptation and extension to cursive recognition pose continuing challenges.

13.1 Introduction

Pen-based hand-held computers are heavily dependent upon fast and accurate handwriting recognition, since the pen serves as the primary means for inputting data to such devices. Some earlier attempts at handwriting recognition have utilized strong, limited language models to maximize accuracy. However, this approach has proven to be unacceptable in real-world applications, generating disturbing and seemingly random word substitutions – known colloquially within Apple and Newton as "The Doonesbury Effect", due to Gary Trudeau's satirical look at first-generation Newton recognition performance. But the original handwriting recognition technology in the Newton, and the current, much-improved "Cursive Recognizer" technology, both of which were licensed from ParaGraph International, Inc., are not the subject of this article.

In Apple's Advanced Technology Group (aka Apple Research Labs), we pursued a different approach, using bottom-up classification techniques based on trainable artificial neural networks (ANNs), in combination with comprehensive but weakly-applied language models. To focus our work on a subproblem that was tractable enough to lead to usable products in a reasonable time, we initially restricted the domain to hand-printing, so that strokes are clearly delineated by pen lifts. By simultaneously providing accurate character-level recognition, dictionaries exhibiting very wide coverage of the language, and the ability to write entirely outside those dictionaries, we have produced a hand-print recognizer that some have called the "first usable" handwriting recognition system. The ANN character classifier required some innovative training techniques to perform its task well. The dictionaries required large word lists, a regular expression grammar (to describe special constructs such as date, time, phone numbers, etc.), and a means of combining all these dictionaries into a comprehensive language model. And well balanced prior probabilities had to be determined for in-dictionary and out-of-dictionary writing. Together with a maximum-likelihood search engine, these elements form the basis of the so-called "Print Recognizer", that was first shipped in Newton OS 2.0 based MessagePad 120 units in December, 1995, and has shipped in all subsequent Newton devices. In the MessagePad 2000 and 2100, despite retaining its label as a "Print Recognizer", it has been extended to handle connected characters (as well as a full Western European character set).

There is ample prior work in combining low-level classifiers with dynamic time warping, hidden Markov models, Viterbi algorithms, and other search strategies to provide integrated segmentation and recognition for writing [15] and speech [11]. And there is a rich background in the use of ANNs as classifiers, including their use as low-level character classifiers in a higher-level word recognition system [2]. But these approaches leave a large number of open-ended questions about how to achieve acceptable (to a real user) levels of performance. In this paper, we survey some of our experiences in exploring refinements and improvements to these techniques.

13.2 System Overview

Apple's print recognizer (APR) consists of three conceptual stages – Tentative Segmentation, Classification, and Context-Driven Search – as indicated in Figure 13.1. The primary data upon which we operate are simple sequences of (x,y) coordinate pairs, plus pen-up/down information, thus defining stroke primitives. The Segmentation stage decides which strokes will be combined to produce *segments* - the tentative groupings of strokes that will be treated as possible characters - and produces a sequence of these segments together with legal transitions between them. This process builds an implicit graph which is then labeled in the Classification stage and examined for a maximum likelihood interpretation in the Search stage. The Classification stage evaluates each segment using the ANN classifier, and produces a vector of output activations that are used as letter-class probabilities. The Search stage then uses these class probabilities together with models of lexical and geometric context to find the N most likely word or sentence hypotheses.

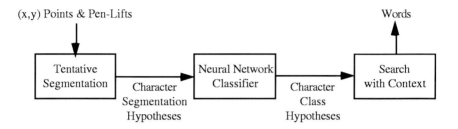

Fig. 13.1. A simplified block diagram of our hand-print recognizer.

13.3 Tentative Segmentation

Character segmentation – the process of deciding which strokes comprise which characters – is inherently ambiguous. Ultimately this decision must be made, but, short of writing in boxes, it is impossible to do so (with any accuracy) in advance, external to the recognition process. Hence the initial segmentation stage in APR produces multiple, tentative groupings of strokes, and defers the final segmentation decisions until the search stage, thus integrating those segmentation decisions with the overall recognition process.

APR uses a potentially exhaustive, sequential enumeration of stroke combinations to generate a sequence of viable character-segmentation hypotheses. These *segments* are subjected to some obvious constraints (such as "all strokes must be used" and "no strokes may be used twice"), and some less obvious filters (to cull "impossible" segmentations for the sake of efficiency). The resulting algorithm produces the actual segments that will be processed as possible characters, along with the legal transitions between these segments.

The legal transitions are defined by *forward* and *reverse delays*. The forward delay indicates the next possible segment in the sequence (though later segments may also be legal), pointing just past the last segment that shares the trailing stroke of the current segment. The reverse delay indicates the start of the current batch of segments, all of which share the same leading stroke. Due to the enumeration scheme, a segment's reverse delay is the same as its stroke count minus one, unless preceeding segments (sharing the same leading stroke) were eliminated by the filters mentioned previously. These two simple delay parameters (per segment) suffice to define an implicit graph of all legal segment transitions. For a transition from segment number i to segment number j to be legal, the sum of segment i's forward delay plus segment j's reverse delay must be equal to $j - i$. Figure 13.2 provides an example of some ambiguous ink and the segments that might be generated from its strokes, supporting interpretations of "dog", "clog", "cbg", or even "%g".

Ink	Segment Number	Segment	Stroke Count	Forward Delay	Reverse Delay
dog	1	c	1	3	0
	2	cl	2	4	1
	3	clo	3	4	2
	4	l	1	2	0
	5	lo	2	2	1
	6	o	1	1	0
	7	g	1	0	0

Fig. 13.2. Segmentation of strokes into tentative characters or segments.

13.4 Character Classification

The output of the segmentation stage is a stream of segments that are then passed to an ANN for classification as characters. Except for the architecture and training specifics detailed below, a fairly standard multi-layer perceptron trained with error back-propagation (BP) provides the ANN character classifier at the heart of APR. A large body of prior work exists to indicate the general applicability of ANN technology as a classifier providing good estimates of *a posteriori* probabilities of each class given the input ([5, 12, 11], and others cited herein). Compelling arguments have been made for why ANNs providing posterior probabilities in a probabilistic recognition formulation should be expected to outperform other recognition approaches [8], and ANNs have performed well as the core of speech recognition systems [10].

13.4.1 Representation

A recurring theme in ANN research is the extreme importance of the representation of the data that is given as input to the network. We experimented with a variety of input representations, including stroke features both anti-aliased (grayscale) and not (binary), and images both anti-aliased and not, and with various schemes for positioning and scaling the ink within the image input window.

In every case, anti-aliasing was a significant win. This is consistent with others' findings, that ANNs perform better when presented with smoothly varying, distributed inputs than they do when presented with binary, localized inputs. Almost the simplest image representation possible, a non-aspect-ratio-preserving, expand-to-fill-the-window image (limited only by a maximum scale factor to keep from blowing dots up to the full window size), together with either a single unit or a thermometer code (some number of units turned on in sequence to represent larger values) for the aspect ratio, proved to be the most effective single-classifier solution. However, the best overall classifier accuracy was ultimately obtained by combining multiple distinct representations into nearly independent, parallel classifiers, joined at a final output layer. Hence representation proved not only to be as important as architecture, but, ultimately, to help define the architecture of our nets. For our final, hand-optimized system, we utilize four distinct inputs, as indicated in Figure 13.3. The stroke count representation was dithered (changed randomly at a small probability), to expand the effective training set, prevent the network from fixating on this simple input, and thereby improve the network's ability to generalize. A schematic of the various input representations can be seen as part of the architecture drawing in Figure 13.4 in the next section.

Input Feature	Resolution	Description
Image	14x14	anti-aliased, scale-to-window, scale-limited
Stroke	20x9	anti-aliased, limited resolution tangent slope, resampled to fixed number of points
Aspect Ratio	1x1	normalized and capped to [0,1]
Stroke Count	5x1	dithered thermometer code

Fig. 13.3. Input representations used in APR.

13.4.2 Architecture

As with representations, we experimented with a variety of architectures, including simple fully-connected layers, receptive fields, shared weights, multiple hidden layers, and, ultimately, multiple nearly independent classifiers tied to a common output layer. The final choice of architecture includes multiple input representations, a first hidden layer (separate for each input representation) using receptive fields, fully connected second hidden layers (again distinct for each representation), and a final, shared, fully-connected output layer. Simple scalar features – aspect ratio and stroke count – connect to both second hidden layers. The final network architecture, for our original English-language system, is shown in Figure 13.4.

Layers are fully connected, except for the inputs to the first hidden layer on the image side. This first hidden layer on the image side consists of 8 separate

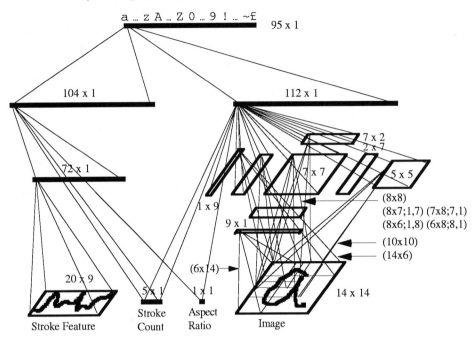

Fig. 13.4. Final English-language net architecture. (See the text for an explanation of the notation.)

grids, each of which accepts inputs from the image input grid with its own receptive field sizes and strides, shown parenthetically in Figure 13.4 as (x-size x y-size; x-stride, y-stride). A stride is the number of units (pixels) in the input image space between sequential positionings of the receptive fields, in a given direction. The 7x2 and 2x7 side panels (surrounding the central 7x7 grid) pay special attention to the edges of the image. The 9x1 and 1x9 side panels specifically examine full-size vertical and horizontal features, respectively. The 5x5 grid observes features at a different spatial scale than the 7x7 grid.

Combining the two classifiers at the output layer, rather than, say, averaging the outputs of completely independent classifiers, allows generic BP to learn the best way to combine them, which is both convenient and powerful. But our *integrated multiple-representations* architecture is conceptually related to and motivated by prior experiments at combining nets such as Steve Nowlan's "mixture of experts" [7].

13.4.3 Normalizing Output Error

Analyzing a class of errors involving words that were misrecognized due to perhaps a single misclassified character, we realized that the net was doing a poor job of representing second and third choice probabilities. Essentially, the net was being forced to attempt unambiguous classification of intrinsically ambiguous patterns due to the nature of the mean squared error minimization in BP, coupled with the typical training vector which consists of all 0's except for the single 1 of the target. Lacking any viable means of encoding legitimate probabilistic ambiguity into the training vectors, we decided to try "normalizing" the "pressure towards 0" vs. the "pressure towards 1" introduced by the output error during training. We refer to this technique as *NormOutErr*, due to its normalizing effect on target versus non-target output error.

We reduce the BP error for non-target classes relative to the target class by a factor that normalizes the total non-target error seen at a given output unit relative to the total target error seen at that unit. Assuming a training set with equal representation of classes, this normalization should then be based on the number of non-target versus target classes in a typical training vector, or, simply, the number of output units (minus one). Hence for non-target output units, we scale the error at each unit by a constant:

$$e' = Ae$$

where e is the error at an output unit, and A is defined to be:

$$A = \frac{1}{d(N_{outputs} - 1)}$$

where $N_{outputs}$ is the number of output units, and d is our tuning parameter, typically ranging from 0.1 to 0.2. Error at the target output unit is unchanged. Overall, this raises the activation values at the output units, due to the reduced pressure towards zero, particularly for low-probability samples. Thus the learning algorithm no longer converges to a least mean-squared error (LMSE) estimate of $p(class|input)$, but to an LMSE estimate of a nonlinear function $f(p(class|input), A)$ depending on the factor A by which we reduced the error pressure toward zero.

Using a simple version of the technique of [3], we worked out what that resulting nonlinear function is. The net will attempt to converge to minimize the modified quadratic error function

$$\langle \hat{E}^2 \rangle = p(1-y)^2 + A(1-p)y^2$$

by setting its output y for a particular class to

$$y = \frac{p}{A - Ap + p}$$

where $p = p(class|input)$, and A is as defined above. For small values of p, the activation y is increased by a factor of nearly $1/A$ relative to the conventional case of $y = p$, and for high values of p the activation is closer to 1 by nearly a factor of A. The inverse function, useful for converting back to a probability, is

$$p = \frac{yA}{yA + 1 - y}$$

We verified the fit of this function by looking at histograms of character-level empirical percentage-correct versus y, as in Figure 13.5.

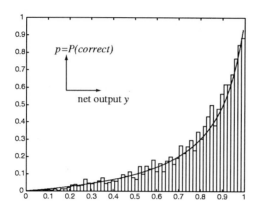

Fig. 13.5. Empirical p vs. y histogram for a net trained with $A = 0.11$ ($d = 0.1$), with the corresponding theoretical curve.

Even for this moderate amount of output error normalization, it is clear that the lower-probability samples have their output activations raised significantly, relative to the 45° line that $A = 1$ yields.

The primary benefit derived from this technique is that the net does a much better job of representing second and third choice probabilities, and low probabilities in general. Despite a small drop in top choice character accuracy when using *NormOutErr*, we obtain a very significant increase in word accuracy by this technique. Figure 13.6 shows an exaggerated example of this effect, for an atypically large value of d (0.8), which overly penalizes character accuracy; however, the 30% decrease in word error rate is normal for this technique. (Note: These data are from a multi-year-old experiment, and are not necessarily representative of current levels of performance on any absolute scale.)

13.4.4 Negative Training

The previously discussed inherent ambiguities in character segmentation necessarily result in the generation and testing of a large number of invalid segments. During recognition, the network must classify these invalid segments just as it would any valid segment, with no knowledge of which are valid or invalid. A significant increase in word-level recognition accuracy was obtained by performing *negative training* with these invalid segments. This consists of presenting invalid segments to the net during training, with all-zero target vectors. We retain control over the degree of negative training in two ways. First is a *negative-training factor* (ranging from 0.2 to 0.5) that modulates the learning rate (equivalently by modulating the error at the output layer) for these negative patterns. This

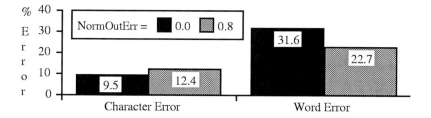

Fig. 13.6. Character and word error rates for two different values of *NormOutErr(d)*. A value of 0.0 disables *NormOutErr*, yielding normal BP. The unusually high value of 0.8 ($A = 0.013$) produces nearly equal pressures towards 0 and 1.

reduces the impact of negative training on positive training, thus modulating the impact on characters that specifically look like elements of multi-stroke characters (e.g., I, 1, l, o, O, 0). Secondly, we control a *negative-training probability* (ranging between 0.05 and 0.3), which determines the probability that a particular negative sample will actually be trained on (for a given presentation). This both reduces the overall impact of negative training, and significantly reduces training time, since invalid segments are more numerous than valid segments. As with *NormOutErr*, this modification hurts character-level accuracy a little bit, but helps word-level accuracy a lot

13.4.5 Stroke Warping

During training (but not during recognition), we produce random variations in stroke data, consisting of small changes in skew, rotation, and x and y linear and quadratic scalings. This produces alternate character forms that are consistent with stylistic variations within and between writers, and induces an explicit aspect ratio and rotation invariance within the framework of standard back-propagation. The amounts of each distortion to apply were chosen through cross-validation experiments, as just the amount needed to yield optimum generalization. (*Cross-validation* is a standard technique for *early stopping* of ANN training, to prevent over-learning of the training set, and thus reduced accuracy on new data outside that training set. The technique consists of keeping aside some subset of the available data, the cross-validation set, and testing on it at some interval, but never training on it, and then stopping the training when accuracy ceases to improve on this cross-validation set, despite the fact that accuracy might continue to improve on the training set.) We chose relative amounts of the various transformations by testing for optimal final, converged accuracy on a cross-validation set. We then increased the amount of all stroke warping being applied to the training set, just to the point at which accuracy on the training set ceased to diverge from accuracy on the cross-validation set.

We also examined a number of such samples by eye to verify that they represent a natural range of variation. A small set of such variations is shown in Figure 13.7.

Fig. 13.7. A few random stroke warpings of the same original "m" data.

Our stroke warping scheme is somewhat related to the ideas of Tangent Dist and Tangent Prop [14, 13] (see chapter 12), in terms of the use of predetermined families of transformations, but we believe it is much easier to implement. It is also somewhat distinct in applying transformations on the original coordinate data, as opposed to using distortions of images. The voice transformation scheme of [4] is also related, but they use a static replication of the training set through a small number of transformations, rather than dynamic random transformations of an essentially infinite variety.

13.4.6 Frequency Balancing

Training data from natural English words and phrases exhibit very non-uniform priors for the various character classes, and ANNs readily model these priors. However, as with *NormOutErr*, we find that reducing the effect of these priors on the net, in a controlled way, and thus forcing the net to allocate more of its resources to low-frequency, low-probability classes is of significant benefit to the overall word recognition process. To this end, we explicitly (partially) balance the frequencies of the classes during training. We do this by probabilistically skipping and repeating patterns, based on a precomputed repetition factor. Each presentation of a repeated pattern is "warped" uniquely, as discussed previously.

To compute the repetition factor for a class i, we first compute a normalized frequency of that class:

$$F_i = \frac{S_i}{\bar{S}}$$

where S_i is the number of samples in class i, and $\bar{S}$ is the average number of samples over all classes, computed in the obvious way:

$$\bar{S} = \frac{1}{C} \sum_{i=1}^{C} S_i$$

with C being the number of classes. Our repetition factor is then defined to be:

$$R_i = \left(\frac{a}{F_i}\right)^b$$

with a and b being adjustable controls over the amount of skipping vs. repeating and the degree of prior normalization, respectively. Typical values of a range from 0.2 to 0.8, while b ranges from 0.5 to 0.9. The factor $a < 1$ lets us do more skipping than repeating; e.g. for $a = 0.5$, classes with relative frequency equal to half the average will neither skip nor repeat; more frequent classes will skip, and less frequent classes will repeat. A value of 0.0 for b would do nothing, giving $R_i = 1.0$ for all classes, while a value of 1.0 would provide "full" normalization. A value of b somewhat less than one seems to be the best choice, letting the net keep some bias in favor of classes with higher prior probabilities.

This explicit prior-bias reduction is conceptually related to Lippmann's [8] and Morgan and Bourlard's [10] recommended method for converting from the net's estimate of posterior probability, $p(class|input)$, to the value needed in an HMM or Viterbi search, $p(input|class)$, which is to divide by $p(class)$ priors. Using that technique, however, should produce noisier estimates for low frequency classes, due to the divisions by low frequencies, resulting in a set of estimates that are not really optimized in a LMSE sense (as the net outputs are). In addition, output activations that are naturally bounded between 0 and 1, due to the sigmoid, convert to potentially very large probability estimates, requiring a re-normalization step. Our method of frequency balancing during training eliminates both of these concerns. Perhaps more significantly, frequency balancing also allows the standard BP training process to dedicate more network resources to the classification of the lower-frequency classes, though we have no current method for characterizing or quantifying this benefit.

13.4.7 Error Emphasis

While frequency balancing corrects for under-represented classes, it cannot account for under-represented writing styles. We utilize a conceptually related probabilistic skipping of patterns, but this time for just those patterns that the net correctly classifies in its forward/recognition pass, as a form of "error emphasis", to address this problem. We define a *correct-train probability* (ranging from 0.1 to 1.0) that is used as a biased coin to determine whether a particular pattern, having been correctly classified, will also be used for the backward/training pass or not. This only applies to correctly segmented, or "positive" patterns, and misclassified patterns are never skipped.

Especially during early stages of training, we set this parameter fairly low (around 0.1), thus concentrating most of the training time and the net's learning capability on patterns that are more difficult to correctly classify. This is the only way we were able to get the net to learn to correctly classify unusual character variants, such as a 3-stroke "5" as written by only one training writer.

Variants of this scheme are possible in which misclassified patterns would be repeated, or different learning rates would apply to correctly and incorrectly classified patterns. It is also related to techniques that use a training subset, from which easily-classified patterns are replaced by randomly selected patterns from the full training set [6].

13.4.8 Annealing

Though some discussions of back-propagation espouse explicit formulae for modulating the learning rate over time, many seem to assume the use of a single, fixed learning rate. We view the stochastic back-propagation process as a kind of simulated annealing, with a learning rate starting very high and decreasing only slowly to a very low value. But rather than using any prespecified formula to decelerate learning, the rate at which the learning rate decreases is determined by the dynamics of the learning process itself. We typically start with a rate near 1.0 and reduce the rate by a multiplicative *decay factor* of 0.9 until it gets down to about 0.001. The rate decay factor is applied following any epoch in which the total squared error increased on the training set, relative to the previous epoch. This "total squared error" is summed over all output units and over all patterns in one full epoch, and normalized by those counts. So even though we are using "online" or stochastic gradient descent, we have a measure of performance over whole epochs that can be used to guide the "annealing" of the learning rate. Repeated tests indicate that this approach yields better results than low (or even moderate) initial learning rates, which we speculate to be related to a better ability to escape local minima.

In addition, we find that we obtain best overall results when we also allow some of our many training parameters to change over the course of a training run. In particular, the correct train probability needs to start out very low to give the net a chance to learn unusual character styles, but it should finish up near 1.0 in order to not introduce a general posterior probability bias in favor of classes with lots of ambiguous examples. We typically train a net in four "phases" according to parameters such as in Figure 13.8.

Phase	Epochs	Learning Rate	Correct Train Prob	Negative Train Prob
1	25	1.0 - 0.5	0.1	0.05
2	25	0.5 - 0.1	0.25	0.1
3	50	0.1 - 0.01	0.5	0.18
4	30	0.01 - 0.001	1.0	0.3

Fig. 13.8. Typical multi-phase schedule of learning rates and other parameters for training a character-classifier net.

13.4.9 Quantized Weights

The work of Asanovic and Morgan [1] shows that two-byte (16-bit) weights are about the smallest that can be tolerated in training large ANNs via back-propagation. But memory is expensive in small devices, and RISC processors, such as the ARM-610 in the first devices in which this technology was deployed, are much more efficient doing one-byte loads and multiplies than two-byte loads and multiplies, so we were motivated to make one-byte weights work.

Running the net for recognition demands significantly less precision than does training the net. It turns out that one-byte weights provide adequate precision for recognition, if the weights are trained appropriately. In particular, a dynamic range should be fixed, and weights limited to that legal range during training, and then rounded to the requisite precision after training. For example, we find that a range of weight values from (almost) -8 to +8 in steps of 1/16 does a good job. Figure 13.9 shows a typical resulting distribution of weight values. If the weight limit is enforced during high-precision training, the resources of the net will be adapted to make up for the limit. Since bias weights are few in number, however, and very important, we allow them to use two bytes with essentially unlimited range. Performing our forward/recognition pass with low-precision, one-byte weights (a ±3.4 fixed-point representation), we find no noticeable degradation relative to floating-point, four-byte, or two-byte weights using this scheme.

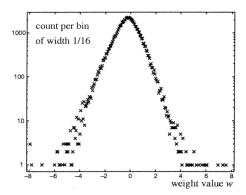

Fig. 13.9. Distribution of weight values in a net with one-byte weights, on a log count scale. Weights with magnitudes greater than 4 are sparse, but important.

We have also developed a scheme for training with augmented one-byte weights. It uses a temporary augmentation of the weight values with two additional low-order bytes to achieve precision in training, but runs the forward pass of the net using only the one-byte high-order part. Thus any cumulative effect of the one-byte rounded weights in the forward pass can be compensated through further training. Small weight changes accumulate in the low-order bytes, and only occasionally carry into a change in the one-byte weights used by the net. In a personal product, this scheme could be used for adaptation to the user, after which the low-order residuals could be discarded and the temporary memory reclaimed.

13.5 Context-Driven Search

The output of the ANN classifier is a stream of probability vectors, one vector for each segmentation hypothesis, with as many potentially nonzero probability elements in each vector as there are characters (that the system is capable of recognizing). In practice, we typically only pass the top ten (or fewer) scored character-class hypotheses, per segment, to the search engine, for the sake of efficiency. The search engine then looks for a minimum-cost path through this vector stream, abiding by the legal transitions between segments, as defined in the tentative-segmentation step discussed previously. This minimum-cost path is the APR system's best interpretation of the ink input by the user, and is returned to the system in which APR is embedded as the recognition result for whole words or sentences of the user's input.

The search is driven by a somewhat *ad hoc*, generative language model, which consists of a set of graphs that are searched in parallel. We use a simple beam search in a negative-log-probability (or *penalty*) space for the best N hypotheses. The beam is based on a fixed maximum number of hypotheses, rather than a particular value. Each possible transition token (character) emitted by one of the graphs is scored not only by the ANN, but by the language model itself, by a simple letter-case model, and by geometric-context models discussed below. The fully integrated search process takes place over a space of character- and word-segmentation hypotheses, as well as character-class hypotheses.

13.5.1 Lexical Context

Context is essential to accurate recognition, even if that context takes the form of a very broad language model. Humans achieve just 90% accuracy on isolated characters from our database. Lacking any context this would translate to a word accuracy of not much more than 60% (0.9^5), assuming an average word length of 5 characters. We obviously need to do much better, with even lower isolated-character accuracy, and we accomplish this by the application of our context models.

A simple model of letter case and adjacency – penalizing case transitions except between the first and second characters, penalizing alphabetic-to-numeric transitions, and so on – together with the geometric-context models discussed later, is sufficient to raise word-level accuracy up to around 77%.

The next large gain in accuracy requires a genuine language model. We provide this model by means of dictionary graphs, and assemblages of those graphs combined into what we refer to as *BiGrammars*. BiGrammars are essentially scored lists of dictionaries, together with specified legal (scored) transitions between those dictionaries. This scheme allows us to use word lists, prefix and suffix lists, and punctuation models, and to enable appropriate transitions between them. Some dictionary graphs are derived from a regular-expression grammar that permits us to easily model phone numbers, dates, times, etc., as shown in Figure 13.10.

```
dig      = [0123456789]
digm01 =    [23456789]

acodenums = (digm01 [01] dig)

acode = { ("1-"?     acodenums "-"):40 ,
          ("1"? "(" acodenums ")"):60 }

phone = (acode? digm01 dig dig "-" dig dig dig dig)
```

Fig. 13.10. Sample of the regular-expression language used to define a simple telephone-number grammar. Symbols are defined by the equal operator; square brackets enclose multiple, alternative characters; parentheses enclose sequences of symbols; curly braces enclose multiple, alternative symbols; an appended colon followed by numbers designates a prior probability of that alternative; an appended question mark means "zero or one occurrence"; and the final symbol definition represents the graph or grammar expressed by this dictionary.

All of these dictionaries can be searched in parallel by combining them into a general-purpose BiGrammar that is suitable for most applications. It is also possible to combine subsets of these dictionaries, or special-purpose dictionaries, into special BiGrammars targeted at more limited contexts. A very simple BiGrammar, which might be useful to specify context for a field that only accepts telephone numbers, is shown in Figure 13.11. A more complex BiGrammar (though still far short of the complexity of our final general-input context) is shown in Figure 13.12.

```
BiGrammar Phone

[Phone.lang 1. 1. 1.]
```

Fig. 13.11. Sample of a simple BiGrammar describing a telephone-only context. The BiGrammar is first named (Phone), and then specified as a list of dictionaries (Phone.lang), together with the probability of starting with this dictionary, ending with this dictionary, and cycling within this dictionary (the three numerical values).

```
BiGrammar FairlyGeneral
(.8
    (.6
        [WordList.dict  .5   .8   1.   EndPunct.lang .2]
        [User.dict      .5   .8   1.   EndPunct.lang .2]
    )
    (.4
        [Phone.lang     .5   .8   1.   EndPunct.lang .2]
        [Date.lang      .5   .8   1.   EndPunct.lang .2]
    )
)

(.2
    [OpenPunct.lang  1.   0.   .5
        (.6
            WordList.dict  .5
            User.dict      .5
        )
        (.4
            Phone.lang     .5
            Date.lang      .5
        )
    ]
)

[EndPunct.lang  0.   .9   .5   EndPunct.lang .1]
```

Fig. 13.12. Sample of a slightly more complex BiGrammar describing a fairly general context. The BiGrammar is first named (FairlyGeneral), and then specified as a list of dictionaries (the *.dict and *.lang entries), together with the probability of starting with this dictionary, ending with this dictionary, and cycling within this dictionary (the first three numerical values following each dictionary name), plus any dictionaries to which this dictionary may legally transition, along with the probability of taking that transition. The parentheses permit easy specification of multiplicative prior probabilities for all dictionaries contained within them. Note that in this simple example, it is not possible (starting probability = 0) to start a string with the EndPunct (end punctuation) dictionary, just as it is not possible to end a string with the OpenPunct dictionary.

We refer to our language model as being "weakly applied" because in parallel with all of the wordlist-based dictionaries and regular-expression grammars, we simultaneously search both an alphabetic-characters grammar ("wordlike") and a completely general, any-character-anywhere grammar ("symbols"). These more flexible models, though given fairly low *a priori* probabilities, permit users to write any unusual character string they might desire. When the prior probabilities for the various dictionaries are properly balanced, the recognizer is able to benefit from the language model, and deliver the desired level of accuracy for common in-dictionary words (and special constructs like phone numbers, etc.), yet can also recognize arbitrary, non-dictionary character strings, especially if they are written neatly enough that the character classifier can be confident of its classifications.

We have also experimented with bi-grams, tri-grams, N-grams, and we are continuing experiments with other, more data-driven language models; so far, however, our generative approach has yielded the best results.

13.5.2 Geometric Context

We have never found a way to reliably estimate a baseline or topline for characters, independent of classifying those characters in a word. Non-recognition-integrated estimates of these line positions, based on strictly geometric features, have too many pathological failure modes, which produce erratic recognition failures. Yet the geometric positioning of characters most certainly bears information important to the recognition process. Our system factors the problem by letting the ANN classify representations that are independent of baseline and size, and then using separate modules to score both the absolute size of individual characters, and the relative size and position of adjacent characters.

The scoring based on absolute size is derived from a set of simple Gaussian models of individual character heights, relative to some running scale parameters computed both during learning and during recognition. This *CharHeight* score directly multiplies the scores emitted by the ANN classifier, and helps significantly in case disambiguation.

We also employ a *GeoContext* module that scores adjacent characters, based on the classification hypotheses for those characters and on their relative size and placement. GeoContext scores each tentative character based on its class and the class of the immediately preceding letter (for the current search hypothesis). The character classes are used to look up expected character sizes and positions in a standardized space (baseline=0.0, topline=1.0). The ink being evaluated provides actual sizes and positions that can be compared directly to the expected values, subject only to a scale factor and offset, which are chosen so as to minimize the estimated error of fit between data and model. This same quadratic error term, computed from the inverse covariance matrix of a full multivariate Gaussian model of these sizes and positions, is used directly as GeoContext's score (or penalty, since it is applied in the -log probability space of the search engine). Figure 13.13 illustrates the bounding boxes derived from the user's ink vs. the table-driven model, with the associated error measures for our GeoContext module.

GeoContext's multivariate Gaussian model is learned directly from data. The problem in doing so was to find a good way to train per-character parameters of top, bottom, width, space, etc., in our standardized space, from data that had no labeled baselines, or other absolute referent points. Since we had a technique for generating an error vector from the table of parameters, we decided to use a back-propagation variant to train the table of parameters to minimize the squared error terms in the error vectors, given all the pairs of adjacent characters and correct class labels from the training set.

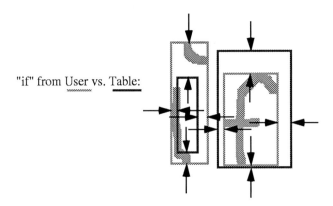

Fig. 13.13. The eight measurements that contribute to the GeoContext error vector and corresponding score for each letter pair.

GeoContext plays a major role in properly recognizing punctuation, in disambiguating case, and in recognition in general. A more extended discussion of GeoContext has been provided by Lyon and Yaeger [9].

13.5.3 Integration with Word Segmentation

Just as it is necessary to integrate character segmentation with recognition via the search process, so is it essential to integrate word segmentation with recognition and search, in order to obtain accurate estimates of word boundaries, and to reduce the large class of errors associated with missegmented words. To perform this integration, we first need a means of estimating the probability of a word break between each pair of tentative characters. We use a simple statistical model of gap sizes and stroke-centroid spacing to compute this probability (*spaceProb*). Gaussian density distributions, based on means and standard deviations computed from a large training corpus, together with a prior probability scale factor, provide the basis for the word-gap and stroke-gap (non-word-gap) models, as illustrated in Figure 13.14. Since any given gap is, by definition, either a word gap or a non-word gap, the simple ratio defined in Figure 13.14 provides a convenient, self-normalizing estimate of the word-gap probability. In practice, that equation further reduces to a simple sigmoid form, thus allowing us to take advantage of a lookup-table-based sigmoid derived for use in the ANN. In a thresholding, non-integrated word-segmentation model, word breaks would be introduced when spaceProb exceeds 0.5; i.e., when a particular gap is more likely to be a word-gap than a non-word-gap. For our integrated system, both word-break and non-word-break hypotheses are generated at each segment transition, and weighted by spaceProb and (1-spaceProb), respectively. The search process then proceeds over this larger hypothesis space to produce best estimates of whole phrases or sentences, thus integrating word segmentation as well as character segmentation.

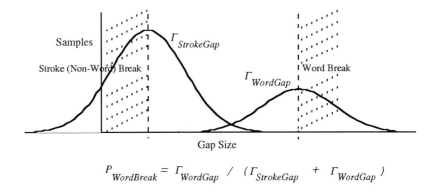

Fig. 13.14. Gaussian density distributions yield a simple statistical model of word-break probability, which is applied in the region between the peaks of the StrokeGap and WordGap distributions. Hashed areas indicate regions of clear cut decisions, where $P_{WordBreak}$ is set to either 0.0 or 1.0, to avoid problems dealing with tails of these simple distributions.

13.6 Discussion

The combination of elements described in the preceeding sections produces a powerful, integrated approach to character segmentation, word segmentation, and recognition. Users' experiences with APR are almost uniformly positive, unlike experiences with previous handwriting recognition systems. Writing within the dictionary is remarkably accurate, yet the ease with which people can write outside the dictionary has fooled many people into thinking that the Newton's "Print Recognizer" does not use dictionaries. As discussed previously, our recognizer certainly does use dictionaries. Indeed, the broad-coverage language model, though weakly applied, is essential for high accuracy recognition. Curiously, there seems to be little problem with dictionary *perplexity* – little difficulty as a result of using very large, very complex language models. We attribute this fortunate behavior to the excellent performance of the neural network character classifier at the heart of the system. One of the side benefits of the weak application of the language model is that even when recognition fails and produces the wrong result, the answer that is returned to the user is typically understandable by the user – perhaps involving substitution of a single character. Two useful phenomena ensue as a result. First, the user learns what works and what doesn't, especially when she refers back to the ink that produced the misrecognition, so the system trains the user gracefully over time. Second, the meaning is not lost the way it can be, all too easily, with whole word substitutions – with that "Doonesbury Effect" found in first-generation, strong-language-model recognizers.

Though we have provided legitimate accuracy statistics for certain comparative tests of some of our algorithms, we have deliberately shied away from claiming specific levels of accuracy in general. Neat printers, who are familiar with the system, can achieve 100% accuracy if they are careful. Testing on data from complete novices, writing for the first time using a metal pen on a glass surface, without any feedback from the recognition system, and with ambiguous instructions about writing with "disconnected characters" (intended to mean printing, but often interpreted to mean writing with otherwise cursive characters but separated by large spaces in a wholly unnatural style), can yield word-level accuracies as low as 80%. Of course, the entire interesting range of recognition accuracies lies between these two extremes. Perhaps a slightly more meaningful statistic comes from common reports on usenet newsgroups, and some personal testing, that suggest accuracies of 97% to 98% in regular use. But for scientific purposes, none of these numbers have any real meaning, since our testing datasets are proprietary, and the only valid tests between different recognizers would have to be based on results obtained by processing the exact same bits, or by analyzing large numbers of experienced users of the systems in the field – a difficult project which has not been undertaken.

One of the key reasons for the success of APR is the suite of innovative neural network training techniques that help the network encode better class probabilities, especially for under-represented classes and writing styles. Many of these techniques – stroke count dithering, normalization of output error, frequency balancing, error emphasis – share a unifying theme: Reducing the effect of *a priori* biases in the training data on network learning significantly improves the network's performance in an integrated recognition system, despite a modest reduction in the network's accuracy for individual characters. Normalization of output error prevents over-represented non-target classes from biasing the net against under-represented target classes. Frequency balancing prevents over-represented classes from biasing the net against under-represented classes. And stroke-count dithering and error emphasis prevent over-represented writing styles from biasing the net against under-represented writing styles. One could even argue that negative training eliminates an absolute bias towards properly segmented characters, and that stroke warping reduces the bias towards those writing styles found in the training data, although these techniques also provide wholly new information to the system.

Though we've offered arguments for why each of these techniques, individually, helps the overall recognition process, it is unclear why prior-bias reduction, in general, should be so consistently valuable. The general effect may be related to the technique of dividing out priors, as is sometimes done to convert from $p(class|input)$ to $p(input|class)$. But we also believe that forcing the net, during learning, to allocate resources to represent less frequent sample types may be directly beneficial. In any event, it is clear that paying attention to such biases and taking steps to modulate them is a vital component of effective training of a neural network serving as a classifier in a maximum-likelihood recognition system.

The majority of this paper describes a sort of snapshot of the system and its architecture as it was deployed in its first commercial release, when it was, indeed, purely a "Print Recognizer". Letters had to be fully "disconnected"; i.e., the pen had to be lifted between each pair of characters. The characters could overlap to some extent, but the ink could not be continuous. Connected characters proved to be the largest remaining class of errors for most of our users, since even a person who normally prints (as opposed to writing in cursive script) may occasionally connect a pair of characters – the cross-bar of a "t" with the "h" in "the", the "o" and "n" in any word ending in "ion", and so on. To address this issue, we experimented with some fairly straightforward modifications to our recognizer, involving the *fragmenting* of user-strokes into multiple system-strokes, or *fragments*. Once the ink representing the connected characters is broken up into fragments, we then allow our standard integrated segmentation and recognition process to stitch them back together into the most likely character and word hypotheses, as always. This technique has proven itself to work quite well, and the version of the "Print Recognizer" in the MessagePad 2000 and 2100 supports recognition of printing with connected characters. This capability was added without significant modification of the main recognition algorithms as presented in this paper. Due to certain assumptions and constraints in the current release of the software, APR is not yet a full cursive recognizer, though that is an obvious next direction to explore.

The net architecture discussed in section 13.4.2 and shown in Figure 13.4 also corresponds to the true printing-only recognizer. The final output layer has 95 elements corresponding to the full printable ASCII character set plus the British Pound sign. Initially for the German market, and now even in English units, we have extended APR to handle diacritical marks and the special symbols needed for most European languages (although there is only very limited coverage of foreign languages in the dictionaries of English units). The main innovation that permitted this extended character set was an explicit handling of any compound character as a *base* plus an *accent*. This way only a few nodes needed to be added to the neural network output layer, representing just the bases and accents, rather than all combinations and permutations of same. And training data for all compound characters sharing a common base or a common accent contributed to the network's ability to learn that base or accent, as opposed to contributing only to the explicit base+accent combination. Here again, however, the fundamental recognizer technology has not changed significantly from that presented in this paper.

13.7 Future Extensions

We are optimistic that our algorithms, having proven themselves to work essentially as well for connected characters as for disconnected characters, may extend gracefully to full cursive.

On a more speculative note, we believe that the technique may extend well to ideographic languages, substituting radicals for characters, and ideographic characters for words.

Finally, a note about learning and user adaptation: For a learning technology such as ANNs, user adaptation is an obvious and natural fit, and was planned as part of the system from its inception. However, due to RAM constraints in the initial shipping product, and the subsequent prioritization of European character sets and connected characters, we have not yet deployed a learning system. We have, however, done some testing of user adaptation, and believe it to be of considerable value. Figure 13.15 shows a comparison of the average performance on an old user-independent net trained on data from 45 writers, and the performance for three individuals using (A) the user-independent net, (B) a net trained on data exclusively from that individual, and (C) a copy of the user-independent net adapted to the specific user by some incremental training. (Note: These data are from a multi-year-old experiment, and are not necessarily representative of current levels of performance on any absolute scale.)

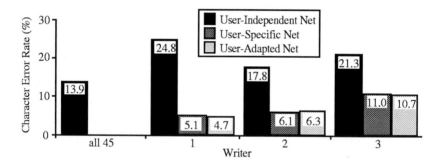

Fig. 13.15. User-adaptation test results for three individual writers with three different nets each, plus the overall results for 45 writers tested on a user-independent net trained on all 45 writers.

An important distinction is being made here between "user-adapted" and "user-specific" nets. "User-specific" nets have been trained with a relatively large corpus of data exclusively from that specific user. "User-adapted" nets were based on the user-independent net, with some additional training using limited data from the user in question. All testing was performed with data held out from all training sets.

One obvious thing to note is the reduction in error rate ranging from a factor of 2 to a factor of 5 that both user-specific and user-adapted nets provide. An equally important thing to note is that the user-adapted net performs essentially as well as a user-specific net – in fact, slightly *better* for two of the three writers. Given ANNs' penchant for local minima, we were concerned that this might not be the case. But it appears that the features learned during the user-independent net training served the user-adapted net well. We believe that a very small amount of training data from an individual will allow us to adapt the user-independent net to that user, and improve the overall accuracy for that user significantly, especially for individuals with more stylized writing, or whose writing style is underrepresented in our user-independent training corpus. And even for writers with common and/or neat writing styles, there is inherently less ambiguity in a single writer's style than in a corpus of data necessarily doing its best to represent essentially all possible writing styles.

These results may be exaggerated somewhat by the limited data in the user-independent training corpus at the time these tests were performed (just 45 writers), and at least two of the three writers in question had particularly problematic writing styles. We have also made significant advances in our user-independent recognition accuracies since these tests were performed. Nonetheless, we believe these results are suggestive of the significant value of user adaptation, even in preference to a user-specific solution.

13.8 Acknowledgements

This work was done in collaboration with Bill Stafford, Apple Computer, and Les Vogel, Angel Island Technologies. We would also like to acknowledge the contributions of Michael Kaplan, Rus Maxham, Kara Hayes, Gene Ciccarelli, and Stuart Crawford. We also received support and assistance from the Newton Recognition, Software Testing, and Hardware Support groups. We are also indebted to our many colleagues in the connectionist community for their advice, help, and encouragement over the years, as well as our many colleagues at Apple who pitched in to help throughout the life of this project.

Some of the techniques described in this paper are the subject of pending U.S. and foreign patent applications.

References

1. K. Asanovic and N. Morgan. Experimental determination of precision requirements for back-propagation training of artificial neural networks. Technical Report TR-91-036, International Computer Science Institute, Berkeley, CA, June 1991.
2. Y. Bengio, Y. LeCun, C. Nohl, and C. Burges. LeRec: A NN/HMM hybrid for on-line handwriting recognition. *Neural Computation*, 7(6):1289–1303, 1995.
3. H. Bourlard and C. J. Wellekens. Links between markov models and multilayer perceptrons. *IEEE Transactions on Pattern Analysis and Machine Intelligence*, PAMI-12(12):1167–1178, December 1990.
4. E. I. Chang and R. P. Lippmann. Using voice transformations to create additional training talkers for word spotting. In G. Tesauro, D. Touretzky, and T. Leen, editors, *Advances in Neural Information Processing Systems*, volume 7, pages 875–882. The MIT Press, 1995.

5. H. Gish. A probabilistic approach to the understanding and training of neural network classifiers. In *Proceedings of the IEEE Conference on Acoustics, Speech and Signal Processing*, pages 1361–1364. IEEE Press, 1990.
6. I. Guyon, D. Henderson, P. Albrecht, Y. LeCun, and P. Denker. Writer independent and writer adaptive neural network for on-line character recognition. In S. Impedovo, editor, *From pixels to features III*, pages 493–506. Elsevier, Amsterdam, 1992.
7. R. A. Jacobs, M. I. Jordan, S. J. Nowlan, and G. E. Hinton. Adaptive mixtures of local experts. *Neural Computation*, 3(1):79–87, 1991.
8. R. P. Lippmann. Neural networks, bayesian a posteriori probabilities, and pattern classification. In V. Cherkassky, J. H. Friedman, and H. Wechsler, editors, *From Statistics to Neural Networks – Theory and Pattern Recognition Applications*, pages 83–104. Springer-Verlag, Berlin, 1994.
9. R. Lyon and L. Yaeger. On-line hand-printing recognition with neural networks. In *Fifth International Conference on Microelectronics for Neural Networks and Fuzzy Systems*, Lausanne, Switzerland, 1996. IEEE Computer Society Press.
10. N. Morgan and H. Bourlard. Continuous speech recognition – an introduction to the hybrid HMM/connectionist approach. *IEEE Signal Processing Mag.*, 13(3):24–42, 1995.
11. S. Renals, N. Morgan, M. Cohen, and H. Franco. Connectionist probability estimation in the Decipher speech recognition system. In *Proceedings of the IEEE International Conference on Acoustics, Speech, and Signal Processing*, volume I, pages 601–604, 1992.
12. M. D. Richard and R. P. Lippmann. Neural Network Classifiers Estimate Bayesian a posteriori probabilities. *Neural Computation*, 3(4):461–483, 1991.
13. P. Simard, Y. LeCun, and J. Denker. Efficient pattern recognition using a new transformation distance. In S. J. Hanson, J. D. Cowan, and C. L. Giles, editors, *Advances in Neural Information Processing Systems 5. Proceedings of the 1992 Conference*, pages 50–58, San Mateo, CA, 1993. Morgan Kaufmann.
14. P. Simard, B. Victorri, Y. LeCun, and J. Denker. Tangent prop—A formalism for specifying selected invariances in an adaptive network. In John E. Moody, Steve J. Hanson, and Richard P. Lippmann, editors, *Advances in Neural Information Processing Systems*, volume 4, pages 895–903. Morgan Kaufmann Publishers, Inc., 1992.
15. C. C. Tappert, C. Y. Suen, and T. Wakahara. The state of the art in on-line handwriting recognition. *IEEE Transactions on Pattern Analysis and Machine Intelligence*, PAMI-12(8):787–808, August 1990.

14
Neural Network Classification and Prior Class Probabilities

Steve Lawrence[1], Ian Burns[2], Andrew Back[3], Ah Chung Tsoi[4], and C. Lee Giles[1]*

[1] NEC Research Institute**, 4 Independence Way, Princeton, NJ 08540
[2] Open Access Pty Ltd, Level 2, 7–9 Albany St, St. Leonards, NSW 2065, Australia
[3] RIKEN Brain Science Institute, 2-1 Hirosawa, Wako-shi, Saitama, 351-0198, Japan
[4] Faculty of Informatics, University of Wollongong, Northfields Ave, Wollongong, NSW 2522, Australia
{lawrence,giles}@research.nj.nec.com, ian.burns@oa.com.au, back@brain.riken.go.jp, Ah_Chung_Tsoi@uow.edu.au
http://www.neci.nj.nec.com/homepages/lawrence/

Abstract. A commonly encountered problem in MLP (multi-layer perceptron) classification problems is related to the prior probabilities of the individual classes – if the number of training examples that correspond to each class varies significantly between the classes, then it may be harder for the network to learn the rarer classes in some cases. Such practical experience does not match theoretical results which show that MLPs approximate Bayesian *a posteriori* probabilities (independent of the prior class probabilities). Our investigation of the problem shows that the difference between the theoretical and practical results lies with the assumptions made in the theory (accurate estimation of Bayesian *a posteriori* probabilities requires the network to be large enough, training to converge to a global minimum, infinite training data, and the *a priori* class probabilities of the test set to be correctly represented in the training set). Specifically, the problem can often be traced to the fact that efficient MLP training mechanisms lead to sub-optimal solutions for most practical problems. In this chapter, we demonstrate the problem, discuss possible methods for alleviating it, and introduce new heuristics which are shown to perform well on a sample ECG classification problem. The heuristics may also be used as a simple means of adjusting for unequal misclassification costs.

14.1 Introduction

It has been shown theoretically that MLPs approximate Bayesian *a posteriori* probabilities when the desired network outputs are *1 of M* and squared-error

---
* Lee Giles is also with the Institute for Advanced Computer Studies, University of Maryland, College Park, MD 20742.
** http://www.neci.nj.nec.com

or cross-entropy cost functions are used [6, 11, 12, 15, 23, 25, 26, 28, 29, 32]. This result relies on a number of assumptions for accurate estimation: the network must be large enough and training must find a global minimum, infinite training data is required, and the *a priori* class probabilities of the test set must be correctly represented in the training set.

In practice, MLPs have also been shown to accurately estimate Bayesian *a posteriori* probabilities for certain experiments [10]. However, a commonly encountered problem in MLP classification is related to the case when the frequency of the classes in the training set varies significantly[1]. If the number of training examples for each class varies significantly between classes then there may be a bias towards predicting the more common classes [3, 4], leading to worse classification performance for the rarer classes. In [5] it was observed that classes with low *a priori* probability in a speech application were "ignored" (no samples were classified as these classes after training). Such problems indicate that either the estimation of Bayesian *a posteriori* probabilities is inaccurate, or that such estimation may not be desired (e.g. due to varying misclassification costs (this is explained further in section 14.4)). Bourlard and Morgan [7] have demonstrated inaccurate estimation of Bayesian *a posteriori* probabilities in speech recognition. This chapter discusses how the problem may occur along with methods of dealing with the problem.

14.2 The Trick

This section describes the tricks for alleviating the aforementioned problem. Motivation for their use and experimental results are provided in the following sections. The methods all consider some kind of scaling which is performed on a class by class basis[2].

14.2.1 Prior Scaling

A method of scaling weight updates on a class by class basis according to the prior class probabilities is proposed in this section. Consider gradient descent weight updates for each pattern: $w_{ki}^l(\text{new}) = w_{ki}^l(\text{old}) + \Delta w_{ki}^l(p)$ where $\Delta w_{ki}^l(p) = -\eta \frac{\partial E(p)}{\partial w_{ki}^l}$, p is the pattern index, and w_{ki} is the weight between neuron k in layer l and neuron i in layer $l-1$. Scaling the weight updates on a pattern by pattern basis is considered such that the total expected update for patterns belonging to each class is equal (i.e. independent of the number of patterns in

---

[1] For the data in general. Others have considered the case of different class probabilities between the training and test sets, e.g. [23].
[2] Anand et al. [2] have also presented an algorithm related to unequal prior class probabilities. However, their algorithm aims only to improve convergence speed. Additionally, their algorithm is only for two class problems and batch update.

the class):

$$\left\langle \sum_{p=1}^{N_p} |s_x \Delta w_{ki}^l(p)|_{p_c=x} \right\rangle = c_1, \forall x \in X \quad (14.1)$$

where p_c is the target classification of pattern p, c_1 is a constant, s_x is a scaling factor, x ranges over all classes X, $<>$ denotes expectation, and the $p_c = x$ subscript indicates that the sum is only over the patterns in a particular class x. This effectively scales the updates for lower frequency classes so that they are higher – the aim is to account for the fact that lower frequency classes tend to be "ignored" in certain situations. We assume that the expected weight update for individual patterns in each class is equal:

$$\left\langle |\Delta w_{ki}^l(p)|_{p_c=x} \right\rangle = c_2, \forall x \in X \quad (14.2)$$

where c_2 is a constant not related to c_1. The scaling factor required is therefore:

$$s_x = \frac{1}{p_x N_c} \quad (14.3)$$

where s_x is the scaling factor for all weight updates associated with a pattern belonging to class x, N_c is the number of classes, and p_x is the prior probability of class x.

Scaling as defined above invalidates the Bayesian *a posteriori* probability proofs (for example, scaling a class by two can be compared with duplicating every pattern in the data for that class – causing changes in probability distributions), i.e. there is no reason to expect that the scaling strategy will be optimal. This, and the empirical result that the scaling may improve performance, leads to the hypothesis that there may be a point between no prior scaling and prior scaling as defined above which produces performance better than either of the two extremes. The following scaling rule can be used to select a degree of scaling between the two extremes:

$$s'_x = 1 - c_s + \frac{c_s}{p_x N_c} \quad (14.4)$$

where $0 \leq c_s \leq 1$ is a constant specifying the amount of prior scaling to use. $c_s = 0$ corresponds to no scaling according to prior probabilities, and $c_s = 1$ corresponds to scaling as above. Prior scaling in this form can be expressed as training with the following alternative cost function[3]:

---

[3] A cost function with similar motivation, the "classification figure-of-merit" (CFM) proposed by Hampshire and Waibel [13], has been suggested as a possible improvement when prior class probabilities vary [3]. In [13], the CFM cost function leads to networks which make different errors to those trained with the MSE criterion, and can therefore be useful for improving performance by combining classifiers trained with the CFM and the MSE. However, networks trained with the CFM criterion do not result in higher classification performance than networks trained with the MSE criterion for the experiments reported in [13].

Definition 1.

$$E = \frac{1}{2} \sum_{k=1}^{N_p} \sum_{j=1}^{N_c} s'_x (d_{kj} - y_{kj})^2 \qquad (14.5)$$

where the network has one output for each of the N_c classes, N_p is the number of patterns, d is the desired or target output, y is the predicted output, and x is the class of pattern k.

When using prior scaling as defined in this section, the individual s'_x values can be large for classes with low prior probability. This may lead to the requirement of decreasing the learning rate in order to prevent the relatively large weight updates interfering with the gradient descent process. Comparing the use of prior scaling and not using prior scaling then becomes problematic because the optimal learning rate is different for each case. An alternative is to normalize the s'_x values so that the maximum is 1. Another possibility is to present patterns repeatedly to the network instead of scaling weight updates, i.e. for a class with a scaling factor of 2 each pattern would be presented twice. This would have the advantage of reducing the range of weight updates in terms of magnitude, e.g. an update of magnitude x might be repeated twice rather than using a single update of magnitude $2x$. This may allow the use of a higher learning rate, and therefore reduce the number of epochs required. However, a disadvantage of repeating patterns is that the effective training set would be larger, resulting in longer training times for the same number of epochs. Such a technique could be done probabilistically, and this is the subject of the next technique.

14.2.2 Probabilistic Sampling

Yaeger et al. [33] (chapter 13) have proposed a method called *frequency balancing* which is similar to the prior scaling method above. In frequency balancing, Yaeger et al. use all training samples in random order for each training epoch and allow each sample to be presented to the network a random number of times, which may be zero or more and is computed probabilistically. A balancing factor is included, which is analogous to the scaling factor above (c_s).

We introduce a very similar method here called *probabilistic sampling* whereby training patterns are chosen randomly in the following manner: the class is chosen randomly with the probability of choosing each class x, being $(1-c_s)p_x + \frac{c_s}{N_c}$. A training sample is then chosen randomly from among all training samples for the chosen class.

14.2.3 Post Scaling

Instead of scaling weight updates or altering the effective class frequencies, it is possible to train the network as usual and then scale the outputs of the network after training. For example, the network could be trained as usual and then the outputs scaled according to the prior probabilities in a similar fashion to the prior scaling method (using equation 14.3 or 14.4). Experiments with this technique alone show that it is not always as successful as prior scaling of the weight updates. This may be because the estimation of the lower frequency classes can be less accurate than that of the higher frequency classes [24] (the deviations of the network outputs from the true values in regions with a higher number of data points influence the squared error cost function more than the deviations in regions with a lower number of points [23]).

The post scaling technique introduced here can also be used to optimize a given criterion, e.g. the outputs may be scaled so that the probability of predicting each class matches the prior probabilities in the training set as closely as possible. Post scaling to minimize a different criterion is demonstrated in the results section. For the results in this chapter, the minimization is performed using a simple hill-climbing algorithm which adjusts a scaling factor associated with each of the outputs of the network.

14.2.4 Equalizing Class Membership

A simple method for alleviating difficulty with unequal prior class probabilities is to adjust (e.g. equalize) the number of patterns in each class, either by subsampling [24] (removing patterns from higher frequency classes), or by duplication (of patterns in lower frequency classes)[4]. For subsampling, patterns can be removed randomly, or heuristics may be used to remove patterns in regions of low ambiguity. Subsampling involves a loss of information which can be detrimental. Duplication involves a larger dataset and longer training times for the same number of training epochs (the convergence time may be longer or shorter).

14.3 Experimental Results

Results on an ECG classification problem are reported in this section after discussing the use of alternative performance measures. Results on a simple artificial problem are also included in the explanation section.

14.3.1 Performance Measures

When the interclass prior probabilities of the classes vary significantly, then the overall classification error may not be the most appropriate performance criterion. For example, a model may always predict the most common class and still provide relatively high performance. Statistics such as the Sensitivity, Positive Predictivity, and False Positive Rate can provide more meaningful results [1]. These are defined on a class by class basis as follows:

The **Sensitivity** of a class is the proportion of events labeled as that class which are correctly detected. For the two class confusion matrix shown in table 14.1 the sensitivity of class 1 is $\frac{c_{11}}{c_{11}+c_{12}}$.

---

[4] The heuristic of adding noise during training [22] could be useful here as with the other techniques in this chapter.

The **Positive Predictivity** of a class is the proportion of events which were predicted to be the class and were labeled as that class. For the two class confusion matrix shown in table 14.1 the positive predictivity of class 1 is $\frac{c_{11}}{c_{11}+c_{21}}$.

The **False Positive Rate** of a class is the proportion of all patterns for other classes which were incorrectly classified as that class. For the two class confusion matrix shown in table 14.1 the false positive rate of class 1 is $\frac{c_{21}}{c_{11}+c_{21}}$.

Class	1	2
1	c_{11}	c_{12}
2	c_{21}	c_{22}

Table 14.1. A sample confusion matrix which is used to illustrate sensitivity, positive predictivity, and false positive rate. Rows correspond to the desired classes and columns correspond to the predicted classes.

No single performance criterion can be labeled as the best for comparing algorithms or models because the best criterion to use is problem dependent. Here, we take the sensitivity as defined above, and create a single performance measure, the mean squared sensitivity error (MSSE). We define the MSSE as follows:

Definition 2.

$$\text{MSSE} = \frac{1}{N_c} \sum_{i=1}^{N_c} (1 - S_i)^2 \qquad (14.6)$$

where N_c = the number of classes and S_i = sensitivity of class i as defined earlier.

Sensitivities range from 0 (no examples of the class correctly classified) to 1 (all examples correctly classified). Thus, a lower MSSE corresponds to better performance. We choose this criterion because each class is given equal importance and the square causes lower individual sensitivities to be penalized more (e.g. for a two class problem, class sensitivities of 100% and 0% produce a higher MSSE than sensitivities of 50% and 50%). Note that this is only one possible criterion, and other criterion could be used in order to reflect different requirements, e.g. specific misclassification costs for each class. The post scaling heuristic can be used with any criterion (and doing so may be simpler than reformulating the neural network training algorithm for the new criterion).

14.3.2 ECG Classification Problem

This section presents results using the beforementioned techniques on an ECG classification problem. The database used is the MIT-BIH Arrhythmia database [21] – a common publicly available ECG database which contains a large number of ECG records that have been carefully annotated by experts. Detection of the following four beat types is considered: Normal (N), Premature Ventricular Contraction (PVC), Supraventricular Contraction (S), and Fusion (F) [21], i.e. there are four output classes. The four classes are denoted 1 (N), 2 (PVC), 3 (S), and 4 (F). An autoregressive model is calculated for a window of 200 samples centered over the peak of the R-wave of each beat. The inputs are the polar coordinates of each pole in the z-plane, i.e. frequency changes are reflected in the angular variation of the poles and damping is reflected in the magnitude variations. The model order was four corresponding to eight input variables. The prior probability of the classes (according to the training data) is (0.737, 0.191, 0.0529, 0.0196) corresponding to beat types (N, PVC, S, F).

MLPs with 20 hidden units were trained with stochastic backpropagation (update after each pattern) using an initial learning rate of 0.02 which was linearly reduced to zero over the training period of 500,000 updates. We used 5,000 points in each of the training, validation and test sets. The validation set was used for early stopping. The following algorithms were used – a) prior scaling with the degree of scaling, c_s, varied from 0 to 1, b) probabilistic sampling with the degree of scaling, c_s, varied from 0 to 1, c) as per a) and b) with the addition of post scaling, and d) equalizing the number of cases in each class by removing cases in more common classes. The post scaling attempted to minimize the MSSE on the training set[5]. 10 trials were performed for each case.

The median test set MSSE for d) was 0.195. The results for probabilistic sampling and probabilistic sampling plus post scaling are shown with box-whiskers plots[6] in figure 14.1. For probabilistic sampling, the best scaling results correspond to a degree of scaling in between no scaling and scaling according to the prior probabilities ($c_s \approx 0.8$). When c_s is larger, the sensitivity of class 1 drops significantly and results in higher false positive rates for the other classes. When c_s is lower, the sensitivity of classes 3 and 4 drops significantly. It can be seen that the addition of post scaling appears to almost always improve performance for this problem. The optimal degree of scaling, $c_s \approx 0.8$, is difficult to determine *a priori*. However, it can be seen that the addition of post scaling makes the selection of c_s far less critical ($c_s = 0.3$ to $c_s = 1.0$ result in similar per-

---

[5] 400 steps were used for the hill climbing algorithm where each step corresponded to either multiplying or dividing an individual output scale factor by a constant which was reduced linearly over time from 1.5 to 1. The time taken was short compared to the overall training time.

[6] The distribution of results is often not Gaussian and alternative means of presenting results other than the mean and standard deviation can be more informative. Box-whiskers plots show the interquartile range (IQR) with a box and the median as a bar across the box. The whiskers extend from the ends of the box to the minimum and maximum values. The median and the IQR are simple statistics which are not as sensitive to outliers as the mean and the standard deviation [31]. The median is the value in the middle when arranging the distribution in order from the smallest to the largest value. If the data is divided into two equal groups about the median, then the IQR is the difference between the medians of these groups. The IQR contains 50% of the points.

formance). Figure 14.2 shows confusion matrices (in graphical form). Without scaling ($c_s = 0$), it can be seen that classes 3 & 4 have low sensitivity. With scaling using $c_s = 1$ all classes are now recognized, however the sensitivity of class 1 is worse and the false positive rate of classes 3 & 4 is significantly worse.

The results for prior scaling and prior scaling combined with post scaling were very similar but slightly worse than the results with probabilistic sampling. The prior scaling results are not plotted in order to make the graph easier to follow, however the qualitative results are as follows: for low c_s, prior scaling and probabilistic sampling perform very similarly. However, for high c_s, probabilistic sampling has a clear advantage for this problem. This is perhaps just as expected – the relatively high variation in prior class probabilities leads to a high variation in weight update magnitudes across the classes when using high c_s. Results for all methods can be seen in table 14.2.

Method	Prior Scaling	Prior Scaling + Post Scaling	Probabilistic Sampling	Probabilistic Sampling + Post Scaling	Equalizing Membership
Average MSSE (for best c_s)	0.10 ($c_s = 0.8$)	0.096 ($c_s = 0.6$)	0.099 ($c_s = 0.8$)	0.089 ($c_s = 0.3$)	0.195
Average MSSE (over all c_s)	0.19	0.10	0.18	0.099	0.195

Table 14.2. Results for the various methods. We show the average results for the best selection of c_s and also an average across all selections of c_s. Note that selection of the optimal value of c_s is less critical when using post scaling in addition to either the prior scaling or probabilistic sampling methods.

14.4 Explanation

This section discusses why the techniques presented can be useful, limitations of the techniques, and how they relate to the theoretical result that MLPs approximate Bayesian *a posteriori* probabilities under certain conditions.

14.4.1 Convergence and Representation Issues

We first list four possible situations:

1. The proofs regarding estimation of Bayesian *a posteriori* probabilities assume networks with an infinite number of hidden nodes in order to obtain accurate approximation. For a given problem, it can be seen that a network which is too small will be unable to estimate the probabilities accurately due to limited resources.

14. Neural Network Classification and Prior Class Probabilities 307

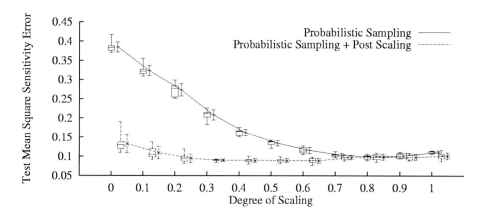

Fig. 14.1. Box-whiskers plots (on the left in each case) along with the usual mean plus and minus one standard deviation plots (on the right in each case) showing the test set MSSE for probabilistic sampling and for probabilistic sampling plus post scaling. Each result is derived from 10 trials with different starting conditions. The probabilistic sampling plus post scaling case is offset by 0.03 to aid viewing. It can be seen that the selection of the scaling degree for the best performance is not as critical when using the combination of probabilistic sampling and post scaling.

2. Training an MLP is NP-complete in general and it is well known that practical training algorithms used for MLPs often result in sub-optimal solutions (e.g. due to local minima). Often, a result of attaining a sub-optimal solution is that not all of the network resources are efficiently used. Experiments with a controlled task have indicated that the sub-optimal solutions often have smaller weights on average [17].

3. Weight decay [16] or weight elimination [30] are often used in MLP training and aim to minimize a cost function which penalizes large weights. These techniques tend to result in networks with smaller weights.

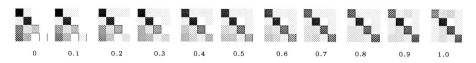

Fig. 14.2. Confusion matrices for the test set as the degree of prior scaling, c_s, is varied from 0 (left) to 1 (right). The columns correspond to the predicted classes and the rows correspond to the desired classes. The classes are (left to right and top to bottom) N, PVC, S, F. For each desired class, the predicted classes are shaded in proportion to the number of examples which are labeled as the desired class. White indicates no predictions. A general trend can be observed where classes S & F are recognized as normal when $c_s = 0$, and progressively more of the normal class examples are recognized as classes PVC, S, & F as c_s approaches 1.

4. A commonly recommended technique with MLP classification is to set the training targets away from the bounds of the activation function (e.g. (-0.8, 0.8) instead of (-1, 1) for the tanh activation function) [14].

These four situations can all lead to a bias towards smaller weights, or "smoother" models[7]. The possibility of such a bias is not taken into account by the proofs regarding posterior probabilities, i.e. the difference between theory and practice may, in part, be explained by violation of the assumption that sufficient convergence is obtained.

When a network is biased towards a "smoother" solution, and accurate fitting of the optimal function is not possible, the result may be a tendency to "ignore" lower frequency classes[8], e.g. if a network has the choice of fitting either a high frequency class or a low frequency class then it can provide a lower MSE by fitting the high frequency class[9]. We demonstrate by example.

We generated artificial training data using the following distributions: class 1: $N(-5, 1, 2) + N(0, 1, 2) + N(5, 1, 2)$, class 2: $N(-2.5, 0.25, 0.5) + N(2.5, 0.25, 0.5)$, where $N(\mu, \sigma, x)$ is a normal distribution with mean μ, standard deviation σ, and is truncated to lie within $(\mu - x, \mu + x)$. We generated 500 training and test examples from these distributions with the probability of selection for classes (1,2) being (0.9,0.1), i.e. the training and test sets have nine times as many samples of class 1 as they do of class 2. Note that there is no overlap between the classes. Figure 14.3 shows typical output probability plots for training an MLP with 10 hidden nodes[10] with and without probabilistic sampling. 10 trials were performed in each case with very similar results (see table 14.3). It can be seen that the network "ignores" class two without the use of probabilistic sampling.

It should be noted that using conjugate gradient training for this simple problem results in relatively accurate estimation of both classes with standard training (alternate parameters with backpropagation may also be successful). Rather than arguing for either backpropagation or conjugate gradient here (neither training algorithm is expected to always find a global minimum in general), we simply note that our experience and the experience of others [7, 18, 19, 27] suggests that conjugate gradient is not superior for many problems – i.e. backpropagation works better on one class of problems and conjugate gradient works

---

[7] In general, smaller weights correspond to smoother functions, however this is not always true. For example, this is not the case when fitting the function $sech(x)$ using two tanh sigmoids [8] (because $sech(x) = \lim_{d \to 0}(\tanh(x+d) - \tanh(x))/d$, i.e. the weights become indefinitely large as the approximation improves).

[8] In relation to the representational capacity (size of the network), Barnard and Botha [3] have observed that MLP networks have a tendency to guess higher probability classes when a network is too small to approximate the decision boundaries reasonably well.

[9] Lyon and Yaeger [20] find that their frequency balancing technique reduces the effect of the prior class probabilities on the network and effectively forces the network to allocate more resources to the lower frequency classes.

[10] 500,000 stochastic training updates with backpropagation, initial learning rate 0.02 reduced linearly to zero.

better on another class. Conjugate gradient resulted in significantly worse performance when tested on the ECG problem. It should be noted that there are many options when implementing a conjugate gradient training algorithm and that poor performance may be attributed to the implementation used. We have used a modified implementation of the algorithm from Fletcher [9].

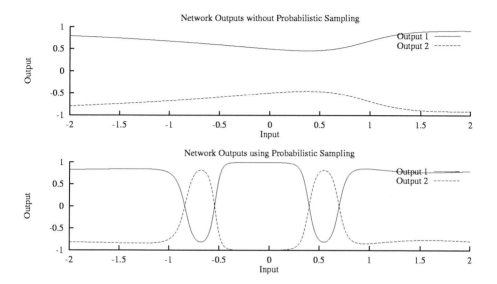

Fig. 14.3. Network outputs for the artificial problem with (below) and without (above) probabilistic sampling. It can be seen that the network "ignores" the lower frequency class without the use of probabilistic sampling. Note that the input has been normalized.

Classification Error	Mean	Standard Deviation
Standard Training	11.4	0.02
With Probabilistic Sampling	0.8	0.004

Table 14.3. Mean and standard deviation of the classification error for the artificial problem both with and without the use of probabilistic sampling.

14.4.2 Overlapping Distributions

Consider figure 14.4. If classes 1 and 2 have distributions differing only by translation (c_1 and c'_2) then the decision threshold between these classes should be chosen at x_1. Equal percentages of each of these classes will be classified as the other class. Now, if the distribution of class 2 is as shown (c_2) then the decision threshold between the classes should be chosen at x_2. In this case, a higher percentage of class 2 will be classified as class 1 than the reverse. If it is desirable to maximize the class by class sensitivity then scaling such that the effective distribution of c_2 is c_2' might be appropriate. Similarly, class 3 (c_3) will be "ignored" without any scaling.

Scaling on a class by class basis may be desired when i) the distribution of samples in the training set does not match the true distribution (e.g. it may be more expensive to collect samples of a particular class)[11], or ii) the distribution of the classes does not represent their relative importance, e.g. in a medical classification problem the cost of misclassifying a diseased case as normal may be much higher than the cost of classifying a normal case as a (possibly) diseased case [24]. The importance of each class may be independent of the class prior probabilities. Note that scaling such that lower frequency classes are made to be artificially more important can be useful when considering a higher level problem. For example, the training data from natural English words and phrases exhibit very non-uniform priors for different characters. Yaeger et al. [33] find that reducing the effect of these priors on the network using frequency balancing improves the performance of the higher level word recognition training.

Observations. a) There is no intrinsic problem if the distributions do not overlap. b) When distributions overlap, it is desirable to preprocess the data in a manner that results in reduced overlap. However, it is often not possible to obtain zero overlap (due to noise, for example).

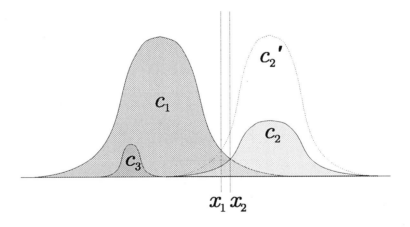

Fig. 14.4. Overlapping distributions.

---

[11] It may be possible to obtain more accurate estimates of class probabilities using data that has class labels without input information. For example, word frequency information can be obtained from text databases and the frequency of various diseases can be obtained from health statistics [23].

14.4.3 Limitations

We note a couple of limitations with the heuristics considered herein:

1. *Local issues.* The heuristics presented counteract biases in the network, training algorithm and/or training data. There is no reason for these biases to be constant throughout the input space, e.g. scaling may be helpful in one region but detrimental in another.
2. *Nonlinear calibration.* There is no reason for the linear scaling heuristics used here to be optimal (in the sense that they best counteract the biases).

14.4.4 *A Posteriori* Proofs

Theoretically it is possible to show that the scaling techniques invalidate the *a posteriori* proofs – when performing scaling on a class by class basis the decision thresholds which are used to determine the winning class should be altered accordingly. This indicates another possible use of the prior scaling and probabilistic sampling techniques when the conditions given above do not exist. This use is related to the problem whereby lower frequency classes may be estimated less accurately than higher frequency classes (see section 14.2.3) – training may be performed with the heuristically altered problem (e.g. so that the class frequencies are effectively equal) and the outputs or decision thresholds can be altered accordingly.

14.5 Conclusions

In practice, training issues or characteristics of a given classification problem can mean that scaling the predicted class probabilities may improve performance in terms of overall classification error and/or in terms of an alternative criterion. We introduced algorithms which a) scale weight updates on a class by class basis according to the prior class probabilities, b) alter class frequencies probabilistically (very similar to the frequency balancing technique of Yaeger et al. [33]), and c) scale outputs after training in order to maximize a given performance criterion. For an electrocardiogram (ECG) classification problem, we found that the prior scaling, probabilistic sampling, and post scaling techniques provided better performance in comparison to a) no heuristics, and b) subsampling in order to equalize the number of cases in each class. The best performance for prior scaling and probabilistic sampling was obtained with a degree of scaling in between no scaling and scaling according to the prior probabilities. The optimal degree was difficult to determine *a priori*. However, it was found that using prior scaling or probabilistic sampling in combination with post scaling made the selection of the optimal degree far less critical.

References

1. AAMI. Testing and reporting performance results of ventricular Arrhythmia detection algorithms. Association for the Advancement of Medical Instrumentation, Arlington, VA, 1987. ECAR-1987.
2. R. Anand, K. G. Mehrotra, C. K. Mohan, and S. Ranka. An improved algorithm for neural network classification of imbalanced training sets. *IEEE Transactions on Neural Networks*, 4(6):962–969, November 1993.
3. E. Barnard and E. C. Botha. Back-propagation uses prior information efficiently. *IEEE Transactions on Neural Networks*, 4(5):794–802, September 1993.
4. E. Barnard and D. Casasent. A comparison between criterion functions for linear classifiers, with an application to neural nets. *IEEE Transactions on Systems, Man, and Cybernetics*, 19(5):1030–1041, 1989.
5. E. Barnard, R.A. Cole, and L. Hou. Location and classification of plosive constants using expert knowledge and neural-net classifiers. *Journal of the Acoustical Society of America*, 84 Supp 1:S60, 1988.
6. H.A. Bourlard and N. Morgan. Links between Markov models and multilayer perceptrons. In D. S. Touretzky, editor, *Advances in Neural Information Processing Systems 1*, volume 1, pages 502–510. Morgan Kaufmann, San Mateo, CA, 1989.
7. H.A. Bourlard and N. Morgan. *Connnectionist Speech Recognition: A Hybrid Approach*. Kluwer Academic Publishers, Boston, MA, 1994.
8. N. Scott Cardell, Wayne Joerding, and Ying Li. Why some feedforward networks cannot learn some polynomials. *Neural Computation*, 6(4):761–766, 1994.
9. R. Fletcher. *Practical Methods of Optimization, Second Edition*. John Wiley & Sons, 1987.
10. S. Geman, E. Bienenstock, and R. Doursat. Neural networks and the bias/variance dilemma. *Neural Computation*, 4(1):1–58, 1992.
11. H. Gish. A probabilistic approach to the understanding and training of neural network classifiers. In *Proceedings of the IEEE Conference on Acoustics, Speech and Signal Processing*, pages 1361–1364. IEEE Press, 1990.
12. J.B. Hampshire and B. Pearlmutter. Equivalence proofs for multilayer perceptron classifiers and the Bayesian discriminant function. In D. S. Touretzky, J. L. Elman, T. J. Sejnowski, and G. E. Hinton, editors, *Proceedings of the 1990 Connectionist Models Summer School*. Morgan Kaufmann, San Mateo, CA, 1990.
13. J.B. Hampshire and A. H. Waibel. A novel objective function for improved phoneme recognition using time delay neural networks. In *International Joint Conference on Neural Networks*, pages 235–241, Washington, DC, June 1989.
14. S. Haykin. *Neural Networks, A Comprehensive Foundation*. Macmillan, New York, NY, 1994.
15. F. Kanaya and S. Miyake. Bayes statistical behavior and valid generalization of pattern classifying neural networks. *IEEE Transactions on Neural Networks*, 2(1):471, 1991.
16. A. Krogh and J.A. Hertz. A simple weight decay can improve generalization. In J.E. Moody, S. J. Hanson, and R. P. Lippmann, editors, *Advances in Neural Information Processing Systems*, volume 4, pages 950–957. Morgan Kaufmann, San Mateo, CA, 1992.
17. S. Lawrence, C. Lee Giles, and A.C. Tsoi. Lessons in neural network training: Overfitting may be harder than expected. In *Proceedings of the Fourteenth National Conference on Artificial Intelligence, AAAI-97*, pages 540–545. AAAI Press, Menlo Park, California, 1997.

18. Y. Le Cun. Efficient learning and second order methods. Tutorial presented at Neural Information Processing Systems 5, 1993.
19. Y. Le Cun and Y. Bengio. Pattern recognition. In Michael A. Arbib, editor, *The Handbook of Brain Theory and Neural Networks*, pages 711–715. MIT Press, Cambridge, Massachusetts, 1995.
20. R. Lyon and L. Yaeger. On-line hand-printing recognition with neural networks. In *Fifth International Conference on Microelectronics for Neural Networks and Fuzzy Systems*, Lausanne, Switzerland, 1996. IEEE Computer Society Press.
21. MIT-BIH. MIT-BIH Arrhythmia database directory. Technical Report BMEC TR010 (Revised), Massachusetts Institute of Technology and Beth Israel Hospital, 1988.
22. A. F. Murray and P. J. Edwards. Enhanced MLP performance and fault tolerance resulting from synaptic weight noise during training. *IEEE Transactions on Neural Networks*, 5(5):792–802, 1994.
23. M.D. Richard and R.P. Lippmann. Neural network classifiers estimate Bayesian *a posteriori* probabilities. *Neural Computation*, 3(4):461–483, 1991.
24. B.D. Ripley. *Pattern Recognition and Neural Networks*. Cambridge University Press, Cambridge, UK, 1996.
25. R. Rojas. A short proof of the posterior probability property of classifier neural networks. *Neural Computation*, 8:41–43, 1996.
26. D.W. Ruck, S.K. Rogers, K. Kabrisky, M.E. Oxley, and B.W. Suter. The multilayer perceptron as an approximation to an optimal Bayes estimator. *IEEE Transactions on Neural Networks*, 1(4):296–298, 1990.
27. W. Schiffman, M. Joost, and R. Werner. Optimization of the backpropagation algorithm for training multilayer perceptrons. Technical report, University of Koblenz, 1994.
28. P.A. Shoemaker. A note on least-squares learning procedures and classification by neural network models. *IEEE Transactions on Neural Networks*, 2(1):158–160, 1991.
29. E. Wan. Neural network classification: A Bayesian interpretation. *IEEE Transactions on Neural Networks*, 1(4):303–305, 1990.
30. A.S. Weigend, D.E. Rumelhart, and B.A. Huberman. Generalization by weight-elimination with application to forecasting. In R. P. Lippmann, J.E. Moody, and D. S. Touretzky, editors, *Advances in Neural Information Processing Systems*, volume 3, pages 875–882. Morgan Kaufmann, San Mateo, CA, 1991.
31. N.A. Weiss and M.J. Hassett. *Introductory Statistics*. Addison-Wesley, Reading, Massachusetts, second edition, 1987.
32. H. White. Learning in artificial neural networks: A statistical perspective. *Neural Computation*, 1(4):425–464, 1989.
33. L. Yaeger, R. Lyon, and B. Webb. Effective training of a neural network character classifier for word recognition. In M.C. Mozer, M.I. Jordan, and T. Petsche, editors, *Advances in Neural Information Processing Systems 9*, Cambridge, MA, 1997. MIT Press.

15
Applying Divide and Conquer to Large Scale Pattern Recognition Tasks

Jürgen Fritsch and Michael Finke

Interactive Systems Laboratories

University of Karlsruhe	Carnegie Mellon University
Am Fasanengarten 5	5000 Forbes Avenue
76128 Karlsruhe, Germany	Pittsburgh, PA 15213, USA
fritsch@ira.uka.de	finkem@cs.cmu.edu
	http://www.cs.cmu.edu/ finkem/

Abstract. Rather than presenting a specific trick, this paper aims at providing a methodology for large scale, real-world classification tasks involving thousands of classes and millions of training patterns. Such problems arise in speech recognition, handwriting recognition and speaker or writer identification, just to name a few. Given the typically very large number of classes to be distinguished, many approaches focus on parametric methods to independently estimate class conditional likelihoods. In contrast, we demonstrate how the principles of modularity and hierarchy can be applied to directly estimate posterior class probabilities in a connectionist framework. Apart from offering better discrimination capability, we argue that a hierarchical classification scheme is crucial in tackling the above mentioned problems. Furthermore, we discuss training issues that have to be addressed when an almost infinite amount of training data is available.

15.1 Introduction

The majority of contributions in the field of neural computation deal with relatively small datasets and, in case of classification tasks, with a relatively small number of classes to be distinguished. Representatives of such problems include the UCI machine learning database [16] and the Proben [20] benchmark set for learning algorithms. Research concentrates on aspects such as missing data, model selection, regularization, overfitting vs. generalization and the bias/variance trade-off. Over the years, many methods and 'tricks' have been developed to optimally learn and generalize when only a limited amount of data is available.

On the other hand, many problems in human computer interaction (HCI) such as speech and handwriting recognition, lipreading and speaker and writer identification require comparably large training databases and also often exhibit a large number of classes to be discriminated, such as (context-dependent) phones, letters and individual speakers or writers. For example, in state-of-the-art large vocabulary continuous speech recognition, we are typically faced with an inventory of several thousand basic acoustic units and training databases consisting of several millions of preprocessed speech patterns. There is only a limited amount of publications available on the sometimes very different problems concerning the choice of learning machines and training algorithms for such tasks and datasets.

This article addresses exactly the latter kind of learning tasks and provides a principled approach to large scale classification problems, exemplifying it on the problem of connectionist speech recognition. Our approach is grounded on the powerful *divide and conquer* paradigm that traditionally has always been applied to problems of rather large size. We argue that a hierarchical approach that modularizes classification tasks is crucial in applying statistical estimators such as artificial neural networks. In that respect, this paper presents not just a single 'trick of the trade', it offers a methodology for large scale classification tasks. Such tasks have traditionally been addressed by building generative models rather than focusing on the prediction of posteriors without making strong assumptions on the distribution of the input.

The remainder of the paper is organized as follows. Section 2 presents the general approach to soft hierarchical classification. Section 3 then discusses methods to design the topology of hierarchical classifiers - a task that is of increasing importance when dealing with large numbers of classes. Finally, section 4 demonstrates in detail the application of hierarchical classification to connectionist statistical speech recognition. Section 5 concludes this paper with a summary.

15.2 Hierarchical Classification

Consider the task of classifying patterns $\mathbf{x}$ as belonging to one of N classes ω_k. Given that we have access to the class conditional probability densities $p(\mathbf{x}|\omega_k)$, Bayes theory states that the optimal decision should be based on the a-posteriori probabilities

$$p(\omega_k|\mathbf{x}) = \frac{p(\mathbf{x}|\omega_k)p(\omega_k)}{\sum_i p(\mathbf{x}|\omega_i)p(\omega_i)}.$$

Given that equal risks are associated with all possible misclassifications, the optimal decision is to choose the class with maximum a-posteriori probability given a specific pattern $\mathbf{x}$. Two distinct approaches have to be considered when applying Bayes theory to a learning from examples task with generally unknown distributions. In the first approach, one tries to estimate class-conditional likelihoods $p(\mathbf{x}|\omega_k)$ and prior probabilities $p(\omega_k)$ from a labeled dataset which are then used to calculate posterior probabilities according to Bayes rule. In principle, this approach can be applied to tasks with an arbitrary large number of classes since the class-conditional likelihoods can be estimated independently. However, such an approach focuses on the modeling of the class-conditional densities. For classification accuracy however, it is more important to model class boundaries.

The second approach accommodates this perspective by directly estimating posterior class probabilities from datasets. It was shown (e. g. [6]) that a large class of artificial neural networks such as multi-layer perceptrons and recurrent neural networks can be trained to approximate posterior class probabilities. The degree of accuracy of the approximation however depends on many factors, among them the plasticity of the network. Comparing the two approaches, the discriminative power of methods that estimate posterior probabilities directly is generally higher, resulting in better classification accuracy especially when the class-conditional distributions are very complex. This fact (among others) explains the success and popularity of neural network classifiers on many learning from examples tasks.

However, when the number of classes to be distinguished increases to say several thousand, neural network estimators of posterior probabilities fail to provide good approximations mainly because of two reasons: First, real-world problems involving such a large number of classes often exhibit an extremely non-uniform distribution of priors, see chapter 14. Many learning algorithms for neural networks (especially stochastic on-line gradient descent) have difficulties with non-uniformly distributed classes. Particularly the distribution of posteriors of infrequent classes tend to be approximated poorly. Second, and more important, one of the prerequisites for training neural networks to estimate posteriors, the 1-out-of-N coding of training targets, implies that the number of output neurons matches the number of classes. It is unfeasible to train a neural network with thousands of output neurons. Also, with increasing number of classes, the complexity of the optimum discriminant functions also increases and the potential for conflicts between classes grows. Thus, from our point of view, typical monolithic neural network classifiers are not applicable because of their limitation to tasks with relatively few classes.

15.2.1 Decomposition of Posterior Probabilities

Applying the principle of *divide and conquer*, we can break down the task of discriminating between thousands of classes into a hierarchical structure of many smaller classification tasks of controlled size. This idea underlies the approaches to decision tree architectures [5, 21, 23]. Decision trees classify input patterns by asking categorical questions at each internal node. Depending on the answer to these questions a single path is followed to one of the child nodes and the process repeats until a leaf node is reached and a (winner) class label is emitted. Therefore, decision tree classifiers can only supply us with hard decisions. No information about the confusability of a specific input pattern is given to us. Rather, we are often interested in the posterior class probabilities because we wish to have a measure of the ambiguity of a decision. Furthermore, we are sometimes required to feed a measure of the degree of membership for all potential classes into a superior decision making process. As we will see in section 4, statistical speech recognition is a typical example for the latter scenario.

Adhering to the *divide and conquer* approach but generalizing the decision tree framework, the statistical method of factoring posteriors can be applied

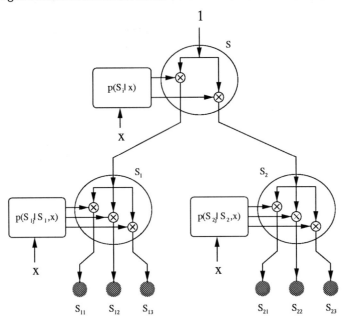

Fig. 15.1. Hierarchical decomposition of posteriors

to design *soft* classification trees [24, 25]. For now, we assume, that optimal posterior probabilities are available. Let S be a (possibly large) set of classes ω_k to be discriminated. Consider we have a method at our disposition which gives us a partitioning of S into M disjoint and non-empty subsets S_i such that members of S_i are almost never confused with members of S_j ($\forall j \neq i$). A particular class ω_k will now be a member of S and exactly one of the subsets S_i. Therefore, we can rewrite the posterior probability of class ω_k as a joint probability of the class and the corresponding subset S_i and factor it according to

$$p(\omega_k|\mathbf{x}) = p(\omega_k, S_i|\mathbf{x}) \quad \text{with} \quad \omega_k \in S_i$$
$$= p(S_i|\mathbf{x})\, p(\omega_k|S_i, \mathbf{x}).$$

Thus, the global task of discriminating between all the classes in S has been converted into (1) discriminating between subsets S_i and (2) independently discriminating between the classes ω_k remaining within each of the subsets S_i. Recursively repeating this process yields a hierarchical tree-organized structure (Fig. 15.1).

Note, that the number of subclasses S_i of each node does not need to be constant throughout the classifier tree and might be subject to optimization during the tree design phase. In order to compute the posterior probability for a specific class, we have to follow the path from root node to the leaf corresponding to the class in question, multiplying all the conditional posteriors along the way. Both the design of the tree structure (*divide*) and the estimation and multiplication (*conquer*) of conditional posteriors at each node are important aspects in this architecture, that have to be considered thoroughly because in practice, only approximations to the conditional posteriors are available.

15.2.2 Hierarchical Interpretation

The presented architecture can be interpreted as a probability mass distribution device. At the root node, an initial probability mass of 1 is fed into the architecture. At each node, the incoming probability mass is multiplied by the respective conditional posterior probabilities and fed into the child nodes. Eventually, the probability mass is distributed among all the leaves (classes) rendering their posterior probabilities. In contrast, classifier trees are mostly used as hard-switching devices, where only a single path from root node to one of the leaves is taken.

A hierarchical decomposition of posterior probabilities through a soft classification tree offers several advantages. If one of the nodes in the tree, for example the root node fails to provide good estimates of conditional posteriors, a hard decision tree will produce many classification errors. In a soft classification tree, such shortcomings will influence the decision process less dramatically. Also, recovery from errors is often possible through a superior decision process.

Another aspect of soft classification trees that can be exploited for various purposes is the sum-to-unity property observable in any horizontal cross-section at any level of the tree. The tree can be cut off at a certain level and still be used as a soft classification tree that computes posterior class probabilities. This is equivalent to creating a new (smaller) set of classes by clustering and merging the original classes according to the tree topology. In general, the resulting classification task will be easier to solve than the original one.

Related to the sum-to-unity property of cross-sections is that the partial posteriors computed on a path from the root node to a leaf are decreasing monotonically. This in turn allows to close paths whenever a suitable threshold is reached, pruning whole subtrees with classes that would otherwise receive posteriors smaller than the threshold. This property yields the possibility to smoothly trade off classification accuracy against computational complexity. In the limit, when only a single path with highest conditional posterior is followed, the soft classification tree transmutes into a hard decision tree.

15.2.3 Estimation of Conditional Node Posteriors

Given a hierarchical decomposition of posterior class probabilities, it remains to instantiate the tree nodes with estimators for the required conditional posteriors. Conditioning a posterior on a subset of classes S_i can be accomplished by restricting the training set of the corresponding learning device to the patterns with a class label from S_i. According to this setting, the available training data in each node is distributed among all its child nodes according to the class partitioning. While the root node receives all available training data, nodes further down the tree receive less data than their predecessors. On the other hand, specialization increases from root node to leaves. This fact has important consequences on learning speed and model selection when training whole hierarchies.

One of the important issues in hierarchical decompositions of posterior probabilities are the unavoidable inaccuracies of practical estimators for the conditional posteriors that have to be provided in each tree node. Neural networks can only be trained to *approximate* the true distribution of posterior class probabilities and the degree of accuracy depends on both the inherent difficulty of the task as given by the training set and the network structure and training schedule being used. Inaccurate approximations to the true distribution of posteriors hurt most in the upper layers of a classification tree - a fact that has to be taken into account by tree design procedures, which we will discuss next.

15.3 Classifier Tree Design

When it comes to the design of soft classifier trees, or equivalently to the design of hierarchical decompositions of class posteriors, the choice of algorithm depends mostly on the number of initial classes. We will first discuss optimal tree structures before we will turn to heuristic design algorithms necessary when dealing with the large number of classes that we have to deal with.

15.3.1 Optimality

The optimal soft classification tree for a given task and given type and structure of estimators for the conditional node posteriors is the one which results in minimum classification error in the Bayes setting. If all the node classifiers would compute the true conditional posteriors, the tree structure would have no influence on the classifier performance because any kind of factoring (through any kind of tree structure) yields an *exact* decomposition of the class posteriors. However, in practice, approximation errors of node classifiers render the choice of tree structure an important issue. For small numbers of classes, the optimal tree can in principle be found by exhaustively training and testing all possible partitionings for a particular node (starting with the root node) and chosing the one that gives the highest recognition accuracy. However, even if restricting the tree structure to binary branching nodes and balanced partitionings, the number K of partitionings that have to be examined at the root node

$$K = \frac{N!}{(\frac{N}{2}!)^2}$$

quickly brings this algorithm to its limits, even for a moderate number of classes N. Therefore, we have to consider heuristics to derive near optimal tree structures. For example, one valid possibility is to assume that the accuracy of achievable approximations to the true posteriors is related to the separability of the corresponding sets of classes.

15.3.2 Prior Knowledge

Following the above mentioned guideline, prior knowledge about the task in question can often be applied to hierarchically partition the global set of classes into reasonable subsets. The goal is to partition the remaining set of classes in a way that intuitively maximizes the separability of the subsets. For example, given a set of phones in a speech recognizer, a reasonable first partitioning would be to build subsets consisting of voiced and unvoiced phones. In larger speech recognition systems where we have to distinguish among multi-state context-dependent phones, prior knowledge such as state and context identity can be used as splitting criterion. In tasks such as speaker or writer identification, features such as gender or age are potential candidates for splitting criteria.

The advantage of such knowledge driven decompositions is a fast tree design phase which is a clear superiority of this approach when dealing with large numbers of classes. However, this method for the design of hierarchical classifiers is subjective and error prone. Two experts in a specific field might disagree strongly on what constitutes a reasonable hierarchy. Furthermore, it is not always the case that *reasonable* partitionings yield good separability of subsets. Expert knowledge can be misleading.

15.3.3 Confusion Matrices

In case the number of classes is small enough to allow the training of a single classifier, the design of a soft classifier tree can be based on the confusion matrix of the trained monolithic classifier. Indicating the confusability of each pair of classes, the confusion matrix yields relatively good measures of the separability of pairs of classes. This information can be exploited for designing a tree structure using a clustering algorithm. For instance, we can define the following (symmetric) distance measure between two disjunct sets of classes S_k and S_l

$$d(S_k, S_l) = - \sum_{\omega_i \in S_k} \sum_{\omega_j \in S_l} C(\omega_i, \omega_j | \mathcal{T}) + C(\omega_j, \omega_i | \mathcal{T})$$

where $C(\omega_i, \omega_j | \mathcal{T})$ denotes the number of times class ω_i is confused with class ω_j as measured on a set of labeled patterns $\mathcal{T}$. The distance $d(S_k, S_l)$ can now be used as a replacement for the usual Euclidean distance measure in a standard bottom-up clustering algorithm. Unfortunately, once the number of classes increases to several thousand, training of a monolithic classifier becomes increasingly difficult.

15.3.4 Agglomerative Clustering

Assuming that separability of classes correlates with approximation accuracy of estimators for the posterior class probabilities, we can go further and assume that separability of classes can be measured by a suitable distance between the class conditional distributions in feature space. We already introduced such a distance measure in form of the elements of a class confusion matrix. Other, computationally less expensive distance measures would be the Euclidean distance between class means or the Mahalanobis distance between the classes second order statistics. Irrespective of the chosen distance measure, the goal always is to group the set of classes in a way that results in maximum inter- and minimum intra-group distances. Solutions to this problem are known as agglomerative clustering algorithms and a large pool of variations of the basic algorithm is available in the literature [7].

15.4 Application to Speech Recognition

In this section, we will demonstrate the main ideas and benefits of the hierarchical classifier approach on the task of large vocabulary continuous speech recognition (LVCSR). More specifically, we will focus on acoustic modeling for statistical speech recognition using hidden Markov models (HMM) [27]. To give an impression of the complexity of such a task: training databases typically consist of tens of millions of speech patterns, the number of acoustic classes being distinguished ranges from ca. 50 (monophones) to over 20000 (context-dependent polyphones).

15.4.1 Statistical Speech Recognition

The basic statistical entity in HMM based speech recognition is the posterior probability of word sequences $W_1, \ldots, W_N$ given a sequence of acoustic observations $\mathbf{X_1}, \ldots, \mathbf{X_M}$ and a set of model parameters Θ

$$P(W_1, \ldots, W_N | \mathbf{X_1}, \ldots, \mathbf{X_M}, \Theta)$$

During training, we are seeking parameters Θ that maximize this probability on the training data

$$\hat{\Theta} = \arg\max\nolimits_{\Theta} \prod_{t=1}^{T} P(W_1, \ldots, W_{N(t)} | \mathbf{X_{1}}, \ldots, \mathbf{X_{M(t)}}, \Theta)$$

and during recognition, we want to find the sequence of words that maximizes this probability for a given acoustic observation and fixed model parameters Θ

$$\hat{W}_1, \ldots, \hat{W}_N = \arg\max\nolimits_{W_1, \ldots, W_N} P(W_1, \ldots, W_N | \mathbf{X_1}, \ldots, \mathbf{X_M}, \Theta)$$

In order to simplify the process of maximizing the posterior probability of word sequences, Bayes rule is usually applied

$$P(W_1, \ldots, W_N | \mathbf{X_1}, \ldots, \mathbf{X_M}) = \frac{P(\mathbf{X_1}, \ldots, \mathbf{X_M} | W_1, \ldots, W_N) \, P(W_1, \ldots, W_N)}{P(\mathbf{X_1}, \ldots, \mathbf{X_M})}$$

This rule separates the estimation process into the so called *acoustic model (AM)* consisting of terms that depend on the acoustic observations $\mathbf{X_1}, \ldots, \mathbf{X_M}$ and the *language model (LM)* consisting of terms that depend only on the sequence of words $W_1, \ldots, W_N$. In this paper we will focus on acoustic modeling using connectionist estimators as a typical example of a task involving the discrimination of thousands of classes. For a review on other important aspects of LVCSR such as pronunciation modeling, language modeling and decoding algorithms we refer the reader to [27].

The task of acoustic modeling (ignoring the denominator) is to estimate parameters Θ^{AM} which maximize

$$P(\mathbf{X_1}, \ldots, \mathbf{X_M} | W_1, \ldots, W_N, \Theta^{AM}).$$

Words W_i are modeled as sequences (or graphs) of phone models. The mapping from words to phone models is usually accomplished by means of a pronunciation dictionary. Phone models in turn are usually modeled as m-state left-to-right hidden Markov models (HMM) to capture the temporal and acoustic variability of speech signals. The following figure shows the process of converting a sequence of words into (1) a pronunciation graph (possibly with pronunciation variants) and (2) an HMM state graph.

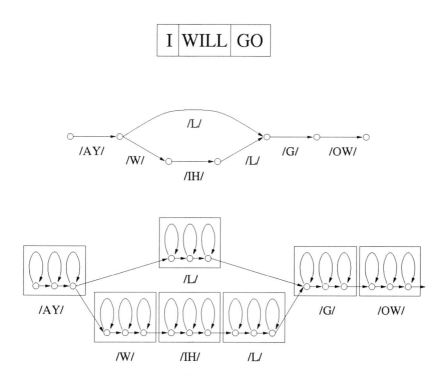

Fig. 15.2. Typical hidden Markov model in speech recognition

In this framework, where word sequences are represented as directed acyclic graphs of HMM states, the likelihood of an acoustic observation can be rewritten as

$$P(\mathbf{X_1},\ldots,\mathbf{X_M}|W_1,\ldots,W_N) = \sum_{s_1,\ldots,s_M} P(\mathbf{X_1},\ldots,\mathbf{X_M}|s_1,\ldots,s_M)\, p(s_1,\ldots,s_M)$$

where the summation extends over all possible state sequences $s_1,\ldots,s_M$ in the HMM model for the word sequence $W_1,\ldots,W_N$. In the Viterbi approximation, the above likelihood is approximated by the probability of the most likely state sequence

$$P(\mathbf{X_1},\ldots,\mathbf{X_M}|W_1,\ldots,W_N) \approx \max_{s_1,\ldots,s_M} P(\mathbf{X_1},\ldots,\mathbf{X_M}|s_1,\ldots,s_M)\, p(s_1,\ldots,s_M).$$

Given a specific state sequence, the likelihood of the acoustic observations given that sequence can be factored as follows

$$P(\mathbf{X_1},\ldots,\mathbf{X_M}|s_1,\ldots,s_M) \approx \prod_{i=1}^{M} p(\mathbf{X_i}|X_1,\ldots,X_{i-1},s_1,\ldots,s_M)\, p(s_1,\ldots,s_M).$$

In the application of first-order hidden Markov models to the estimation of such likelihoods one usually makes two simplifying assumptions:

- Independence of Observations:

$$P(\mathbf{X_1},\ldots,\mathbf{X_M}|s_1,\ldots,s_M) \approx \prod_{i=1}^{M} p(\mathbf{X_i}|s_1,\ldots,s_M)\, p(s_1,\ldots,s_M)$$

- First-order Assumption:

$$P(\mathbf{X_1},\ldots,\mathbf{X_M}|s_1,\ldots,s_M) \approx \prod_{i=1}^{M} p(\mathbf{X_i}|s_i)\, p(s_i|s_{i-1})$$

15.4.2 Emission and Transition Modeling

Mainstream LVCSR systems follow the above approach by modeling emission probability distributions $p(\mathbf{X_i}|s_i)$ and transition probabilities $p(s_i|s_{i-1})$ separately and independently. Emission probability distributions are usually modeled using mixture densities from the exponential family, such as the mixture of Gaussians

$$p(\mathbf{X_i}|s_i) = \sum_{k=1}^{n} \gamma_k N_k(\mathbf{X_i}|s_i)$$

where the γ_k denote *mixture coefficients* and the N_k *mixture component densities* (here: normal distributions). Transition probabilities on the other hand are modeled by simple multinomial probabilities since they are conditioned on a discrete variable only (not on the input vector).

The advantage of this approach is a decoupled estimation process that separates temporal and acoustic modeling. As such, it allows to easily vary HMM state topologies after training in order to modify temporal behavior. For instance, state duplication is a popular technique to increase the minimum duration constraint in phone models. Having separated emission and transition probability estimation, state duplication consists of simply sharing acoustic models among multiple states and adapting the transition probabilities.

However, the disadvantage of the above approach is a mismatch in the dynamic range of emission and transition probabilities. The reason is that transition probabilities are modeled separately as multinomial probabilities, constrained by the requirement to sum to one. This leads to a dominant role of emission probabilities with transition probabilities hardly influencing overall system performance.

15.4.3 Phonetic Context Modeling

So far we have assumed that only one HMM is required per modeled monophone (see Fig. 15.2). Since the English language can be modeled by approximately 45 monophones, one might get the impression that only that number of HMM models need to be trained. In practice however, one observes an effect called coarticulation that causes large variations in the way specific monophones actually sound, depending on their phonetic context.

Usually, explicit modeling of phones in phonetic context yields great gains in recognition accuracy. However, it is not immediately clear how to achieve robust context-dependent modeling. Consider, for example, so called *triphone* models. A triphone essentially represents the realization of a specific monophone in a specific context spanning one phone to the left and right. Assuming an inventory of 45 monophones, the number of (theoretically) possible triphones is $45^3 = 91125$. Many of these triphones will occur rarely or never in actual speech due to the linguistic constraints in the language. Using triphones therefore results in a system which has too many parameters to train. To avoid this problem, one has to introduce a mechanism for sharing parameters across different triphone models.

Typically, a CART like decision tree is adopted to cluster triphones into *generalized triphones* based on both their a-priori probability and their acoustic similarity. Such a top-down clustering requires the specification of viable attributes to be used as questions on phonetic context in order to split tree nodes. Mostly, linguistic classes such as vowel, consonant, fricative, plosive, etc. are being employed. Furthermore, one can generalize triphones to *polyphones* by allowing dependence on a wider context (and not just the immediate left and right neighboring phones). Fig. 15.3 shows a typical decision tree for the clustering of polyphonic variations of a particular monophone state.

The collection of all leaf nodes of decision trees for each monophone state in a given system represents a robust and general set of context-dependent sub-

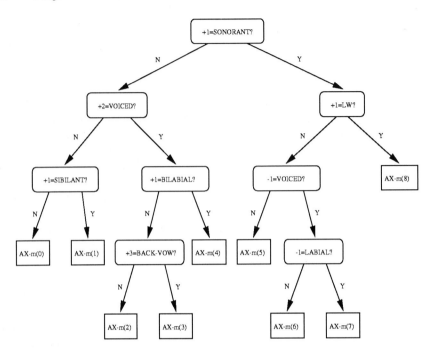

Fig. 15.3. Phonetic Context Modeling using Decision Trees. Shown is a decision tree modeling phonetic contexts of middle state (3-state HMM) of monophone /AX/.

phonetic units. Since each of these units corresponds to several triphone HMM states, they are often called *tied states*. Typically, a large vocabulary continuous speech recognizer models between 3000 and 24000 such tied states. Mainstream LVCSR systems scale to any number of context-dependent modeling units since emission and transition models are independently estimated for each tied state.

15.4.4 Connectionist Acoustic Modeling

Locally discriminant connectionist acoustic modeling is the most popular approach to integrate neural networks into an HMM framework [3, 4, 18]. It is based on converting estimates of local posterior class probabilities to scaled likelihoods using Bayes rule. These scaled likelihoods can then be used as observation probability estimates in standard HMMs. For a moderately small number N of HMM states, a neural network can be trained to jointly estimate posterior probabilities $p(s_i|\mathbf{X}_i)$ for each state s_i given an input vector $\mathbf{X}_i$. Bayes rule yields the corresponding scaled [1] class conditional likelihoods

---

[1] The missing additional term consisting of the probability of the input vector $p(\mathbf{X}_i)$ is usually omitted because it is independent of the class/state identity and therefore does not influence a Viterbi style search for the most likely state sequence.

$$\hat{p}(\mathbf{X_i}|s_i) = \frac{p(s_i|\mathbf{X_i})}{p(s_i)}.$$

While $p(s_i|\mathbf{X_i})$ is estimated using a neural network, prior probabilities $p(s_i)$ can be estimated by relative frequencies as observed in the training data. Several researchers (e. g. [3, 14]) have reported improvements with connectionist acoustic modeling when the technique for the estimation of emission probabilities was the only difference in comparison. Since mainstream HMMs for speech recognizers are mostly trained in a maximum likelihood framework using the Expectation-Maximization (EM) algorithm, incorporation of discriminatively trained neural networks that focus on modeling of class boundaries instead of class distributions is often observed to be beneficial. Also, compared to mixtures of Gaussians based acoustic models, connectionist acoustic models are often reported to achieve the same accuracy with far less parameters.

However, when the number of HMM states is increased to model context-dependent polyphones (triphones,quintphones), a single neural network can no longer be applied to estimate posteriors. It becomes necessary to factor the posterior state probabilities [17] and modularize the process of estimating those posteriors. In most approaches, the posteriors are factored on phonetic context or monophone identity (e.g. [4, 9, 15]). Viewing factoring as a hierarchical decomposition of posteriors, we generalized the approaches to context-dependent connectionist acoustic modeling by introducing a tree structured hierarchy of neural networks (HNN) [12, 13] corresponding to a multi-level factoring of posteriors based on a-priori knowledge such as broad sound classes (silence, noises, phones), phonetic context and HMM state identity. Fig. 15.4 shows the topology of such a structure.

At the top of this hierarchy, we discriminate silence, noise and speech sounds by means of two networks (SIL-Net, SPEECH-Net). The motivation for this specific partitioning comes from the observation that these three classes are easy to distinguish acoustically. The remainder of the tree structure decomposes the posterior of speech, conditioning on monophone, context and state identity as these are convenient sound classes modeled by any phone based HMM speech recognizer. The hierarchy of Fig. 15.4 can be decomposed even further, for instance by factoring conditional monophone posteriors (estimated by the MONO-Net) based on linguistic features (e.g. voiced/unvoiced, vowel/consonantal, fricative etc.). The motivation behind such a decomposition is twofold. First, it reduces the number of local classes in each node, improving approximation accuracy and second, it yields a decoupled and specialized set of expert networks having to handle a smaller amount of phonetic variation.

However, as mentioned in section 3, the use of prior knowledge for the design of a hierarchy of neural networks does not take into account dissimilarity of the observed classes in feature space. We therefore developed an agglomerative clustering algorithm to automatically design such hierarchies for the estimation of

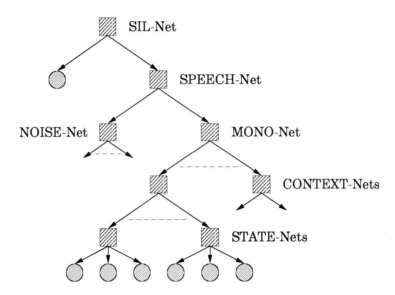

Fig. 15.4. Topology of a Hierarchy of Neural Networks (HNN) to estimate context-dependent posteriors, factored based on a-priori phonetic knowledge

posteriors for a large number of classes. We termed this framework ACID/HNN [11].

15.4.5 ACID Clustering

ACID (**A**gglomerative **C**lustering based on **I**nformation **D**ivergence) is a bottom-up clustering algorithm for the design of tree-structured soft classifiers such as a hierarchy of neural networks (HNN) [10, 11]. Although developed for connectionist acoustic modeling, the algorithm can in principle be used for any kind of classification task. Starting from a typically very large set of initial classes, for example the set of decision tree clustered HMM states in a speech recognizer [2], the ACID algorithm constructs a binary hierarchy. The nodes in the resulting tree are then instantiated with estimators for the respective conditional posterior probabilities, for instance in form of an HNN. The clustering metric in the ACID algorithm is the symmetric information divergence [26]

$$d(s_i, s_j) = \int_{\mathbf{x}} (p(\mathbf{x}|s_i) - p(\mathbf{x}|s_j)) \log \frac{p(\mathbf{x}|s_i)}{p(\mathbf{x}|s_j)} \, d\mathbf{x}$$

between class conditional densities of clusters. In contrast to standard agglomerative clustering algorithms which mostly represent clusters by their means and employ the Euclidean distance metric, we chose to represent clusters by

---

[2] In our case, we experimented with up to 24000 initial classes

15. Applying Divide and Conquer to Large Scale Pattern Recognition Tasks

parametric mixture densities (mixtures of Gaussians) in the ACID algorithm. Modeling clusters with mixture densities is much more adequate than just using the mean and it still allows to cluster large amounts of classes in a reasonable time. The symmetric information divergence (also called Kullback-Leibler distance) measures the dissimilarity of two distributions and was therefore chosen as the clustering metric. Typically, each initial class (state) is modeled using a single full covariance multivariate Gaussian density

$$p(\mathbf{x}|s_i) = \frac{1}{\sqrt{(2\pi)^n |\Sigma_i|}} \exp\{-\frac{1}{2}(\mathbf{x}-\mu_i)^t \Sigma_i^{-1}(\mathbf{x}-\mu_i)\}$$

with mean vector μ_i and covariance matrix Σ_i. Clustering then continuously merges initial classes which corresponds to building mixture densities based on the Gaussians. The symmetric information divergence between two states s_i and s_j with Gaussian distributions amounts to

$$d(s_i, s_j) = \frac{1}{2} tr\{(\Sigma_i - \Sigma_j)(\Sigma_j^{-1} - \Sigma_i^{-1})\}$$
$$+ \frac{1}{2} tr\{(\Sigma_i^{-1} + \Sigma_j^{-1})(\mu_i - \mu_j)(\mu_i - \mu_j)^t\}$$

The computation of this distance measure requires $O(n^2)$ multiplications and additions (assuming pre-computed inverse covariance matrices), where n is the dimensionality of the input feature space. To reduce the computational load of the ACID clustering algorithm, one can model the class conditional likelihoods with diagonal covariance matrices only. Feature space transformations such as principal component analysis and linear discriminant analysis can be used to approximate such distributions. When using diagonal covariance Gaussians, the symmetric information divergence simplifies to

$$d(s_i, s_j) = \frac{1}{2} \sum_{k=1}^{n} \frac{(\sigma_{jk}^2 - \sigma_{ik}^2)^2 + (\sigma_{ik}^2 + \sigma_{jk}^2)(\mu_{ik} - \mu_{jk})^2}{\sigma_{ik}^2 \sigma_{jk}^2}$$

where σ_{ik}^2 and μ_{ik} denote the k-th coefficient of the variance and mean vectors of state s_i, respectively. The evaluation of the latter distance measure requires only $O(n)$ multiplications and additions.

Making the simplifying assumption of linearity of information divergence, we can define the following distance measure between clusters of Gaussians S_k and S_l

$$D(S_k, S_l) = \sum_{s_i \in S_k} p(s_i|S_k) \sum_{s_j \in S_l} p(s_j|S_l) d(s_i, s_j)$$

This distance measure is used in the **ACID** clustering algorithm:

> 1. Initialize algorithm with N clusters S_i, each containing
>   (1) a parametric model of the class-conditional likelihood and
>   (2) a count C_i, indicating the frequency of class s_i in the training set.
> 2. Compute within cluster priors $p(s_i|S_k)$ for each cluster S_k, using the counts C_i
> 3. Compute the symmetric divergence measure $D(S_k, S_l)$ between all pairs of clusters S_k and S_l.
> 4. Find the pair of clusters with minimum divergence, S_k^* and S_l^*
> 5. Create a new cluster $S = S_k^* \cup S_l^*$ containing all Gaussians of S_k^* and S_l^* plus their respective class counts. The resulting parametric model is a mixture of Gaussians where the mixture coefficients are the class priors
> 6. Delete clusters S_k^* and S_l^*
> 7. While there are at least 2 clusters remaining, continue with 2.

ACID Initialization Initialization requires to estimate class conditional likelihoods for all (tied) states modeled by the recognizer. The number N of initial classes therefore is determined by other parts of the speech recognizer, namely by the phonetic decision tree that is typically applied to cluster phonetic contexts, or equivalently to tie HMM states [27]. Initial class conditional densities for these classes can be computed from state alignments using either the Viterbi or the Forward-Backward algorithm on training data and corresponding HMM state graphs generated from training transcriptions. Estimation of initial parametric models for the ACID algorithm therefore requires a single pass through the training data. After initial models have been estimated, the actual ACID clustering does not require any further passes through the training data. Furthermore, note that this algorithm clusters HMM states without knowledge of their phonetic identity solemnly based on acoustic dissimilarity.

ACID Dendrograms For illustration purposes, Fig. 15.5 shows a dendrogram of a typical ACID clustering run on a relatively small set of only 56 initial classes corresponding to the set of single-state monophone HMMs in a context-independent speech recognizer. The set of classes consists of 44 standard English phones along with 7 noise sounds (marked with a plus), 4 phones modeling interjections (marked with an ampersand) and silence (SIL).

Already the top level split separates silence, breathing and noise sounds (lower subtree) almost perfectly from phonetic sounds (upper subtree). Furthermore, clusters of acoustically similar phones can be observed in the ACID tree, for instance

- IX,IH,IY,Y
- JH,CH,SH,ZH
- Z,S,F
- ER,AXR,R

15. Applying Divide and Conquer to Large Scale Pattern Recognition Tasks

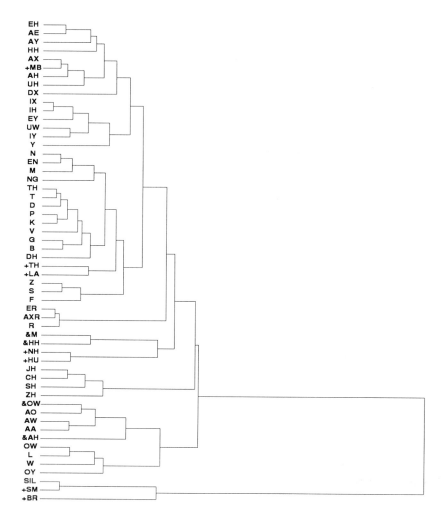

Fig. 15.5. Typical dendrogram of ACID clustering

ACID clustering was found to be quite effective in generating a hierarchical decomposition of a classification task into subtasks of increasing difficulty (when traversing the tree from root node to leaves). In the case of connectionist acoustic modeling for speech recognition, we observed that nodes in the upper layers of an ACID clustered HNN tree distinguish between broad phonetic classes, whereas nodes further down the tree begin to distinguish the particular phones within a broad phonetic class. Thus, ACID clustering constitutes an effective algorithm for discovering inherent hierarchical structure and exploiting it for modular classification.

Model Selection The choice of model size and topology becomes very important in the application of hierarchical soft classifiers to tasks such as connectionist speech recognition. While the global tree topology is determined by the outcome of the ACID clustering (or any other tree design procedure), it remains to decide on local (node-internal) classifier topology. The task of a local classifier is to estimate conditional posterior probabilities based on the available training data. Since a particular local estimator is conditioned on all predecessor nodes in the tree, it only receives training data from all the classes (leaves) that can be reached from the respective node. This amounts to a gradually diminishing training set when going from root node to nodes further down the tree. Fig. 15.6 shows this property of HNNs with a plot of the amount of available training patterns vs. node depth for a binary hierarchy with 6000 leaves. Note the logscale on the ordinate.

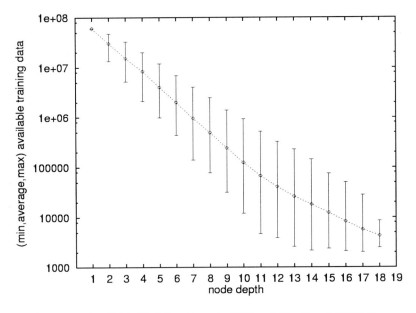

Fig. 15.6. Available Training Data in Different Depths of HNN Tree

When deciding on the local model complexity, we consider tree nodes as lying in a continuum between the following two extrema:

Top of the Hierarchy
 – large amounts of training data available
 – allows for large node classifiers
 – relatively easy, general classification tasks

Bottom of the Hierarchy
 – only small amounts of training data available
 – requires relatively small node classifiers

15. Applying Divide and Conquer to Large Scale Pattern Recognition Tasks 333

- comparably hard classification tasks
- high degree of specialization

Ideally, the complexity of local node classifiers should be selected so as to maximize generalization ability of the complete hierarchy. Generalization, on the other hand, is influenced by three factors: (1) size and distribution of the training set, (2) model complexity and (3) classification complexity of the specific task at hand. Obviously, we can not alter the latter of these factors. Furthermore, in the context of our architecture, we assume that the size of the training set for each node is fixed by the tree topology, once the hierarchy has been designed. Therefore, we have to choose model complexity based on available training data and difficulty of classification task.

In our experiments in connectionist acoustic modeling, we typically use multi layer perceptron (MLP) nodes with a single hidden layer and control model complexity by varying the number of hidden units. We use standard projective kernels with tanh activations for the hidden units and a task dependent non-linearity for the output units (sigmoid for binary and softmax for multiway classification). The overall number of weights in such a network depends linearly on the number of hidden units. According to [1] and with some approximations, a rule of thumb is to choose the number of hidden units M to satisfy

$$N > \frac{M}{\epsilon}$$

where N is the size of the training set and ϵ is the expected error rate on the test set. In our case, the variation in the number of training patterns in the different nodes dominates the above formula. Therefore, we set the number of hidden units proportional to b^{-n}, where b is the branching factor of the classification tree and n is the node depth. As long as the tree is approximately balanced in terms of the prior distribution of child nodes, this strategy leads to hidden layers with size proportional to the number of available training patterns. A more fundamental treatment of model complexity using multiple training runs and cross validation is desirable. However, in case of large-scale applications such as speech recognition, such a strategy is not realizable because of the high computational cost resulting from very large training databases. Less heuristic approaches to select model complexity still have to be explored.

15.4.6 Training Hierarchies of Neural Networks on Large Datasets

For the demonstration of various aspects of training large and complex structures such as hierarchies of neural networks on typical datasets, we report on experiments on the Switchboard [19] speech recognition database. Switchboard is a large corpus of conversational American English dialogs, recorded in telephone quality all over the US. It consists of about 170 hours of speech which typically corresponds to about 60 million training samples. The corpus currently serves as a benchmark for the official evaluation of state-of-the-art speech recognition systems. Switchboard is a comparably hard task, current best systems achieve word error rates in the vicinity of 30-40%. Fig. 15.7 shows the structure of an HNN based connectionist acoustic model for an HMM based recognizer, in our case the Janus recognition toolkit (JanusRTk) [8].

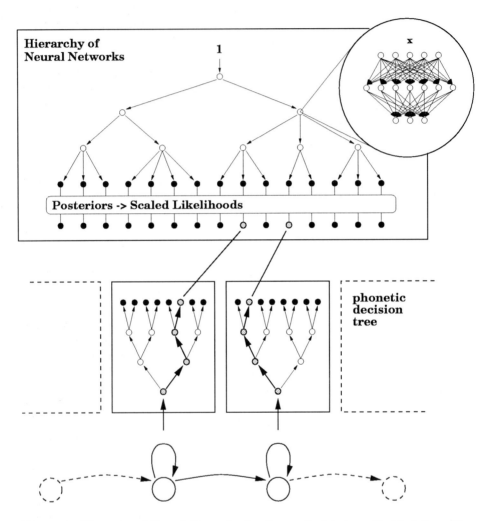

Fig. 15.7. Hierarchy of Neural Networks for Connectionist Acoustic Modeling: The upper part shows an ACID clustered HNN after node merging. This architecture computes posterior probabilities for a set of generalized polyphones. To allow for integration into the HMM framework, these posteriors are converted to scaled likelihoods. The correspondence to actual HMM states is accomplished by means of phonetic decision trees.

Due to the inherent variability and complexity of the task and the large amount of training data, typical speech recognition systems model several thousand distinct subphonetic units (HMM states) as base classes. This requires to train an estimator for posterior probabilities of thousands of distinct acoustic classes based on millions of training samples, in order to take advantage of the full modeling granularity of the speech recognizer.

In the following, we will discuss several aspects of training a hierarchical soft classifier on large datasets such as Switchboard. Due to the modular structure of the classifier, the size of the model inventory and the training database, the following discussion leads to rather unique problems and solutions. However, it is important to emphasize that these properties stem from the structure of the classifier and the size of the task - not from the specific task of acoustic modeling for speech recognition. Thus, they are transferable to comparably large tasks, e. g. handwriting, speaker or face recognition.

Classifier Tree Topology Depending on the number of classes to be modeled, tree design algorithm, branching factor and size and structure of local node classifiers have to be chosen. For Switchboard, we were experimenting with three systems consisting of 6000, 10000 and 24000 distinct classes, respectively. We used the ACID clustering algorithm to design an initial tree structure from the set of base classes for the 6k and 24k systems. As a second step of the tree design phase, we applied a greedy node merging algorithm on the ACID clustered hierarchy. Node merging decreases the number of internal nodes while increasing the average branching factor (arity) of the tree. Training of such hierarchies is less problematic than training of the original binary tree structure since the difference among nodes (in terms of the number of available training patterns) is somewhat extenuated and the overall number of networks is reduced. However, local classification tasks change from 2-way (binomial) to more complex multi-way (multinomial) problems which might have an impact on the accuracy of estimating conditional posteriors. Therefore, we constrain the node merging algorithm to produce nodes with a maximum branching factor of 8-12. This value was found to improve training speed while not affecting overall classifier accuracy. Considerably larger branching factors are not reasonable in our case as we would gradually loose the advantage of the hierarchical structure by flattening the tree.

For the 10k system, we were using the architecture of Fig. 15.4 that was designed by prior knowledge, not taking into account any measure of class similarity. This structure exhibits a larger average branching factor and less depth than the ACID clustered trees. Although we could decrease the branching factor at the MONO node by introducing linguistic classes as mentioned earlier, we still have large branching factors at the context nodes which are much harder to resolve with prior knowledge only.

The resulting tree nodes were instantiated with MLPs of varying size of the (single) hidden layer. The local MLPs output layer were parameterized with the softmax non-linearity for two reasons. First, it complies to the property of the modeled probability distribution to sum up to one, and second, the softmax function implements the expected value of the multinomial probability density. Fig.

level	# nodes = # networks	# hidden units/network
1	1	256
2	1	256
3	1	256
4	3	192
5	19	128
6	121	64
7	816	32
total	962	

level	# nodes = # networks	# hidden units/network
1	1	128
2	10	128
3	77	64
4	524	32
5	3434	16
total	4046	

Fig. 15.8. Overview of ACID clustered HNNs for 6k (left) and 24k (right) classes

15.8 gives an overview of the structure of the ACID/HNN systems. Tree compactification reduced the number of internal nodes of the 24k system from 24k to about 4k by increasing the average number of local classes (average branching factor) from 2 to about 8. Especially when dealing with large numbers of classes, we found that moderate tree compactification improved classifier performance. The overall numbers of parameters of the tree classifiers were 2M for the 6k system, 2.4M for the 10k system and 3.1M for the 24k system.

Training Algorithm and Parameters Training a distributed, hierarchically organized collection of neural networks on different amounts of training data is a challenging task. Our training criterion is maximum likelihood, assuming a multinomial probability model (1-out-of-N) over all base classes. A target class label is associated with each training pattern, indicating the correct base class. All networks in nodes along the path from root node to the current target class' leaf receive the current pattern for training. Because of the large amount of training data, we use on-line (stochastic) gradient ascent in log-likelihood with small batches (10-100 patterns) to train the individual networks. More elaborate training algorithms which apply second order methods in optimizing the objective function are too costly in our scenario - a single epoch of training, processing all 60 million patterns in the training database takes 3-5 days on a Sparc Ultra workstation. A practical training algorithm therefore must not take longer than 1-4 epochs to converge. Furthermore, because of the large number of networks that have to be trained, a potential training algorithm can not be allowed to use large amounts of memory - which could be the case with second order methods. Of course, training of individual node classifiers is independent and can therefore easily be parallelized for shared memory multi-processors which alleviates the latter requirement.

Since we are relying on *stochastic* gradient ascent in our training method, we additionally use a simple momentum term to smooth gradients. Also, we use local learning rates for each network that are initialized with a global learning rate and adapted individually to the specific learning task. The global learning rate is annealed in an exponentially decaying scheme:

$$\eta_G^{n+1} = \eta_G^n * \gamma_G.$$

Typically, we use an initial global learning rate η_G between 0.001 and 0.01, a momentum constant of 0.5 and a global annealing factor γ_G of $0.999\ldots0.9999$ applied after each batch update.

In order to accommodate the different learning speeds of the node classifiers due to the different amount of available training data, we control individual learning rates using the following measure of correlation between successive gradient vectors g_{n-1} and g_n:

$$\alpha_n = \arccos\left(\frac{g_n^t g_{n-1}}{\|g_n\|\|g_{n-1}\|}\right)$$

α_n measures the angle between the gradients g_{n-1} and g_n. Small angles indicate high correlation and therefore steady movement in weight space. Therefore, we increase the learning rate linearly up to the current maximum (as determined by initial learning rate, annealing factor and number of updates performed) whenever $\alpha_n < 90°$ for several batch updates M. Large angles, on the other hand, indicate random jumps in weight space. We therefore decrease the learning rate exponentially whenever $\alpha_n > 90°$ for several batch updates M. In summary, we obtain the following update rule for local learning rate η_i of network i:

$$\eta_i^{n+1} = \min\left\{\eta_G^{n+1}, \begin{cases}\eta_i^n + \delta \\ \eta_i^n * \gamma \\ \eta_i^n\end{cases}\right\} \quad \text{if} \quad \begin{cases}\frac{1}{M}\left(\sum_{k=0}^M \alpha_{n-k}\right) < 90° - \epsilon \\ \frac{1}{M}\left(\sum_{k=0}^M \alpha_{n-k}\right) > 90° + \epsilon \\ \text{else}\end{cases}$$

with linear increase $\delta = 0.001\ldots0.01$ and exponential annealing factor $\gamma = 0.5\ldots0.9$. The number of batch updates M controls smoothing of α whereas ϵ controls the influence of the global learning rate. For $\epsilon \to 90°$, local learning rates are forced to follow the global learning rate, whereas low values of ϵ allow local learning rates to develop individually. Typical values that have been used in our experiments are $M = 10$ and $\epsilon = 20°$.

Adapting individual learning rates to the training speed is a critical issue in hierarchical classifiers. Networks at the top of the tree have to be trained on very large amounts of training data. Therefore, learning rates must be allowed to become relatively small in order to benefit from all the data and not reach the point of saturation too early. On the other hand, networks at the bottom of the tree have to be trained with comparably small amounts of data. In order to train these networks within the same small number of passes through the overall data, we have to apply comparably large learning rates to reach a maximum in local likelihood as fast as possible. However, unconstrained adaptation of learning rates with aggressive optimization of learning speed may result in failure to converge. In our experiments with global initialization of all networks using the same maximum learning rate, global annealing of the maximum learning rate and local adaptation of individual learning rates that are constrained to never become larger than the global learning rate gives best results.

Generalization/Overfitting Simply speaking, we did not observe any overfitting in our experiments. Taking a look at the training of a large hierarchy in terms of performance on an independent cross-validation set (Fig. 15.9), we can see that the likelihood on this data levels off, but never starts to decrease again, as is often observed on smaller tasks. In the plots of Fig. 15.9, the vertical lines indicate multiple epochs (passes) through the training data (consisting of 87000 utterances). Obviously, the large amount of available training data allows for excellent generalization, early stopping was not necessary. This behavior is surprising at first sight, because we did not use any kind of explicit regularization of the local MLPs. At second sight, however, we can identify several reasons for the good generalization of HNNs on this task:

- Training data can be considered very noisy in our case, since samples come from a large variety of different speakers and recording conditions. Training with noisy data is similar to regularization and therefore improves generalization [2].
- Consider the hierarchy for the 6k classes system (Fig. 15.8). Some of the 816 networks at the bottom of the tree probably have not seen enough training patterns to generalize well to new data. Although all of these networks together constitute 85% of the total number of networks, they contribute just as one of 7 networks to any particular posterior probability. The networks in the upper part of the hierarchy have the largest influence on the evaluation of posterior probabilities. For those networks, the amount of available training data can be considered so abundant that test set error approaches training set error rate. In other words, optimal generalization can be achieved.

Results We evaluate the proposed hierarchical classifiers as connectionist acoustic models in a speech recognition system. Performance of speech recognizers is usually measured in terms of the word error rate on a reasonably large set of test utterances. In our case, we test the different acoustic classifiers with the Janus [8] recognizer on the first 30 seconds of each speaker in the official 1996 Switchboard evaluation test set, consisting of 366 utterances not present in the training set.

acoustic classifier	# classes	# parameters	word error rate
HNN	10000	2.0 M	37.3 %
ACID/HNN	6000	2.4 M	36.7 %
ACID/HNN	24000	3.1 M	33.3 %

15. Applying Divide and Conquer to Large Scale Pattern Recognition Tasks 339

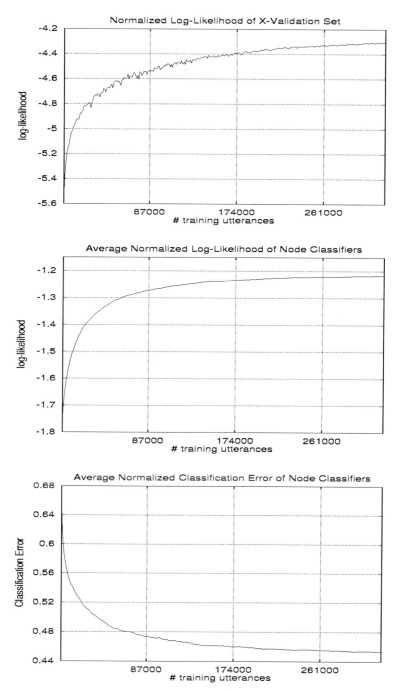

Fig. 15.9. Cross-Validation during training of 24k ACID/HNN architecture

The above results are competitive with those of state-of-the-art systems and indicate a clear advantage of the ACID clustered over the pre-determined hierarchical classifiers. We suspect, that the reason for the better performance of automatically clustered hierarchies of neural networks is the difference in tree topology. Automatically clustered HNNs such as the presented ACID/HNN trees exhibit small and comparably uniform average branching factors that allow to robustly train estimators of conditional posterior probabilities. In contrast, hand-crafted hierarchies such as the 10k HNN tree contain nodes with rather large branching factors. Fig. 15.10 shows the branching factors for all the networks in the 10k tree structure.

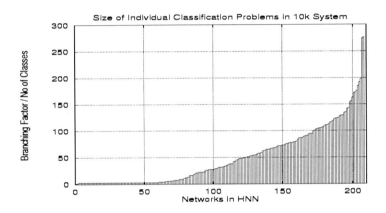

Fig. 15.10. Branching Factors of Individual Nodes in 10k HNN

The largest observed branching factor in this tree was 276. This requires the joint estimation of conditional posterior probabilities for as many as 276 classes which may result in rather poor approximations to the true posterior probabilities for some of the networks in the tree.

Furthermore, the superior performance of both ACID/HNN classifiers over the hand-crafted 10k tree, demonstrates the full scalability of the hierarchical approach and justifies the increase in the number of parameters. Earlier attempts to train hand-crafted hierarchies for 24k classes failed to provide classifiers that could be used as acoustic models in a speech recognizer. Poor approximations to the real posterior probabilities led to instabilities in decoding when dividing by priors in this case. Apart from that, we do not know of any other non-parametric approach capable of directly and discriminatively estimating posterior probabilities for such a large amount of classes.

15.5 Conclusions

We have presented and discussed a methodology for the estimation of posterior probabilities for large numbers of classes using a hierarchical connectionist framework. The aim of the paper is to demonstrate the necessity of hierarchical approaches to modularize classification tasks in large-scale application domains such as speech recognition, where thousands of classes have to be considered and millions of training samples are available. The *divide and conquer* approach proves to be a versatile tool in breaking down the complexity of the original problem into many smaller tasks. Furthermore, agglomerative clustering techniques can be applied to automatically impose a suitable hierarchical structure on a given set of classes, even in the case this set contains tens of thousands of classes. In contrast to the relatively small standard benchmarks for learning machines, aspects such as choice of training method, model selection and generalization ability appear in different light when tackling large-scale probability estimation problems.

References

1. E. B. Baum, D. Haussler (1989) *What Size Net Gives Valid Generalization?*, Neural Computation 1, pp 151-160.
2. C. M. Bishop (1995) *Training with Noise is Equivalent to Tikhonov Regularization*, Neural Computation 7, issue 1, Jan 1995, pp 108-116.
3. H. Bourlard, N. Morgan (1994) *Connectionist Speech Recognition – A Hybrid Approach*, Kluwer Academic Press, 1994.
4. H. Bourlard, N. Morgan (1992) *A Context Dependent Neural Network for Continuous Speech Recognition*, IEEE Proc. Intl. Conf. on Acoustics, Speech and Signal Processing, volume 2, pp 349-352, San Francisco, CA.
5. L. Breiman, J. H. Friedman, R. A. Olshen & C. J. Stone (1984) *Classification and Regression Trees*, Wadsworth International Group, Belmont, CA.
6. J. Bridle (1990) *Probabilistic Interpretation of Feed Forward Classification Network Outputs, with Relationships to Statistical Pattern Recognition*, In Neurocomputing: Algorithms, Architectures, and Applications, F. Fogelman-Soulie and J. Hérault, eds. Springer Verlag, New York.
7. R. Duda, P. Hart (1973) *Pattern Classification and Scene Analysis*, John Wiley & Sons, Inc.
8. M. Finke, J. Fritsch, P. Geutner, K. Ries & T. Zeppenfeld (1997) *The JanusRTk Switchboard/Callhome 1997 Evaluation System*, Proceedings of LVCSR Hub5-e Workshop, May 13-15, Baltimore, Maryland.
9. H. Franco, M. Cohen, N. Morgan, D. Rumelhart & V. Abrash (1994) *Context-Dependent Connectionist Probability Estimation in a Hybrid Hidden Markov Model – Neural Net Speech Recognition System*, Computer Speech and Language, Vol. 8, No 3, pp 211-222, July 1994.
10. J. Fritsch, M. Finke (1997) *ACID/HNN: Clustering Hierarchies of Neural Networks for Context-Dependent Connectionist Acoustic Modeling*, In Proceedings of International Conference on Acoustics, Speech and Signal Processing, May 1998, Seattle, Wa.
11. J. Fritsch (1997) *ACID/HNN: A Framework for Hierarchical Connectionist Acoustic Modeling*, In Proceedings of IEEE Workshop on Automatic Speech Recognition and Understanding, December 1997, Santa Barbara, Ca.
12. J. Fritsch, M. Finke & A. Waibel (1997) *Context-Dependent Hybrid HME/HMM Speech Recognition using Polyphone Clustering Decision Trees*, Intl. Conf. on Acoustics, Speech and Signal Processing, volume 3, pp 1759, Munich, Germany.

13. J. Fritsch (1996) *Modular Neural Networks for Speech Recognition*, Tech. Report CMU-CS-96-203, Carnegie Mellon University, Pittsburgh, PA.
14. M. M. Hochberg, G. D. Cook, S. J. Renals, A. J. Robinson, & R. S. Schechtman (1995) *The 1994 ABBOT Hybrid Connectionist-HMM Large-Vocabulary Recognition System*, In Spoken Language Systems Technology Workshop, pp 170-176, ARPA, Jan. 1995.
15. D. J. Kershaw, M. M. Hochberg & A. J. Robinson (1995) *Context-Dependent Classes in a Hybrid Recurrent Network-HMM Speech Recognition System*, Tech. Rep. CUED/F-INFENG/TR217, Cambridge University Engineering Department, Cambridge, England.
16. C. J. Merz, P. M. Murphy (1996) *UCI Repository of Machine Learning Databases*, http://www.ics.uci.edu/ mlearn/MLRepository.html, University of California, Department of Information and Computer Science.
17. N. Morgan, H. Bourlard (1992) *Factoring Networks by a Statistical Method*, Neural Computation 4, No. 6, pp 835-838, 1992.
18. N. Morgan, H. Bourlard (1995) *An Introduction to Hybrid HMM/Connectionist Continuous Speech Recognition*, Signal Processing Magazine, pp 25-42, May 1995.
19. NIST (1997) *Conversational Speech Recognition Workshop, DARPA Hub-5E Evaluation*, May 13-15/1997, Baltimore, Maryland.
20. L. Prechelt (1994) *Proben1 – A Set of Neural Network Benchmark Problems and Benchmarking Rules*, Technical Report 21/94, University of Karlsruhe, Germany.
21. J. R. Quinlan (1986) *Induction of Decision Trees*, Machine Learn. 1, pp 81-106.
22. L. R. Rabiner (1989) *A Tutorial on Hidden Markov Models and Selected Applications in Speech Recognition*, Proceedings of the IEEE 77, pp 257-285.
23. S. R. Safavian, D. Landgrebe (1991) *A Survey of Decision Tree Classifier Methodology*, IEEE Transactions on Systems, Man and Cybernetics, Vol.21, No.3, pp 660-674.
24. J. Schürmann, W. Doster (1984) *A Decision Theoretic Approach to Hierarchical Classifier Design*, Pattern Recognition 17 (3), pp 359-369.
25. J. Schürmann (1996) *Pattern Classification: A Unified View of Statistical and Neural Approaches*, John Wiley & Sons, Inc., New York, 1996.
26. J. T. Tou, R. C. Ganzales (1974) *Pattern Recognition Principles*, Addison Wesley, Reading, Massachusettes.
27. S. Young (1996) *Large Vocabulary Continuous Speech Recognition: a Review*, CUED Technical Report, Cambridge University.

Tricks for Time Series

Preface

In the last section we focus on tricks related to time series analysis and economic forecasting. In chapter 16, John Moody opens with a survey of the challenges of macroeconomic forecasting including problems such as noise, nonstationarities, nonlinearities, and the lack of good *a priori* models. Lest one be discouraged, descriptions of many possible neural network solutions are next presented including hyperparameter selection (e.g. for regularization, training window length), input variable selection, model selection (size and topology of network), better regularizers, committee forecasts, and model visualization.

The survey is followed by a more detailed description of **smoothing regularizers, model selection methods** (e.g. prediction risk, nonlinear cross-validation (NCV) (p. 361)) and **sensitivity-based pruning (SBP)** (p. 363) for input selection. The goal of using regularizers is to introduce bias into the model. But what is the "right bias"? Weight decay may be too *ad hoc* in that it does not consider the nature of the function being learned. As an alternative, the author presents several new smoothing regularizers for both feedforward and recurrent networks that empirically are found to work better.

In model selection, **prediction risk** is used as the criterion for determining "best fits". Several methods for estimating prediction risk are discussed and compared: generalized cross-validation (GCV), Akaike's final prediction error (FPE), predicted squared error (PSE), and generalized prediction error (GPE) which can be expressed in terms of the effective number of parameters.

In cross validation, separate networks are trained on different subsets of the data to obtain estimates of the prediction risk. With nonlinear loss functions, however, each network may converge to distinct local minima making comparison difficult. NCV alleviates this problem by initializing each network to be trained on a CV subset in the same minimum w_0 (obtained on the full training set). This way the CV errors computed on the different subsets estimate the prediction risk locally around this minimum w_0 and not by using some remote local minima. SBP is used to select the "best subset" of input variables to use. Here a **sensitivity measure** (e.g. delta error, average gradient, average absolute gradient, RMS gradient) (p. 364) is used to measure the change in the training error that would result if a given input is removed. Input variables are ranked based on importance and, beginning with the least important, are pruned one at a time, retraining in between. Finally, John Moody shows how sensitivity measures can be examined visually over time to better understand the role of each input (p. 365). Throughout, empirical results are presented for forecasting the U.S. Index of Industrial Production.

In chapter 17, Ralph Neuneier and Hans Georg Zimmermann describe in detail their impressive integrated system – with a lot of different tricks – for neural network training in the context of economic forecasting. The authors discuss all design steps of their system, e.g. input preprocessing, cost functions, handling of outliers, architecture, regularization techniques, learning techniques, robust estimation, estimation of error bars, pruning techniques. As in chapter 1, the tricks here are also highly interleaved, i.e. many tricks will not retain their full efficiency if they are used individually and not en bloc. Let us start with the preprocessing for which the authors use: **squared inputs** (see chapter 7), **scaled relative differences, scaled forces** in form of scaled curvatures or mean reverting that can characterize turning points in time series (p. 374). To limit the influence of **outliers** the previously mentioned inputs are transformed by

$$x' = \tanh(wx),$$

where the parameters w are learned as one layer in the training process (p. 375). Subsequently, a **bottleneck network** reduces the number of inputs to the relevant ones (p. 376). Since networks for time series prediction are in general bottom heavy, i.e. the input dimension is large while the output dimension is very small, it is important to increase the number of output targets, so that more useful error signals can be backpropagated (see also [3,2]). For this the authors introduce two output layers: (1) a **point prediction layer**, where not only the value to be predicted, i.e. y_{t+n}, but also a number of neighboring values in time are used and (2) an **interaction layer**, where differences between these values, i.e. $y_{t+n+1} - y_{t+n}$, corresponding to curvature are employed (p. 379). Following this interesting trick, several point predictions are **averaged** to reduce the variance of the prediction, similar to bagging [1] (p. 381). This overall design gives rise to an 11-layer neural network architecture, where the available prior knowledge from financial forecasting is coded.

For training this large – but rather constrained – architecture, **cost functions** are defined (p. 387) and a method based on the **CDEN approach** (p. 388) is proposed for estimating the error bars of a forecast.

To train the network the so-called **vario-η learning rule** is introduced, which is essentially a stochastic approximation of a Quasi-Newton algorithm with individual learning rates for each weight (p. 396). The authors discuss how the simple pattern-by-pattern rule has structural consequences that improve generalization behavior; or to put it differently: the stochasticity of learning implicitly includes a curvature penalty on unreliable network parts.

Then the authors raise the provoking question of the Observer-Observer dilemma: to create a model based on observed data while, at the same time, using this model to judge the correctness of new incoming data (p. 395). This leads to (1) clearning and (2) training with noise on input data. The rationale behind **clearning** is that a very noisy environment, such as financial data, will spoil a good prediction if the data is taken too seriously. That is, we are allowed to move the data point a little bit in input space in order to get smoother predictions (p. 399). Similarly, different additive noise levels for each input are chosen by an algorithm in order to **adapt the noise level** to the (estimated) importance of the input: a small noise level is used for perfectly described or unimportant inputs whereas a large noise level should be chosen for a poorly described (likely noise corrupted) input (p. 401).

In the next step, pruning methods for optimizing the architecture are described. They are: (1) **node-pruning** (p. 405) and (2) several types of **weight-pruning**: (a) **stochastic** pruning (p. 405), (b) **early-brain-damage** pruning (p. 406), (c) **inverse-kurtosis** pruning (p. 407), and (d) **instability** pruning (p. 409). The authors use a combination of instability pruning and early-brain-damage in their application; the first gives stable models and the latter generates very sparse networks.

Finally, the whole set of tricks is combined into an integrated training process and monitored on a validation set: early/late stopping, pruning and weight decay regularization are alternated (p. 418) to obtain an excellent estimate of the German bond rate from June 1994 to May 1996 (p. 419).

Jenny & Klaus

References

1. L. Breiman. Bagging predictors. *Machine Learning*, 26(2):123–140, 1996.
2. R. Caruana and V. R. de Sa. Promoting poor features to supervisors: Some inputs work better as outputs. In Michael C. Mozer, Michael I. Jordan, and Thomas Petsche, editors, *Advances in Neural Information Processing Systems*, volume 9, page 389. The MIT Press, 1997.
3. R. Caruana, L. Pratt, and S. Thrun. Multitask learning. *Machine Learning*, 28:41, 1997.

16
Forecasting the Economy with Neural Nets: A Survey of Challenges and Solutions

John Moody

Department of Computer Science & Engineering
Oregon Graduate Institute of Science & Technology
PO Box 91000, Portland, OR 97291, USA
moody@cse.ogi.edu   http://www.cse.ogi.edu/~moody/

Abstract. Macroeconomic forecasting is a very difficult task due to the lack of an accurate, convincing model of the economy. The most accurate models for economic forecasting, "black box" time series models, assume little about the structure of the economy. Constructing reliable time series models is challenging due to short data series, high noise levels, nonstationarities, and nonlinear effects. This chapter describes these challenges and presents some neural network solutions to them. Important issues include balancing the *bias/variance tradeoff* and the *noise/nonstationarity tradeoff*. A brief survey of methods includes hyperparameter selection (regularization parameter and training window length), input variable selection and pruning, network architecture selection and pruning, new smoothing regularizers, committee forecasts and model visualization. Separate sections present more in-depth descriptions of smoothing regularizers, architecture selection via the *generalized prediction error (GPE)* and *nonlinear cross-validation (NCV)*, input selection via *sensitivity based pruning (SBP)*, and model interpretation and visualization. Throughout, empirical results are presented for forecasting the U.S. Index of Industrial Production. These demonstrate that, relative to conventional linear time series and regression methods, superior performance can be obtained using state-of-the-art neural network models.

16.1 Challenges of Macroeconomic Forecasting

Of great interest to forecasters of the economy is predicting the "business cycle", or the overall level of economic activity. The business cycle affects society as a whole by its fluctuations in economic quantities such as the unemployment rate (the misery index), corporate profits (which affect stock market prices), the demand for manufactured goods and new housing units, bankruptcy rates, investment in research and development, investment in capital equipment, savings rates, and so on. The business cycle also affects important socio-political factors such as the general mood of the people and the outcomes of elections.
The standard measures of economic activity used by economists to track the business cycle include the Gross Domestic Product (GDP) and the Index of

Industrial Production (IP). GDP is a broader measure of economic activity than is IP. However, GDP is computed by the U.S. Department of Commerce on only a quarterly basis, while Industrial Production is more timely, as it is computed and published monthly. IP exhibits stronger cycles than GDP, and is therefore more interesting and challenging to forecast. (See figure 16.1.) In this chapter, all empirical results presented are for forecasting the U.S. Index of Industrial Production.

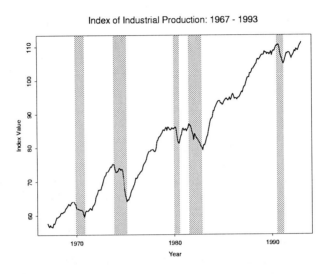

Fig. 16.1. U.S. Index of Industrial Production (IP) for the period 1967 to 1993. Shaded regions denote official recessions, while unshaded regions denote official expansions. The boundaries for recessions and expansions are determined by the National Bureau of Economic Research based on several macroeconomic series. As is evident for IP, business cycles are irregular in magnitude, duration, and structure.

Macroeconomic modeling and forecasting is challenging for several reasons:

Non-Experimental Science: Like evolutionary biology and cosmology, macroeconomics is largely a non-experimental science. There is only one instance of the world economy, and the economy of each country is not a closed system. Observing the state of an economy in aggregate is difficult, and it is generally not possible to do controlled experiments.

No *a priori* Models: A convincing and accurate scientific model of business cycle dynamics is not yet available due to the complexities of the economic system, the impossibility of doing controlled experiments on the economy, and non-quantifiable factors such as mass psychology and sociology that influence economic activity. There are two main approaches that economists have used to model the macroeconomy, econometric models and linear time series models:

Econometric Models: These models attempt to model the macroeconomy at a relatively fine scale and typically contain hundreds or thousands of equations and variables. The model structures are chosen by hand, but model parameters are estimated from the data. While econometric models are of some use in understanding the workings of the economy qualitatively, they are notoriously bad at making quantitative predictions.

Linear Time Series Models: Given the poor forecasting performance of econometric models, many economists have resorted to analyzing and forecasting economic activity by using the empirical "black box" techniques of standard linear time series analysis. Such time series models typically have perhaps half a dozen to a dozen input series. The most reliable and popular of these models during the past decade or so have been bayesian vector autoregressive (BVAR) models [22]. As we have found in our own work, however, neural networks can often outperform standard linear time series models. The lack of an *a priori* model of the economy makes input variable selection, the selection of lag structures, and network model selection critical issues.

Noise: Macroeconomic time series are intrinsically very noisy and generally have poor signal to noise ratios. (See figures 16.2 and 16.3.) The noise is due both to the many unobserved variables in the economy and to the survey techniques used to collect data for those variables that are measured. The noise distributions are typically heavy tailed and include outliers. The combination of short data series and significant noise levels makes controlling model variance, model complexity, and the *bias / variance tradeoff* important issues [9]. One measure of complexity for nonlinear models is P_{eff}, the *effective number of parameters* [24, 25]. P_{eff} can be controlled to balance bias and variance by using regularization and model selection techniques.

Nonstationarity: Due to the evolution of the world's economies over time, macroeconomic series are intrinsically nonstationary. To confound matters, the definitions of many macroeconomic series are changed periodically as are the techniques employed in measuring them. Moreover, estimates of key series are periodically revised retroactively as better data are collected or definitions are changed. Not only do the underlying dynamics of the economy change with time, but the noise distributions for the measured series vary with time also. In many cases, such nonstationarity shortens the usable length of the data series, since training on older data will induce biases in predictions. The combination of noise and nonstationarity gives rise to a *noise / nonstationarity tradeoff* [23], where using a short training window results in too much model variance or *estimation error* due to noise in limited training data, while using a long training window results in too much model bias or *approximation error* due to nonstationarity.

Nonlinearity: Traditional macroeconomic time series models are linear [12, 14]. However, recent work by several investigators have suggested that nonlinearities can improve macroeconomic forecasting models in some cases [13, 27, 39, 35, 40]. (See table 16.1 and figures 16.2 and 16.3.) Based upon our own experience, the degree of nonlinearity captured by neural network models of macroeconomic series tends to be mild [27, 20, 38, 42, 28, 45]. Due to the high noise levels and

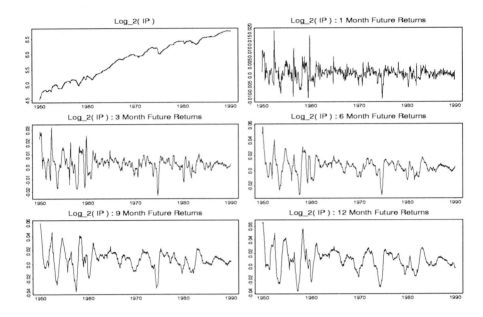

Fig. 16.2. The U.S. Index of Industrial Production and five return series (rates of change measured as log differences) for time scales of 1, 3, 6, 9, and 12 months. These return series served as the prediction targets for the standard Jan 1950 - Dec 1979 / Jan 1980 - Jan 1990 benchmark results reported in [27]. The difficulty of the prediction task is evidenced by the poor signal to noise ratios and erratic behavior of the target series. For the one month returns, the performance of our neural network predictor in table 1 suggests that the SNR is around 0.2. For all returns series, significant nonstationarities and deviations from normality of the noise distributions are present.

limited data, simpler models are favored. This makes reliable estimation of nonlinearities more difficult.

16.2 A Survey of Neural Network Solutions

We have been investigating a variety of algorithms for neural network model selection that go beyond the *vanilla* neural network approach.[1] The goal of this work is to construct models with minimal prediction risk (expected test set error). The techniques that we are developing and testing are described below. Given the brief nature of this survey, I have not attempted to provide an exhaustive list of the many relevant references in the literature.

---

[1] We define a *vanilla* neural network to be a fully connected, two-layer sigmoidal network with a full set of input variables and a fixed number of hidden units that is trained on a data window of fixed length with backprop and early stopping using a validation set. No variable selection, pruning, regularization, or committee techniques are used.

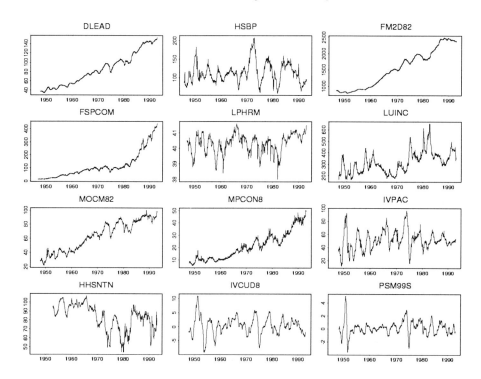

Fig. 16.3. The U.S. Index of Leading Indicators (DLEAD) and its 11 component series as currently defined. The Leading Index is a key tool for forecasting business cycles. The input variables for the IP forecasting models included transformed versions of DLEAD and several of its components [27]. The difficulty of macroeconomic forecasting is again evident, due to the high noise levels and erratic behaviors of DLEAD and its components. (Note that the component series included in DLEAD have been changed several times during the past 47 years. The labels for the various series are those defined in Citibase: HSBP denotes housing starts, FM2D82 is M2 money supply, FSPCOM is the Standard & Poors 500 stock index, and so on.)

Hyperparameter Selection: Hyperparameters are parameters that appear in the training objective function, but not in the network itself. Examples include the regularization parameter, the training window length, and robust scale parameters. Examples of varying the regularization parameter and the training window length for a 12 month IP forecasting model are shown in figures 16.4 and 16.5. Varying the regularization parameter trades off bias and variance, while varying the training window length trades off noise and nonstationarity.

Input Variable Selection and Pruning: Selecting an informative set of input variables and an appropriate representation for them ("features") is critical to the solution of any forecasting problem. The variable selection and representation problem is part of the overall model selection problem. Variable selection procedures can be either model-independent or model-dependent. The Delta

Prediction Horizon (Months)	Trivial (Average of Training Set)	Univariate AR(14) Model Iterated Pred.	Multivariate Linear Reg. Direct Pred.	Sigmoidal Nets w/ PC Pruning Direct Pred.
1	1.04	0.90	0.87	0.81
2	1.07	0.97	0.85	0.77
3	1.09	1.07	0.96	0.75
6	1.10	1.07	1.38	0.73
9	1.10	0.96	1.38	0.67
12	1.12	1.23	1.20	0.64

Table 16.1. Comparative summary of normalized prediction errors for rates of return on Industrial Production for the period January 1980 to January 1990 as presented in [27]. The four model types were trained on data from January 1950 to December 1979. The neural network models significantly outperform the trivial predictors and linear models. For each forecast horizon, the normalization factor is the variance of the target variable for the training period. Nonstationarity in the IP series makes the test errors for the trivial predictors larger than 1.0. In subsequent work, we have obtained substantially better results for the IP problem [20, 38, 42, 28, 45].

Test, a model independent procedure, is a nonparametric statistical algorithm that selects meaningful predictor variables by direct examination of the data set [36]. Other model-independent techniques make use of the mutual information [4, 5, 46] or joint mutual information [46]. Sensitivity-based pruning (SBP) techniques are model-dependent algorithms that prune unnecessary or harmful input variables from a trained network [33, 30, 25, 42, 21]. Sensitivity based pruning methods are described in greater detail in section 16.4.6.

Model Selection and Pruning: A key technique for controlling the bias / variance tradeoff for noisy problems is to select the size and architecture of the network. For two-layer networks, this includes selecting the number of internal units, choosing a connectivity structure, and pruning unneeded nodes, weights, or weight matrix eigennodes. A constructive algorithm for selecting the number of internal units is sequential network construction (SNC) [2, 30, 25]. Techniques for pruning weights and internal nodes include sensitivity-based pruning methods like optimal brain damage (OBD) [19] and optimal brain surgeon (OBS) [15]. Our recently-proposed supervised principal components pruning (PCP) method [20] prunes weight matrix eigennodes, rather than weights. Since PCP does not require training to a local minimum, it can be used with early stopping. It has computational advantages over OBS, and can outperform OBD when input variables or hidden node activities are noisy and correlated. Figure 16.6 shows reductions in prediction errors obtained by using PCP on a set of IP forecasting models. Section 16.4 describes the model selection problem and the use of estimates of prediction risk such as *nonlinear cross-validation* (NCV) to guide the selection process in greater detail.

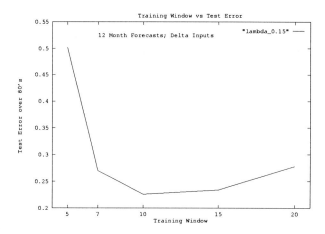

Fig. 16.4. Example of the *Noise / Nonstationary Tradeoff* and selection of the best training window, in this case 10 years [38, 28]. The longer training windows of 15 and 20 years yield higher test set error due to the model bias induced by nonstationarity. The shorter training windows of 5 and 7 years have significantly higher errors due to model variance resulting from noise in the data series and smaller data sets. The test errors correspond to models trained with the best regularization parameter 0.15 indicated in figure 16.5.

Better Regularizers: Introducing biases in a model via regularization or pruning reduces model variance and can thus reduce prediction risk. Prediction risk can be best minimized by choosing appropriate biases. One such set of biases are smoothing constraints. We have proposed new classes of smoothing regularizers for both feedforward and recurrent networks [29, 45] that often yield better performance than the standard weight decay approach. These are described in greater detail in section 16.3.

Committee Forecasts: Due to the extremely noisy nature of economic time series, the control of forecast variance is a critical issue. One approach for reducing forecast variance is to average the forecasts of a committee of models. Researchers in economics have studied and used combined estimators for a long time, and generally find that they outperform their component estimators and that unweighted averages tend to outperform weighted averages, for a variety of weighting methods [12, 44, 6]. Reductions of prediction error variances obtained by unweighted committee averaging for a selection of different IP forecasting models are shown in figure 16.7.

Model Interpretation and Visualization: It is important not only to be able to make accurate forecasts, but to also understand what factors influence the forecasts that are made. This can be accomplished via the sensitivity analyses described in sections 16.4.6 and 16.4.8 and the visualization tool presented in section 16.4.8.

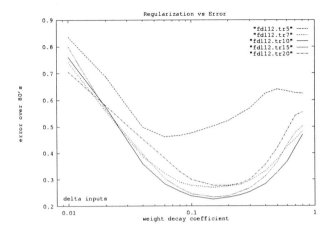

Fig. 16.5. Example of the effect of regularization (weight decay) parameter on test error [38, 28]. The five curves are for training windows of length 5, 7, 10, 15, and 20 years. The *Bias / Variance Tradeoff* is clearly evident in all the curves; the minimum test set errors occur for weight decay parameters of order 0.1. Larger errors due to bias occur for larger weight decay coefficients, while larger errors due to model variance occur for smaller values of the coefficient.

16.3 Smoothing Regularizers for Better Generalization

Introducing biases in a model via regularization or pruning reduces model variance and can thus reduce prediction risk (see also chapters 2-6). Prediction risk can be best minimized by choosing appropriate biases. Quadratic weight decay [37, 18, 17], the standard approach to regularization used in the neural nets community, is an *ad hoc* function of the network weights. Weight decay is *ad hoc* in the sense that it imposes direct constraints on the weights independent of the nature of the function being learned or the parametrization of the network model. A more principled approach is to require that the function $f(W, x)$ learned by the network be smooth. This can be accomplished by penalizing the mth order curvature of $f(W, x)$. The regularization or penalty functional is then the smoothing integral

$$S(W, m) = \int d^d x\, \Omega(x) \left\| \frac{d^m f(W, x)}{dx^m} \right\|^2 , \qquad (16.1)$$

where $\Omega(x)$ is a weighting function and $\| \, \|$ denotes the Euclidean tensor norm.[2] Since numerical computation of (16.1) generally requires expensive Monte Carlo

---

[2] The relation of this type of smoothing functional to radial basis functions has been studied by [10]. However, the approach developed in that work does not extend to standard feedforward sigmoidal networks, which are a special case of projective basis function networks (PBF's).

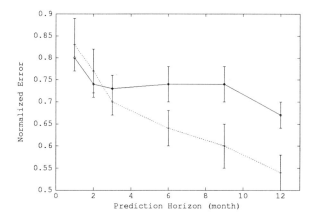

Fig. 16.6. Prediction errors for two sets of neural network models for 12 month returns for IP, with (dotted line) and without (solid line) Supervised Principal Components Pruning (PCP) [20]. Each data point is the mean error for 11 nets, while the error bars represent one standard deviation. Statistically significant improvements in prediction performance are obtained for the 6, 9, and 12 month prediction horizons by using the PCP algorithm to reduce the network complexities. While techniques like optimal brain damage and optimal brain surgeon prune weights from the network, PCP reduces network complexity and hence model variance by pruning eigennodes of the weight matrices. Unlike the unsupervised use of principal components, PCP removes those eigennodes that yield the greatest reduction in estimated prediction error.

integrations and is therefore impractical during training, we have derived algebraically simple approximations and bounds to $S(W, m)$ for feedforward networks that can be easily evaluated at each training step [29]. For these new classes of algebraically simple m^{th}-order smoothing regularizers for networks of projective basis functions (PBF's) $f(W, x) = \sum_{j=1}^{N} u_j g \left[x^T v_j + v_{j0} \right] + u_0$, $W = (u, v)$ with general transfer functions $g[\cdot]$, the regularizers are:

$$R_G(W, m) = \sum_{j=1}^{N} u_j^2 \|v_j\|^{2m-1} \quad \text{Global Form}$$

$$R_L(W, m) = \sum_{j=1}^{N} u_j^2 \|v_j\|^{2m} \quad \text{Local Form.}$$

Our empirical experience shows that these new smoothing regularizers typically yield better prediction accuracies than standard weight decay.

In related work, we have derived an algebraically-simple regularizer for recurrent nets [45]. This regularizer can be viewed as a generalization of the first order Tikhonov stabilizer (the $m = 1$ *local form* above) to dynamic models. For

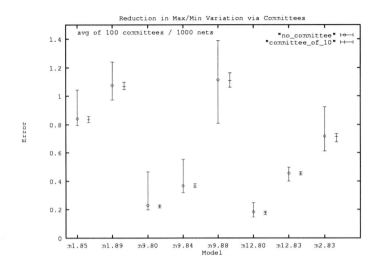

Fig. 16.7. Reduction in error variance for prediction of the U.S.Index of Industrial Production by use of combining forecasts (or committees) [38, 28]. Abscissa points are various combinations of prediction horizon and test period. For example, "m12.80" denotes networks trained to make 12 month forecasts on the ten years prior to 1979 and tested by making true *ex ante* forecasts on the year 1980. Performance metric is normalized mean square error (NMSE) computed over the particular year. All training sets have length 10 years. For each point, bars show range of values for either 1000 individual models, or 100 committees of 10. The individual networks each have three sigmoidal internal units, one linear output, and typically a dozen or so input variables selected by the δ-test from an initial set of 48 candidate variables.

two layer networks with recurrent connections described by

$$Y(t) = \mathbf{g}\left(AY(t-\tau) + VX(t)\right) \, , \, \hat{Z}(t) = UY(t) \, ,$$

the training criterion with the regularizer is

$$\mathcal{E} = \frac{1}{N}\sum_{t=1}^{N} \|Z(t) - \hat{Z}(W, I(t))\|^2 + \lambda \rho_\tau^2(W) \, ,$$

where $W = \{U, V, A\}$ is the network parameter set, $Z(t)$ are the targets, $I(t) = \{X(s), s = 1, 2, \cdots, t\}$ represents the current and all historical input information, N is the size of the training data set, $\rho_\tau^2(W)$ is the regularizer, and λ is a regularization parameter. The closed-form expression for the regularizer for time-lagged recurrent networks is:

$$\rho_\tau(W) = \frac{\gamma\|U\|\|V\|}{1-\gamma\|A\|}\left[1 - e^{\frac{\gamma\|A\|-1}{\tau}}\right] \, ,$$

where $\|\ \|$ is the Euclidean matrix norm and γ is a factor which depends upon the maximal value of the first derivatives of the internal unit activations $g(\ )$.

Simplifications of the regularizer can be obtained for simultaneous recurrent nets ($\tau \mapsto 0$), two-layer feedforward nets, and one layer linear nets. We have successfully tested this regularizer in a number of case studies and found that it performs better than standard quadratic weight decay. A comparison of this recurrent regularizer to quadratic weight decay for 1 month forecasts of IP is shown in figure 16.8.

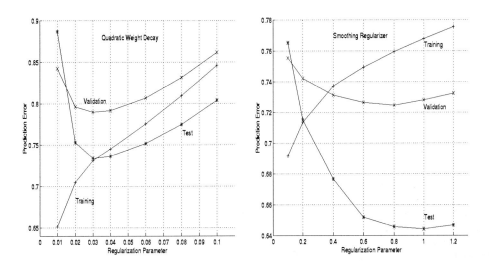

Fig. 16.8. Regularization parameter vs. normalized prediction errors for the task of predicting the one month rates of change of the U.S. Index of Industrial Production [45]. The example given is for a recurrent network trained with standard weight decay (left) or with the new recurrent smoothing regularizer (right). For standard weight decay, the optimal regularization parameter is 0.03 corresponding to a test error of 0.734. For the new smoothing regularizer, the optimal regularization parameter which leads to the least validation error is 0.8 corresponding to a test error of 0.646. The new recurrent regularizer thus yields a 12% reduction in test error relative to that obtained using quadratic weight decay.

16.4 Model Selection and Interpretation

In this section, we provide a more in-depth discussion of several issues and techniques for neural network model selection, including the problem of selecting inputs. We describe techniques for selecting architectures via estimates of the prediction risk, especially the *generalized prediction error (GPE)* and *nonlinear cross-validation (NCV)*. We present *sensitivity-based pruning (SBP)* methods for selecting input variables, and demonstrate the use of these methods for predicting the U.S. Index of Industrial Production. Finally, we discuss some approaches to model interpretation and visualization that enable an understanding of economic relationships.

16.4.1 Improving Forecasts via Architecture and Input Selection

For the discussion of architecture selection in this paper, we focus on the most widely used neural network architecture, the two-layer *perceptron* (or *backpropagation*) network. The response function for such a network with architecture λ having I_λ input variables, H_λ internal (hidden) neurons, and a single output is:

$$\hat{\mu}_\lambda(x) = h(u_0 + \sum_{j=1}^{H_\lambda} u_j\, g(v_{j0} + \sum_{i=1}^{I_\lambda} v_{ji}\, x_i)) \,. \tag{16.2}$$

Here, h and g are typically sigmoidal nonlinearities, the v_{ji} and v_{j0} are input weights and thresholds, the u_j and u_0 are the output weights and threshold, and the index λ is an abstract label for the specific two layer perceptron network architecture. While we consider for simplicity this restricted class of perceptron networks in this section, our approach can be easily generalized to networks with multiple outputs and multiple layers.

For two layer perceptrons, the *architecture selection problem* is to find a good, near-optimal architecture λ for modeling a given data set. The architecture λ is characterized by the number of hidden units H_λ, the subset of input variables I_λ, and the subset of weights u_j and v_{ji} that are non-zero. If all of the u_j and v_{ji} are non-zero, the network is referred to as *fully connected*. Since an exhaustive search over the space of possible architectures is impossible, the procedure for selecting this architecture requires a heuristic search. See Figure 16.9 for examples of heuristic search strategies and [25] and [30] for additional discussion.

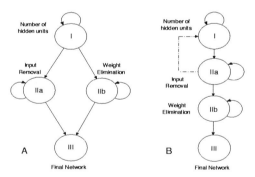

Fig. 16.9. Heuristic Search Strategies: After selecting the number of hidden units H_λ, the input removal and weight elimination can be carried out in parallel (A) or sequentially (B). In (B), the selection of the number of hidden units and removal of inputs may be iterated (dashed line).

In this section, we focus on selecting the "best subset" of input variables for predicting the U.S. Index of Industrial Production. In order to avoid an exhaustive search over the exponentially-large space of architectures obtained by considering all possible combinations of inputs, we employ a directed search strategy using the *sensitivity-based input pruning (SBP)* algorithm (see section 16.4.6).

16.4.2 Architecture Selection via the Prediction Risk

The notion of "best fits" can be captured via an objective criterion; such as *maximum a posteriori probability (MAP)*, minimum *Bayesian information criterion (BIC)*, minimum *description length (MDL)*, or *generalization ability*. The generalization ability can be defined precisely as the *prediction risk* P_λ, the expected performance of an estimator in predicting new observations. In this section, we use the prediction risk as our selection criterion for two reasons. First, it is straightforward to compute, and second, it provides more information than MAP, BIC, or MDL, since it tells us how much confidence to put in predictions produced by our best model.

Consider a set of observations $D = \{(x_j, t_j); j = 1 \ldots N\}$ that are assumed to be generated as $t_j = \mu(x_j) + \epsilon_j$ where $\mu(x)$ is an unknown function, the inputs x_j are drawn independently with an unknown stationary probability density function $p(x)$, the ϵ_j are independent random variables with zero mean ($\bar{\epsilon} = 0$) and variance σ_ϵ^2, and the t_j are the observed target values. The learning or regression problem is to find an estimate $\hat{\mu}_\lambda(x; D)$ of $\mu(x)$ given the data set D from a class of predictors or models $\mu_\lambda(x)$ indexed by λ. In general, $\lambda \in \Lambda = (S, A, W)$, where $S \subset X$ denotes a chosen subset of the set of available input variables X, A is a selected architecture within a class of model architectures $\mathcal{A}$, and W are the adjustable parameters (weights) of architecture A.

The *prediction risk* P_λ (defined above) can be approximated by the expected performance on a finite test set. P_λ can be defined for a variety of loss functions. For the special case of squared error, it is:

$$P_\lambda = \int dx\, p(x)[\mu(x) - \hat{\mu}(x)]^2 + \sigma_\epsilon^2 \qquad (16.3)$$

$$\approx E\{\frac{1}{N}\sum_{j=1}^{N}(t_j^* - \hat{\mu}_\lambda(x_j^*))^2\} \qquad (16.4)$$

where (x_j^*, t_j^*) are new observations that were not used in constructing $\hat{\mu}_\lambda(x)$. In what follows, we shall use P_λ as a measure of the generalization ability of a model. Our strategy is to choose an architecture λ in the model space Λ which minimizes an estimate of the prediction risk P_λ.

16.4.3 Estimation of Prediction Risk

Since it is not possible to exactly calculate the prediction risk P_λ given only a finite sample of data, we have to estimate it. The restriction of limited data makes the model selection and prediction risk estimation problems more difficult. This is the typical situation in economic forecasting, where the time series are short.

A limited training set results in a more severe bias/variance (or underfitting vs overfitting) tradeoff (see e.g. [9]), so the model selection problem is both more challenging and more crucial. In particular, it is easier to overfit a small training set, so care must be taken to select a model that is not too large. Also, limited data sets make prediction risk estimation more difficult if there is not enough data available to hold out a sufficiently large independent test sample. In such situations, one must use alternative approaches which enable the estimation of prediction risk from the training data, such as data resampling and algebraic estimation techniques. Data resampling methods include nonlinear refinements of ν–fold cross–validation (NCV) and bootstrap estimation, while algebraic estimates (in the regression context) include Akaike's *final prediction error (FPE)* [1], for linear models, and the recently proposed *generalized prediction error (GPE)* for nonlinear models [31, 24, 25], which is identical to the independently-derived network information criterion [34]. For comprehensive discussions of prediction risk estimation, see [8, 16, 43, 25].

16.4.4 Algebraic Estimates of Prediction Risk

Predicted Squared Error for Linear Models. For linear regression models with the squared error loss function, a number of useful algebraic estimates for the prediction risk have been derived. These include the well known *generalized cross–validation (GCV)* [7, 11] and Akaike's *final prediction error (FPE)* [1] formulas:

$$GCV_\lambda = ASE_\lambda \frac{1}{\left(1 - \frac{Q_\lambda}{N}\right)^2} \qquad FPE_\lambda = ASE_\lambda \left(\frac{1 + \frac{Q_\lambda}{N}}{1 - \frac{Q_\lambda}{N}}\right). \qquad (16.5)$$

Q_λ denotes the number of weights of model λ (ASE_λ denotes the average squared error). Note that although GCV and FPE are slightly different for small sample sizes, they are asymptotically equivalent for large N:

$$GCV_\lambda \approx FPE_\lambda \approx ASE_\lambda \left(1 + 2\frac{Q_\lambda}{N}\right) \qquad (16.6)$$

A more general expression of *predicted squared error (PSE)* is:

$$PSE_\lambda = ASE_\lambda + 2\hat{\sigma}^2 \frac{Q_\lambda}{N}, \qquad (16.7)$$

where $\hat{\sigma}^2$ is an estimate of the noise variance in the data. Estimation strategies for (16.7) and its statistical properties have been analyzed by [3]. FPE is obtained as special case of PSE by setting $\hat{\sigma}^2 \equiv ASE_\lambda/(N - Q_\lambda)$. See [8, 16, 43] for tutorial treatments.

It should be noted that PSE, FPE and GCV are asymptotically unbiased estimates of the prediction risk for the neural network models considered here under certain conditions. These are: (1) the noise ϵ_j in the observed targets t_j is independent and identically distributed, (2) the resulting model is unbiased, (3) weight decay is not used, and (4) the nonlinearity in the model can be neglected. For PSE, we further require that an asymptotically unbiased estimate of $\widehat{\sigma}^2$ is used. In practice, however, essentially all neural network fits to data will be biased and/or have significant nonlinearity.

Although PSE, FPE and GCV are asymptotically unbiased only under the above assumptions, they are much cheaper to compute than NCV since no re-training is required.

Generalized Prediction Error (GPE) for Nonlinear Models. The predicted squared error PSE, and therefore the final prediction error FPE, are special cases of the *generalized prediction error GPE* [31, 24, 25]. We present an abbreviated description here.

GPE estimates the prediction risk for biased nonlinear models which may use general loss functions and include regularizers such as weight decay. The algebraic form is

$$GPE_\lambda \equiv \mathcal{E}_{\lambda\text{train}} + \frac{2}{N} \operatorname{tr} \widehat{V}\widehat{G}_\lambda \ , \qquad (16.8)$$

where $\mathcal{E}_{\lambda\text{train}}$ is the training set error (average value of loss function on training set), $\widehat{V}$ is a nonlinear generalization of the estimated *noise covariance matrix* of the observed targets, and $\widehat{G}_\lambda$ is the estimated *generalized influence matrix*, a nonlinear analog of the standard influence or hat matrix.

GPE can be expressed in an equivalent form as:

$$GPE_\lambda = \mathcal{E}_{\lambda\text{train}} + 2\,\widehat{\sigma}^2_{\text{eff}} \frac{\widehat{Q}_{\lambda\text{eff}}}{N}, \qquad (16.9)$$

where $\widehat{Q}_{\text{eff}} \equiv \operatorname{tr}\widehat{G}$ is the estimated *effective* number of model parameters, and $\widehat{\sigma}^2_{\text{eff}} \equiv (\operatorname{tr}\widehat{V}\widehat{G})/(\operatorname{tr}\widehat{G})$ is the estimated effective noise variance in the data. For nonlinear and/or regularized models, $\widehat{Q}_{\lambda\text{eff}}$ is generally not equal to the number of weights Q_λ.

When the noise in the target variables is assumed to be independent with uniform variance and the squared error loss function is used, (16.9) simplifies to:

$$GPE_\lambda = ASE_\lambda + 2\widehat{\sigma}^2 \frac{\widehat{Q}_{\lambda\text{eff}}}{N} \ . \qquad (16.10)$$

Note that replacing $\widehat{Q}_{\lambda\text{eff}}$ with Q_λ gives the expression for PSE. Various other special cases of (16.8) and (16.10) have been derived by other authors and can be found in [8, 16, 43]. *GPE* was independently derived by [34], who called it the Network Information Criterion (NIC).

16.4.5 NCV: Cross-Validation for Nonlinear Models

Cross-validation (CV) is a sample re–use method for estimating prediction risk; it makes maximally efficient use of the available data. A perturbative refinement of CV for nonlinear models is called *nonlinear cross-validation (NCV)* [25, 30].

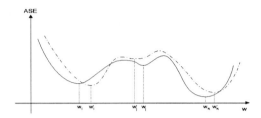

Fig. 16.10. A nonlinear model can have many local minima in the error function. Each local minimum w_i, w_j and w_k (solid curve) corresponds to a different set of parameters and thus to a different model. Training on a different finite sample of data or retraining on a subsample, as in nonlinear cross-validation, gives rise to a slightly different error curve (dashed) and perturbed minima w'_i, w'_j and w'_k. Variations due to data sampling in error curves and their minima are termed *model variance*.

Let the data D be divided into ν randomly selected disjoint subsets D_j of roughly equal size: $\cup_{j=1}^{\nu} D_j = D$ and $\forall i \neq j$, $D_i \cap D_j = \emptyset$. Let N_j denote the number of observations in subset D_j. Let $\hat{\mu}_{\lambda(D_j)}(x)$ be an estimator trained on all data except for $(x,t) \in D_j$. Then, the cross-validation average squared error for subset j is defined as

$$CV_{D_j}(\lambda) = \frac{1}{N_j} \sum_{(x_k,t_k) \in D_j} \left(t_k - \hat{\mu}_{\lambda(D_j)}(x_k)\right)^2 . \tag{16.11}$$

These are averaged over j to obtain the ν-fold cross-validation estimate of prediction risk:

$$CV(\lambda) = \frac{1}{\nu} \sum_j CV_{D_j}(\lambda) . \tag{16.12}$$

Typical choices for ν are 5 and 10. Leave–one–out CV is obtained in the limit $\nu = N$. CV is a nonparametric estimate of the prediction risk that relies only on the available data.

The frequent occurrence of multiple minima in nonlinear models (see Figure 16.10), each of which represents a different predictor, requires a refinement of the cross-validation procedure. This refinement, *nonlinear cross-validation (NCV)*, is illustrated in Figure 16.11 for $\nu = 5$.

A network is trained on the entire data set D to obtain a model $\hat{\mu}_\lambda(x)$ with weights W_0. These weights are used as the starting point for the ν-fold cross–validation procedure. Each subset D_j is removed from the training data in turn. The network is re-trained using the remaining data starting at W_0 (rather than using random initial weights). Under the assumption that deleting a subset from the training data does not lead to a large difference in the locally-optimal weights, the retraining from W_0 "perturbs" the weights to obtain $W_i, i = 1\ldots\nu$. The Cross-Validation error computed for the "perturbed models" $\hat{\mu}_{\lambda(D_j)}(x)$ thus estimates the prediction risk for the model with locally-optimal weights W_0 as desired, and not the performance of other predictors at other local minima.

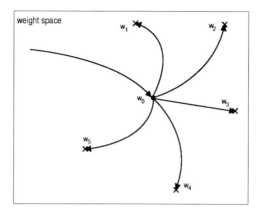

Fig. 16.11. Illustration of the computation of 5-fold *nonlinear cross-validation (NCV)*. First, the network is trained on all data to obtain weights w_0 which are used as starting point for the cross-validation. Each data subset $D_i, i = 1\ldots 5$ is removed from the training data D in turn. The network is trained, starting at W_0, using the remaining data. This "perturbs" the weights to obtain w_i. The test error of the "perturbed model" w_i is computed on the hold-out sample D_i. The average of these errors is the 5-fold CV estimate of the prediction risk for the model with weights w_0.

If the network would be trained from random initial weights for each subset, it could converge to a different minimum corresponding to W_i different from the one corresponding to W_0. This would correspond to a different model. Thus, starting from W_0 assures us that the cross-validation estimates the prediction risk for a particular model in question corresponding to $W \approx W_0$.

16.4.6 Pruning Inputs via Directed Search and Sensitivity Analysis

Selecting a "best subset" of input variables is a critical part of model selection for forecasting. This is especially true when the number of available input series is large, and exhaustive search through all combinations of variables is computationally infeasible. Inclusion of irrelevant variables not only does not help prediction, but can reduce forecast accuracy through added noise or systematic bias. A model-dependent method for input variable selection is *sensitivity-based pruning (SBP)* [41, 32, 30, 25]. Extensions to this approach are presented in [21]. With this algorithm, candidate architectures are constructed by evaluating the effect of removing an input variable from the fully connected network. These are ranked in order of increasing training error. Inputs are then removed following a "Best First" strategy, i.e. selecting the input that, when removed, increases the training error least.

The SBP algorithm computes a *sensitivity measure* S_i to evaluate the change in training error that would result if input x_i were removed from the network.

One such sensitivity measure is the *delta error*, defined as:

$$\text{Delta Error (DE)} \quad S_i = \frac{1}{N} \sum_{j=1}^{N} S_{ij} \quad (16.13)$$

where S_{ij} is the sensitivity computed for exemplar x_j. Since there are usually many fewer inputs than weights, a direct evaluation of S_i is feasible:

$$S_{ij} = SE(\bar{x}_i, w_\lambda) - SE(x_{ij}, w_\lambda) \quad (16.14)$$

$$\bar{x}_i = \frac{1}{N} \sum_{j=1}^{N} x_{ij}$$

S_i measures the effect on the training squared error (SE) of replacing the ith input x_i by its average $\bar{x}_i$ for all exemplars (replacement of a variable by its average value removes its influence on the network output).

Note that in computing S_i, no retraining is done in evaluating $SE(\bar{x}_i, w_\lambda)$. Also note that it is not sufficient to just set $x_{ij} = 0 \ \forall \ j$, because the value of the bias of each hidden unit was determined during training and would not be offset properly by setting the input arbitrarily to zero. Of course, if the inputs are normalized to have zero mean prior to training, then setting an input variable to zero is equivalent to replacing it by its mean.

If S_i is large, the network error will be significantly increased by replacing the ith input variable with its mean. If S_i is small, we need the help of other measures to decide whether the ith input variable is useful. Three additional sensitivity measures [21] can be computed based on perturbating an input or a hidden variable and monitoring network output variations:

$$\text{Average Gradient (AG)} \quad S_i = \frac{1}{N} \sum_{j=1}^{N} \frac{\partial f^{(j)}}{\partial x_i} ,$$

$$\text{Average Absolute Gradient (AAG)} \quad S_i = \frac{1}{N} \sum_{j=1}^{N} |\frac{\partial f^{(j)}}{\partial x_i}| ,$$

$$\text{RMS Gradient (RMSG)} \quad S_i = \sqrt{\frac{1}{N} \sum_{j=1}^{N} \left[\frac{\partial f^{(j)}}{\partial x_i}\right]^2} ,$$

where for ease of notation we define $f(x) \equiv \hat{\mu}_\lambda(x)$ and $f^{(j)} \equiv f(x_{1j}, \ldots, x_{ij}, \ldots, x_{dj})$ is the network output given the jth input data pattern.

These three sensitivity measures together offer useful information. If S_i^{AG} is positive and large, then on average the change of the direction of the network output f is the same as that of the ith input variable. If S_i^{AG} is negative and has large magnitude, the change of the direction of f on average is opposite to that of the ith input variable. When S_i^{AG} is close to zero, we can not get much information from this measure. If S_i^{AAG} is large, the output f is sensitive to the ith input variable; if S_i^{AAG} is small, f is not sensitive to the ith input variable. If S_i^{RMSG} is very different from S_i^{AAG}, the ith input series could be very noisy and have a lot of outliers.

16.4.7 Empirical Example

As described in [42], we construct neural network models for predicting the rate of change of the U.S. Index of Industrial Production (IP). The prediction horizon for the IP results presented here is 12 months. Following previous work [27, 20], the results reported here use networks with three sigmoidal units and a single linear output unit.

The data set consists of monthly observations of IP and other macroeconomic and financial series for the period from January 1950 to December 1989. The data set thus has a total of 480 exemplars. Input series are derived from around ten raw time series, including IP, the Index of Leading Indicators, the Standard & Poors 500 Index, and so on. Both the "unfiltered" series and various "filtered" versions are considered for inclusion in the model, for a total of 48 possible input variables. The target series and all 48 candidate input series are normalized to zero mean and unit standard deviation. Figures 16.12 and 16.13 show the results of the sensitivity analysis for the case where the training-set consists of 360 exemplars randomly chosen from the 40 year period; the remaining 120 monthly observations constitute the test-set.

Local optima for the number of inputs are found at 15 on the FPE curve and 13 on the NCV curve. Due to the variability in the FPE and NCV estimates (readily apparent in figure 16.13 for NCV), we favor choosing the first good local minimum for these curves rather than a slightly better global minimum. This local minimum for NCV corresponds to a global minimum for the test error. Choosing it leads to a reduction of 35 (from 48 to 13) in the number of input series and a reduction in the number of network weights from 151 to 46. Inclusion of additional input variables, while decreasing the training error, does not improve the test-set performance.

This empirical example demonstrates the effectiveness of the *sensitivity-based pruning (SBP)* algorithm guided by an estimate of prediction risk, such as the *nonlinear cross-validation (NCV)* algorithm, for selecting a small subset of input variables from a large number of available inputs. The resulting network models exhibit better prediction performances, as measured by either estimates of prediction risk or errors on actual test sets, than models that make use of all 48 input series.

16.4.8 Gaining Economic Understanding through Model Visualization

Although this chapter has focussed on data-driven time series models for economic forecasting, it is possible to extract information from these models about

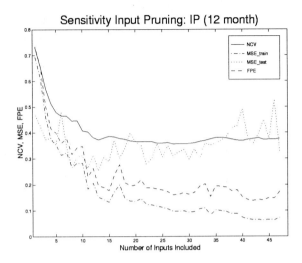

Fig. 16.12. Sensitivity-Based Pruning (SBP) method for selecting a subset of input variables for a neural net forecasting model [42]. The original network was trained on all 48 input variables to predict the 12 month percentage changes in Industrial Production (IP). The variables have been ranked in order of decreasing importance according to a sensitivity measure. The input variables are pruned one-by-one from the network; at each stage, the network is retrained. The figure shows four curves: the Training Error, Akaike Final Prediction Error (FPE), *Nonlinear Cross-Validation Error (NCV)* [30, 25], and the actual Test Error. NCV is used as a selection criterion and suggests that only 13 of the variables should be included. NCV predicts the actual test error quite well relative to FPE.

the structure of the economy. The sensitivity analyses presented above in section 16.4.6 provide a global understanding about which inputs are important for predicting quantities of interest, such as the business cycle. Further information can be gained, however, by examining the evolution of sensitivities over time [21].

Sensitivity analysis performed for an individual exemplar provides information about which input features play an important role in producing the current prediction. Two sensitivity measures for individual exemplars can be defined as:

$$\text{Delta Output (DO)} \quad S_i = \Delta f_i^{(j)}$$
$$\equiv f(x_{1j}, \ldots, x_{ij}, \ldots, x_{dj})$$
$$- f(x_{1j}, \ldots, \overline{x_i}, \ldots, x_{dj}),$$
$$\text{Output Gradient (OG)} \quad S_i = \frac{\partial f^{(j)}}{\partial x_{ij}},$$

where as above we define $f^{(j)} \equiv f(x_{1j}, \ldots, x_{ij}, \ldots, x_{dj})$. If S_i^{DO} or S_i^{OG} is large, then the ith variable plays an important role in making the current prediction,

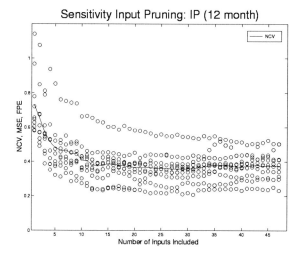

Fig. 16.13. Sensitivity Input Pruning for IP (12 month prediction horizon). The figure illustrates the spread in test-set error for each of the 10 subsets used to calculate NCV (denoted by circles). The NCV error is the average of these test-set errors.

and slightly changing the value of the variable may cause a large change in the network output. Figure 16.14 gives an example of a graphical display of the individual exemplar sensitivities.

Using this graphical display, we can observe which input variables play important roles in producing the current forecast, or which input variables, when we change them, can significantly increase or decrease the forecast error. We can also observe how the roles of different input variables change through time. For example, in Figure 16.14, for exemplars with indices from 56 to 60, the third input variable has large negative sensitivity measures. Starting with the 65th exemplar, the sixth input variable starts to play an important role, and this lasts until the 71th exemplar. This kind of display provides insight into the dynamics of the economy, as learned from the data by the trained neural network model.

16.5 Discussion

In concluding this brief survey of the algorithms for improving forecast accuracy with neural networks, it is important to note that many other potentially useful techniques have been proposed (see also chapter 17). Also, the empirical results presented herein are intended to be illustrative, rather than definitive. Further work on both the algorithms and forecasting models may yield additional improvements. As a final comment, I would like to emphasize that given the difficulty of macroeconomic forecasting, no single technique for reducing prediction risk is sufficient for obtaining optimal performance. Rather, a combination of techniques is required.

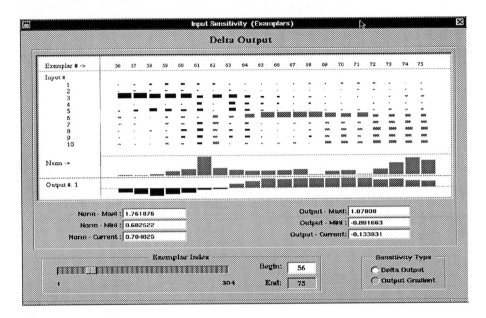

Fig. 16.14. Sensitivity analysis results for individual exemplars for a 10 input model for predicting the U.S. Index of Industrial Production [21]. Black and gray represent negative and positive respectively. The size of a rectangle represents the magnitude of a value. The monthly time index changes along the horizontal direction. The indices of input variables are plotted vertically. This type of graph shows which input variables are important for making forecasts at various points in time. This enables an understanding of economic relationships.

Acknowledgements

The author wishes to thank Todd Leen, Asriel Levin, Yuansong Liao, Hong Pi, Steve Rehfuss, Thorsteinn Rögnvaldsson, Matthew Saffell, Joachim Utans, Lizhong Wu and Howard Yang for their many contributions to this research. This chapter is an expanded version of [26]. This work was supported at OGI by ONR/ARPA grants N00014-92-J-4062 and N00014-94-1-0071, NSF grants CDA-9309728 and CDA-9503968, and at Nonlinear Prediction Systems by DARPA contracts DAAH01-92-CR361 and DAAH01-96-CR026.

References

1. H. Akaike. Statistical predictor identification. *Ann. Inst. Statist. Math.*, 22:203–217, 1970.
2. T. Ash. Dynamic node creation in backpropagation neural networks. *Connection Science*, 1(4):365–375, 1989.
3. A. Barron. Predicted squared error: a criterion for automatic model selection. In S. Farlow, editor, *Self-Organizing Methods in Modeling*. Marcel Dekker, New York, 1984.

4. R. Battiti. Using mutual information for selecting features in supervised neural net learning. *IEEE Trans. on Neural Networks*, 5(4):537–550, July 1994.
5. B. Bonnlander. Nonparametric selection of input variables for connectionist learning. Technical report, PhD Thesis. Department of Computer Science, University of Colorado, 1996.
6. R.T. Clemen. Combining forecasts: A review and annotated bibliography. *International Journal of Forecasting*, (5):559–583, 1989.
7. P. Craven and G. Wahba. Smoothing noisy data with spline functions: Estimating the correct degree of smoothing by the method of generalized cross-validation. *Numer. Math.*, 31:377–403, 1979.
8. R.L. Eubank. *Spline Smoothing and Nonparametric Regression*. Marcel Dekker, Inc., 1988.
9. S. Geman, E. Bienenstock, and R. Doursat. Neural networks and the bias/variance dilemma. *Neural Computation*, 4(1):1–58, 1992.
10. F. Girosi, M. Jones, and T. Poggio. Regularization theory and neural network architectures. *Neural Computation*, 7:219–269, 1995.
11. G. Golub, H. Heath, and G. Wahba. Generalized cross validation as a method for choosing a good ridge parameter. *Technometrics*, 21:215–224, 1979.
12. C.W.J. Granger and P. Newbold. *Forecasting Economic Time Series*. Academic Press, San Diego, California, 2nd edition, 1986.
13. C.W.J. Granger and T. Terasvirta. *Modelling Nonlinear Economic Relationships*. Oxford University Press, 1993.
14. J.D. Hamilton. *Time Series Analysis*. Princeton University Press, 1994.
15. B. Hassibi and D.G. Stork. Second order derivatives for network pruning: Optimal brain surgeon. In Stephen Jose Hanson, Jack D. Cowan, and C. Lee Giles, editors, *Advances in Neural Information Processing Systems 5*, pages 164–171. Morgan Kaufmann Publishers, San Mateo, CA, 1993.
16. T.J. Hastie and R.J. Tibshirani. *Generalized Additive Models*, volume 43 of *Monographs on Statistics and Applied Probability*. Chapman and Hall, 1990.
17. A.E. Hoerl and R.W. Kennard. Ridge regression: applications to nonorthogonal problems. *Technometrics*, 12:69–82, 1970.
18. A.E. Hoerl and R.W. Kennard. Ridge regression: biased estimation for nonorthogonal problems. *Technometrics*, 12:55–67, 1970.
19. Y. LeCun, J.S. Denker, and S.A. Solla. Optimal brain damage. In D.S. Touretzky, editor, *Advances in Neural Information Processing Systems 2*. Morgan Kaufmann Publishers, 1990.
20. A.U. Levin, T.K. Leen, and J.E. Moody. Fast pruning using principal components. In J. Cowan, G. Tesauro, and J. Alspector, editors, *Advances in Neural Information Processing Systems 6*. Morgan Kaufmann Publishers, San Francisco, CA, 1994.
21. Y. Liao and J.E. Moody. A neural network visualization and sensitivity analysis toolkit. In Shun ichi Amari, Lei Xu, Lai-Wan Chan, Irwin King, and Kwong-Sak Leung, editors, *Proceedings of the International Conference on Neural Information Processing*, pages 1069–74. Springer-Verlag Singapore Pte. Ltd., 1996.
22. R.B. Litterman. Forecasting with Bayesian vector autoregressions – five years of experience. *Journal of Business and Economic Statistics*, 4(1):25–38, 1986.
23. J. Moody. Challenges of Economic Forecasting: Noise, Nonstationarity, and Nonlinearity. Invited talk presented at Machines that Learn, Snowbird Utah, April 1994.
24. J. Moody. The effective number of parameters: an analysis of generalization and regularization in nonlinear learning systems. In J.E. Moody, S.J. Hanson, and R.P.

Lippmann, editors, *Advances in Neural Information Processing Systems 4*, pages 847–854. Morgan Kaufmann Publishers, San Mateo, CA, 1992.

25. J. Moody. Prediction risk and neural network architecture selection. In V. Cherkassky, J.H. Friedman, and H. Wechsler, editors, *From Statistics to Neural Networks: Theory and Pattern Recognition Applications*. Springer-Verlag, 1994.

26. J. Moody. Macroeconomic Forecasting: Challenges and Neural Network Solutions. In *Proceedings of the International Symposium on Artificial Neural Networks*. Hsinchu, Taiwan, 1995. Invited keynote address.

27. J. Moody, A. Levin, and S. Rehfuss. Predicting the U.S. index of industrial production. *Neural Network World*, 3(6):791–794, 1993. Special Issue: Proceedings of Parallel Applications in Statistics and Economics '93.

28. J. Moody, S. Rehfuss, and M. Saffell. Macroeconomic forecasting with neural networks. Manuscript in preparation., 1999.

29. J. Moody and T. Rögnvaldsson. Smoothing regularizers for projective basis function networks. In *Advances in Neural Information Processing Systems 9 (Proceedings of NIPS*96)*. MIT Press, Cambridge, 1997.

30. J. Moody and J. Utans. Architecture selection strategies for neural networks: Application to corporate bond rating prediction. In A.N. Refenes, editor, *Neural Networks in the Captial Markets*. John Wiley & Sons, 1994.

31. J.E. Moody. Note on generalization, regularization and architecture selection in nonlinear learning systems. In B.H. Juang, S.Y. Kung, and C.A. Kamm, editors, *Neural Networks for Signal Processing*, pages 1–10. IEEE Signal Processing Society, 1991.

32. J.E. Moody and J. Utans. Principled architecture selection for neural networks: Application to corporate bond rating prediction. In J.E. Moody, S.J. Hanson, and R.P. Lippmann, editors, *Advances in Neural Information Processing Systems 4*, pages 683–690. Morgan Kaufmann Publishers, San Mateo, CA, 1992.

33. M.C. Mozer and P. Smolensky. Skeletonization: A technique for trimming the fat from a network via relevance assessment. In David S. Touretzky, editor, *Advances in Neural Information Processing Systems 1*. Morgan Kaufmann Publishers, San Mateo, CA, 1990.

34. N. Murata, S. Yoshizawa, and S. Amari. Network information criterion – determining the number of hidden units for an artificial neural network model. *IEEE Transactions on Neural Networks*, 5(6):865–872, 1994.

35. M. Natter, C. Haefke, T. Soni, and H. Otruba. Macroeconomic forecasting using neural networks. In *Neural Networks in the Capital Markets 1994*, 1994.

36. H. Pi and C. Peterson. Finding the embedding dimension and variable dependencies in time series. *Neural Computation*, pages 509–520, 1994.

37. D. Plaut, S. Nowlan, and G. Hinton. Experiments on learning by back propagation. Technical Report CMU-CS-86-126, Dept. of Computer Science, Carnegie-Mellon University, Pittsburgh, Pennsylvania, 1986.

38. S. Rehfuss. Macroeconomic forecasting with neural networks. Unpublished simulations., 1994.

39. H. Rehkugler and H.G. Zimmermann, editors. *Neuronale Netze in der Ökonomie*. Verlag Vahlen, 1994.

40. N.R. Swanson and H. White. A model selection approach to real-time macroeconomic forecasting using linear models and artificial neural networks. Discussion paper, Department of Economics, Pennsylvania State University, 1995.

41. J. Utans and J. Moody. Selecting neural network architectures via the prediction risk: Application to corporate bond rating prediction. In *Proceedings of the First*

International Conference on Artificial Intelligence Applications on Wall Street. IEEE Computer Society Press, Los Alamitos, CA, 1991.
42. J. Utans, J. Moody, and S. Rehfuss. Selecting input variables via sensitivity analysis: Application to predicting the U.S. business cycle. In *Proceedings of Computational Intelligence in Financial Engineering*. IEEE Press, 1995.
43. G. Wahba. *Spline models for observational data*. CBMS-NSF Regional Conference Series in Applied Mathematics, 1990.
44. R.L. Winkler and S. Makridakis. The combination of forecasts. *Journal of Royal Statistical Society*, (146), 1983.
45. L. Wu and J. Moody. A smoothing regularizer for feedforward and recurrent networks. *Neural Computation*, 8(2), 1996.
46. H. Yang and J. Moody. Input variable selection based on joing mutual information. Technical report, Department of Computer Science, Oregon Graduate Institute, 1998.

17
How to Train Neural Networks

Ralph Neuneier and Hans Georg Zimmermann

Siemens AG, Corporate Technology, D-81730 München, Germany
{Ralph.Neuneier,Georg.Zimmermann}@mchp.siemens.de
http://w2.siemens.de/zfe_nn/homepage.html

Abstract. The purpose of this paper is to give a guidance in neural network modeling. Starting with the preprocessing of the data, we discuss different types of network architecture and show how these can be combined effectively. We analyze several cost functions to avoid unstable learning due to outliers and heteroscedasticity. The Observer - Observation Dilemma is solved by forcing the network to construct smooth approximation functions. Furthermore, we propose some pruning algorithms to optimize the network architecture. All these features and techniques are linked up to a complete and consistent training procedure (see figure 17.25 for an overview), such that the synergy of the methods is maximized.

17.1 Introduction

The use of neural networks in system identification or regression tasks is often motivated by the theoretical result that in principle a three layer network can approximate any structure contained in a data set [14]. Consequently, the characteristics of the available data determine the quality of the resulting model. The authors believe that this is a misleading point of view, especially, if the amount of useful information that can be extracted from the data is small. This situation arises typically for problems with a low signal to noise ratio and a relative small training data set at hand. Neural networks are such a rich class of functions, that the control of the optimization process, i. e. the learning algorithm, pruning, architecture, cost functions and so forth, is a central part of the modeling process. The statement that "The neural network solution is not better than [any classical] method" has been used too often in order to describe the results of neural network modeling. At any rate, the assessment of this evaluation presupposes a precise knowledge of the procedure involved in the neural network solution that has been achieved. This is the case because a great variety of additional features and techniques can be applied at the different stages of the process to prevent all the known problems like overfitting and sensitivity to outliers concerning neural networks. Due to the lack of a general recipe, one can often find a statement declaring that the quality of the neural network model depends strongly on the person who generated the model, which is usually perceived as negative. In contrast, we consider the additional features an outstanding advantage of neural networks compared to classical methods, which typically do not allow such a sophisticated control of their optimization. The aim of our article is to provide a set of techniques to efficiently exploit the capabilities of neural networks. More important, these features will be combined in such a way that we will achieve a maximal effect of synergy by their application.

First, we begin with the preprocessing of the data and define a network architecture. Then, we analyze the interaction between the data and the architecture (Observer - Observation Dilemma) and discuss several pruning techniques to optimize the network topology. Finally, we conclude by combining the proposed features into a unified training procedure (see fig. 17.25 in section 17.8).

Most of the paper is relevant to nonlinear regression in general. Some considerations are focussed on time series modeling and forecasting. All the proposed approaches have been tested on diverse tasks we have to solve for our clients, e. g. forecasting financial markets. The typical problem can be characterized by a relative small set of very noisy data and a high dimensional input vector to cover the complexity of the underlying dynamical system. The paper gives an overview of the unified training procedure we have developed to solve such problems.

17.2 Preprocessing

Besides the obvious scaling of the data (in the following abbreviated by scale(·)), which transforms the different time series such that each series has a mean value of zero and a statistical variance of one, some authors have proposed complicated preprocessing functions. In the field of financial forecasting, these functions are often derived from technical analysis in order to capture some of the underlying dynamics of the financial markets (see [37] and [25] for some examples).

After many experiments with real data, we have settled with the following simple transformations. If the original time series which has been selected as an input, is changing very slowly with respect to the prediction horizon, i. e. there is no clearly identifiable mean reverting equilibrium, then an indicator for the inertia and an information of the driving force has been proven to be very informative. The inertia can be described by a momentum (relative change, eq. 17.1) and the force by the acceleration of the time series (normalized curvature, eq. 17.2). If we have a prediction horizon of n steps into the future the original time series x_t is transformed in the following way:

$$\text{momentum:} \quad \tilde{x}_t = \text{scale}\left(\frac{x_t - x_{t-n}}{x_{t-n}}\right), \qquad (17.1)$$

$$\text{force:} \quad \hat{x}_t = \text{scale}\left(\frac{x_t - 2x_{t-n} + x_{t-2n}}{x_{t-n}}\right). \qquad (17.2)$$

In eq. 17.1, the relative difference is computed to eliminate exponential trends which, for example, may be caused by inflationary influences. Using only the preprocessing functions of eq. 17.1 typically leads to poor models which only follow obvious trends. The forces, i. e. the transformations by eq. 17.2, are important to characterize the turning points of the time series.

A time series may be fast in returning to it's equilibrium state after new information has entered the market, as is the case for most prices of goods and stock rates. In this case, we substitute eq. 17.2 by a description of the forces which drive the price back to the estimated equilibrium. A simple way to estimate the underlying price equilibrium is to take the average over some past values of the time series. Instead of using the relative difference between the estimate and the current value, we look at the difference between the equilibrium and the past value, which lies in the middle of the averaging window. In our example, this is formulated as

$$\hat{x}_t = \text{scale}\left(\frac{x_{t-n} - \frac{1}{2n+1}\sum_{\tau=0}^{2n} x_{t-\tau}}{x_{t-n}}\right). \qquad (17.3)$$

Note that in eq. 17.3 we use x_{t-n} instead of the present value x_t leading to an estimation of the equilibrium centered around x_{t-n}. Since, in fact, we are interested in measuring the tension between a price level and the underlying market equilibrium, this estimation offers a more appropriate description at the cost of ignoring the newest possible point information. This concept, known as *mean reverting* dynamics in economics, is analog to the behavior of a pendulum in physics.

17.3 Architectures

We will present several separate architectural building blocks (figures 17.1 to 17.7), which will finally be combined into a unified neural network architecture for time series analysis (figure 17.8). Most of the structural elements can be used in general regression.

17.3.1 Net Internal Preprocessing by a Diagonal Connector

In order to have potentially very different inputs on a comparable scale, we standardize the input data to zero mean and unit variance. A problem with this common approach is that outliers in the input can have a large impact. This is particularly serious for data that contain large shocks, as in finance and economics. To avoid this problem, we propose an additional, *net internal* (short for *network internal*) preprocessing of the inputs by scaling them according to

$$x' = \tanh(wx). \qquad (17.4)$$

The implementation of this preprocessing layer is shown in fig. 17.1. This layer has the same number of hidden units as the input layer and uses standard tanh squashing functions. The particular weight matrix between the input layer and the preprocessing layer is only a square diagonal matrix.

For short term forecasts (e. g. modeling daily returns of stock markets), we typically initialize the weights with the value of 0.1 to ensure that the tanh is in its linear range, which in turn ensures that the external inputs pass through essentially unchanged. If interested in long term models (e. g. six month forecast horizon), we prefer to start with an initial value of 1. The reason is that monthly data are typically more contaminated by "non-economical" effects like political shocks. The large initial weight values eliminate such outliers from the beginning.

The weights in the diagonal connector are restricted to be positive to avoid fluctuations of the sign of x' during training. This constraint keeps eq. 17.4 as a monotonic transformation with the ability to limit outliers. No bias should be used for the preprocessing layer to prevent numerical ambiguities. We found that these additional constraints improve the training stability of the preprocessing layer.

These weights will be adapted during training in the same way as all the other weights within the network. In practice, we observe both growing and shrinking values for the weights. Growing values cause a larger proportion of the input range to be compressed by the squashing function of the tanh. Very small values of diagonal elements indicate the option to prune the corresponding input.

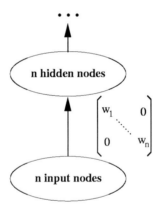

Fig. 17.1. Net internal preprocessing to limit the influence of outliers and to eliminate unimportant inputs.

17.3.2 Net Internal Preprocessing by a Bottleneck Network

It is an old idea to use a bottleneck network to shrink the dimensionality of the input vector. This technique was not only used to build encoders [1], but also to perform a principle component analysis [26]. Using a bottleneck as an internal preprocessing building block within a larger network offers the opportunity for us to compress the input information. In addition, the tanh squashing function of the bottleneck layer acts as a limiter of the outliers in the data.

In a first attempt, one may implement a bottleneck architecture as an input-hidden-output-layer sub-network using the inputs also as targets on the output level. An additional connector from the hidden layer is connected to the remaining parts of the network. This design allows the compression of input signals, but there are two major disadvantages implied. If we apply input pruning to the original inputs the decompression becomes disordered. Furthermore, adding noise to the inputs becomes more complex because the disturbed inputs have also to be used as targets at the output layer.

A distinct approach of the bottleneck sub-networks which avoids these difficulties is indicated in fig. 17.2. On the left side of the diagram 17.2 we have the typical compressor-decompressor network. The additional connection is a frozen identity matrix, which duplicates the input layer to the output layer. Since the target for this layer is set to zero, the output of the compressor-decompressor network will adopt the negative values of the inputs in order to compensate for the identical copy of the inputs. Even though the sign of the input signals has been reversed, this circumstance does not imply any consequence for the information flow through the bottleneck layer. This architecture allows an appropriate elimination of inputs as well as the disturbance of the input signals by artificial noise.

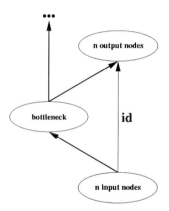

Fig. 17.2. Net internal preprocessing cluster using a bottleneck.

In our experiments, we observed that it is possible to train the bottleneck in parallel with the rest of the network. As a consequence, the bottleneck can be shrunk as long as the error on the training set remains at the same level as it would without a compression. As will be described in section 17.6 it is important to use a dynamically controlled input noise as a regularization method during learning. The architecture of fig. 17.2 allows the combination of noise and net internal preprocessing by data compression.

17.3.3 Squared Inputs

Following the suggestion of G. Flake in [10], we also supply the network with the squared values of the inputs which leads to an integration of global and local decision making. The processing of the original inputs and the squared inputs within a tanh activation function enables the network to act as a combination of widely-known neural networks using sigmoidal activation function (MLP) and radial basis function networks (RBF). The output of a typical three-layer MLP is

$$y = \sum_j v_j \tanh\left(\sum_i w_{ji} x_i - \theta_j\right), \qquad (17.5)$$

whereas the output of a RBF computes to

$$y = \sum_j v_j\, e^{-\frac{1}{2}\sum_i \left(\frac{x_i - \mu_{ji}}{\sigma_{ji}}\right)^2}. \qquad (17.6)$$

Some research papers typically deal with the comparison of these two types of basis functions whereas Flake proposes the following combining approach

$$y = \sum_j v_j \tanh\left(\sum_i w_{ji} x_i + u_{ji} x_i^2 - \theta_j\right), \qquad (17.7)$$

which covers the MLP of eq. 17.5 and simultaneously approximates the RBF output eq. 17.6 to a sufficient level.

Nevertheless, we propose some minor changes to Flake's approach by taking the square of the internally preprocessed inputs instead of supplying the squared original inputs separately. That way, we take advantage of the preprocessing cluster because it limits the outliers. Furthermore, pruning inputs leads to a simultaneous elimination of its linear and squared transformation. A possible implementation is shown in fig. 17.3. The connector id is a fixed identity matrix. The cluster above this connector uses square as the nonlinear activation function. The next hidden cluster is able to create MLP and RBF structures with its tanh squashing function.

One might think that the addition of a new connector between the square and the hidden layer with all its additional weights would boost the overfitting. Our experience indicates just the opposite. Furthermore, when optimizing a usual MLP, one typically stops training before the error on the training set gets too small (*early stopping*) in order to increase the probability of good results on the test set. We have observed, that the error on the training set can be very small while at the same time achieving good test results, even if one does not use weight pruning. Our experiences can be understood on the basis of the local modeling features of the architecture. That is, local artifacts in the input space can be encapsulated, so that they have no more global influence which may lead to bad generalization performance. This is especially true for high dimensional input spaces.

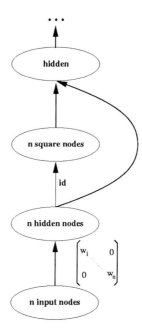

Fig. 17.3. Net internal preprocessing cluster and the square cluster, which produces the input signals for following clusters. Note, that the connector from bias to "hidden" is required but suppressed for visual clarity.

17.3.4 Interaction Layer

This section summarizes the articles of Weigend and Zimmermann in [36, 32]. In most applications of neural networks in financial engineering the number of inputs is huge (of the order of a hundred), but only a single output is used. This situation can be viewed as a large "inverted" funnel; the hard problem is that the only information available for learning is that of a single output. This is one of the reasons for the "data-hungriness" of single-output neural networks, i. e. a large set of training data is often required in order to distinguish nonlinearities from noise. The flip-side of the coin is that small data sets need to be regularized, thus only allowing an identification of a simple model, which implies a bias towards linear models. This circumstance is not to be confused with the bias towards linear models which is a consequence of some techniques for avoiding overfitting such as early stopping or weight decay.

One approach for increasing the information flow from the output side to the input side is to increase the number of output units. In the simplest case, two output units can be used, one to predict the return, the other one to predict the sign of the return [34].

Since we have to model dynamical systems, we would like to provide enough information to characterize the state of the autonomous part of the dynamics on the output side, similar to Takens' theorem for the notion of state on the input side [11]. The idea of multiple output units has recently been popularized in the connectionist community in a non-time series context by Caruana [6].

The present paper is concerned with time series, so the embedding of the output can be done analogously to the input side of a tapped "delay" line, indicated in fig. 17.4 as the *point prediction layer*.

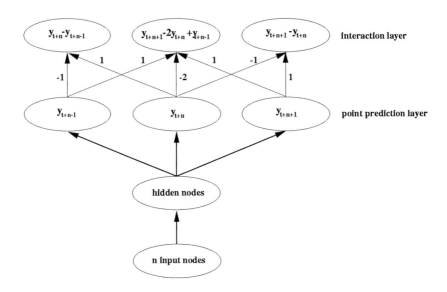

Fig. 17.4. Point predictions followed by the interaction layer.

For the forecast we are interested in y_{t+n}, the n-step ahead forecast of variable y. Additionally, we also want to predict y_{t+n-1} and y_{t+n+1}. However, after experimenting with this architecture, we do not consider it to be much of a benefit to our aim, because there seems only very little interaction between the behavior of the outputs as reflected through the implicit transfer by sharing hidden units.

This prompted us to introduce an explicit second output layer, the *interaction layer*. The additional layer computes the next-neighbor derivatives $(y_{t+n} - y_{t+n-1})$ and $(y_{t+n+1} - y_{t+n})$, as well as the curvature $(y_{t+n+1} - 2y_{t+n} + y_{t+n-1})$.

Differences between neighbors are encoded as fixed weights between point prediction and interaction layer, so they do not enlarge the number of parameters to be estimated. The overall cost function is the sum of all six contributions where all individual contributions are weighted evenly. If the target values are not properly scaled, it may be useful to give equal contribution to the error and to scale each error output by the average error of that output unit.

If the point forecasts were perfect, the interaction layer would have no effect at all. To explain the effect of non-zero errors, consider fig. 17.5. Both predictions of the three points have the same pointwise errors at each of the three neighboring points. However, both the slopes and the curvature are correct in Model 1 (they don't give additional errors), but do add to the errors for Model 2.[1]

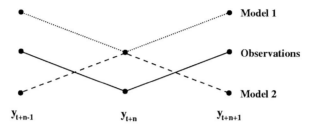

Fig. 17.5. Geometric interpretation of the effect of the interaction layer on the cost function. Given are three curves connecting points at three adjacent steps in time. Whereas Model 1 (connected by dashed line) and Model 2 (connected by the dotted line) have identical pointwise errors to the observations (connected by the solid line), taking derivatives and curvatures into account favors Model 1.

The principle of the interaction layer can also be used to model further relationships on the output side. Let us take as an example a forecasting model of several exchange rates. Between all these forecasts we should guarantee that there is no arbitrage between the assets at the forecast horizon. In other words, at y_{t+n} there is no way to change money in a closed loop and have a positive return. These intermarket relationships can be realized in form of an interaction layer.

In general, the control of intermarket relationships becomes more important if we proceed from forecasting models to portfolio analysis models. In the latter the correct forecast of the interrelationships of the assets is more important than a perfect result on one of the titles.

17.3.5 Averaging

Let us assume, that we have m sub-networks for the same learning task. Different solutions of the sub-networks may be caused by instabilities of the learning or by a different design of the sub-networks. It is a well known principle that averaging the output of several networks may give us a better and more stable result [24, 4].

These advantages can be clearly seen if we define as average error function

$$E_{\text{average}} = \frac{1}{T}\sum_{t=1}^{T}\left[\left(\frac{1}{m}\sum_{i=1}^{m} y_{i,t+n}\right) - y_{t+n}^d\right]^2 \qquad (17.8)$$

---

[1] Note that in the case of a quadratic error function, the interaction layer can be substituted by a single output layer of point predictions, combined with a positive definite non-diagonal quadratic form as target function.

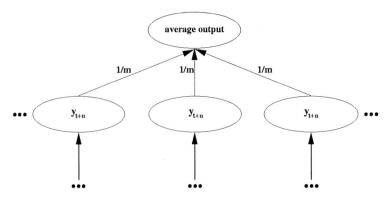

Fig. 17.6. Averaging of several point predictions for the same forecast horizon.

with y_{t+n}^d as the target pattern, $y_{i,t+n}$ as the output of sub-network i, m as the number of subnetworks, and T as the number of training patterns. Assuming that the errors of the sub-networks are uncorrelated,

$$\frac{1}{T}\sum_{t=1}^{T}(y_{i,t+n} - y_{t+n}^d)(y_{j,t+n} - y_{t+n}^d) = 0, \quad \forall\, i \neq j \qquad (17.9)$$

leads to

$$E_{\text{average}} = \frac{1}{m}\left(\frac{1}{m}\sum_{i=1}^{m}\left(\frac{1}{T}\sum_{t=1}^{T}(y_{i,t+n} - y_{t+n}^d)^2\right)\right) \qquad (17.10)$$

$$= \frac{1}{m}\,\text{average}\,(E_{\text{sub-networks}}) \,. \qquad (17.11)$$

According to this argument, the error due to the uncertainty of the training can be reduced. Finally, it is to be noted that averaging adds no additional information about the specific application of our network.

17.3.6 Regularization by Random Targets

It is known in the neural network community that the addition of random targets can improve the learning behavior[2]. An architectural realization of this idea is shown fig. 17.7.

In economical applications we typically use large input vectors in order to capture all probably relevant indicators. The resulting large number of parameters if connecting the input layer to a hidden layer of an appropriate size is the source of overfitting. To partly remedy this effect, one can extend the neural

---

[2] The authors would appreciate any useful citation.

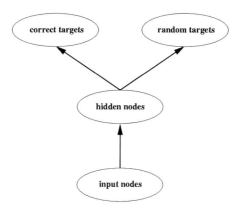

Fig. 17.7. A simple architecture using the original targets (upper left branch) and the random targets for regularization (upper right branch).

network with additional but random targets. As a decreasing error of this outputs can only be achieved by memorizing these random events, this technique absorbs a part of the overparametrization of the network. It should not be confused with artificial noise on the output side because the additional patterns are randomly selected according a probability distribution before the training and are held fixed during the learning.

The question if enough or too many additional random targets are supplied can be answered by observing the learning behavior. During training the error with respect to these random targets is steadily decreasing until convergence. If the number of additional targets is too small one may observe overfitting effects (e. g. using a validation set) on the real targets after this convergence. On the other hand, if the the number of additional targets is too large the learning slows down.

One may suspect the parameters focusing on the random targets may have a unpredictable effect on the generalization set but we could not observe such a behavior. If using squared inputs (see sec. 17.3.3), their local learning possibilities supports this by encapsulating local artifacts.

17.3.7 An Integrated Network Architecture for Forecasting Problems

As a next step in our investigation, we suppose our task to be an economic forecasting model. In the following section we integrate the architectural building blocks discussed above into an eleven layer network specialized to fulfill our purpose (see fig. 17.8). As indicated in section 17.2, the net external preprocessing contains at least two inputs per original time series: the momentum in the form of the relative difference (eq. 17.1) and a force indicator in the form of a curvature or mean reverting description (eq. 17.2).

The lowest part of the network shows the net internal preprocessing by a diagonal connector. Alternatively, we could also use the bottleneck network. By the square layer, we allow our network to cover the difference and similarity analysis of MLP- and RBF- networks. The signals from the internally preprocessed inputs and their squared values are used as inputs, weighted by the associated parameters, to the hidden embedding and hidden force layers (see fig. 17.8).

In contrast to typical neural networks, the upper part of the net is organized in a new way. The underlying dynamical system is supposed to be characterized by an estimation of different features around the forecast horizon. We distinguish these indicators by two particular characteristics, embeddings and forces. The forecast of these different features has been separated in two branches of the network in order to avoid interferences during training. Instead of directly forecasting our final target which has the form:

$$\frac{y_{t+n} - y_t}{y_t}, \qquad (17.12)$$

we use the following indicators as targets in these two output layers:

embeddings:

$$u_i = \frac{y_{t+n+i} + y_{t+n} + y_{t+n-i}}{3y_t} - 1 \quad i = 1, \ldots m$$

$$v_i = \frac{\frac{1}{2i+1}\sum_{j=-i}^{i} y_{t+n+j} - y_t}{y_t} \quad i = 1, \ldots m$$

$$(17.13)$$

forces:

$$r_i = \frac{-y_{t+n+i} + 2y_{t+n} - y_{t+n-i}}{3y_t} \quad i = 1, \ldots m$$

$$s_i = \frac{y_{t+n} - \frac{1}{2i+1}\sum_{j=-i}^{i} y_{t+n+j}}{y_t} \quad i = 1, \ldots m$$

Target u_i describes a normalized 3-point embedding with respect to our forecast horizon $t+n$ while v_i represents a complete average around $t+n$ versus present time. The forces r_i and s_i are formulated as curvatures or mean reverting (see also section 17.2). Similar to the embeddings they describe features of a dynamical system with an increasing width. The different widths in turn allow a complete characterization of the time series, analogous to the characterization by points in Takens' Theorem [29].

The motivation of this design is contained in the step from the embedding and force output layers to the multiforecast output layer. We have

$$\frac{y_{t+n} - y_t}{y_t} = u_i + r_i \; i = 1, \ldots m$$
$$\frac{y_{t+n} - y_t}{y_t} = v_i + s_i \; i = 1, \ldots m \, .$$
(17.14)

That means, we get $2m$ estimations of our final target by simply adding up the embeddings and their associated forces in pairs. In the network this is easy to realize by two identity connectors. After the multiforecast layer we can add an averaging connector to get our final output. This averaging can be done by fixed weights $1/2m$ or by learning after finishing the training of the lower network.

The control-embedding and control-force clusters are motivated by the following observation. On the embedding / force output level the network has to estimate only slightly different features depending on the width parameter. Neural networks have a tendency to estimate these features too similarly. To counteract this behavior we have to add two additional clusters which control the difference between the individual outputs inside the embedding and the force cluster. Thus, the network has not only to learn u_i and u_{i+1} on the embedding output level but also the difference $u_i - u_{i+1}$ on the control embedding level. The same is valid for v_i, r_i, s_i.

From a formal viewpoint the multiforecast cluster, the control-embedding cluster and control-force cluster are interaction layers supporting the identification of the underlying dynamical system in form of embeddings and forces. Keep in mind, that although the full network seems to be relatively complex, most of the connectors are fixed during training. Those are only used to produce the appropriate information flows whose design is the real focus of this section.

Our proposed network design allows for an intuitive evaluation of the dimension of the target indicators *embeddings* and *forces*: how to choose m in eq. 17.13? Start with a relative large m and train the network to the minimal error on the training set as will be described in section 17.8. Then train only the weights of the up to now fixed connector between cluster multi-forecast and average forecast (fig. 17.8). If the dimension m has been chosen too large training may lead to such weights which suppress the long range embeddings and forces. Thus, it is possible to achieve an optimal dimension m of embedding and forces. For the case of a six month forecast horizon we were successful with a value of $m = 6$.

The eleven layer network automatically integrates aspects of section 17.3.6 concerning random targets. If the dimension m has been chosen too large, then the extreme target indicators act as random targets. On the other hand, if the forecast problem is characterized by high noise in the short term, the indicators for smaller m values generate random targets. Thus, choosing m too large does not harm our network design as discussed in section 17.3.6, but can improve generalization by partly adsorbing the overparametrization of the network.

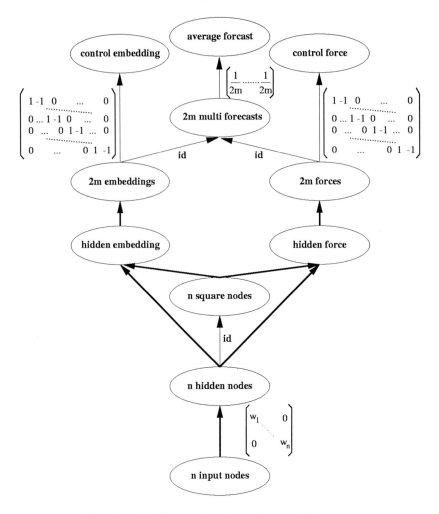

Fig. 17.8. The integrating eleven layer architecture.

The different forecasts o_i of the multiforecast cluster in fig. 17.8 can be used to estimate structural instability s of the network model by

$$s = \sum_{i=1}^{2m} |o_i - \bar{o}| \quad \text{with} \quad \bar{o} = \frac{1}{2m} \sum_{i=1}^{2m} o_i \, . \tag{17.15}$$

A subsequent decision support system using this network can interpret the measurements s as indicators how much one can trust the model. Note that these values of uncertainty must not be identified with error bars as described in section 17.5 because the s merely quantify the instability of the learning.

17.4 Cost Functions

Typical error functions can be written as a sum of individual terms over all T training patterns,

$$E = \frac{1}{T} \sum_{t=1}^{T} E_t , \qquad (17.16)$$

with the individual error E_t depending on the network output $y(x_t, w)$ and the given target data y_t^d. The often used square error,

$$E_t = \frac{1}{2} \left(y(x_t, w) - y_t^d \right)^2 , \qquad (17.17)$$

can be derived from the maximum-likelihood principle and a Gaussian noise model. Eq. 17.17 yields relatively simple error derivatives and results in asymptotically best estimators under certain distribution assumptions, i. e. homoscedasticity. In practical applications, however, several of these assumptions are commonly violated, which may dramatically reduce the prediction reliability of the neural network. A problem arising from this violation is the large impact outliers in the target data can have on the learning, which is also a result of scaling the original time series to zero mean and unit variance. This effect is particularly serious for data in finance and economics which contain large shocks. Another cause of difficulties in financial time series analysis is heteroscedasticity, i.e. situations where the variance of target variable changes over time. We will especially consider cases where the variance is input-dependent: $\sigma_t^2 = \sigma^2(x_t)$.

We propose two approaches to reduce the problems resulting from outliers and heteroscedasticity.

17.4.1 Robust Estimation with LnCosh

Outliers are common in financial time series and are usually caused by "information shocks" such as announcements of government data or dividends paid by companies that are out of line with market expectations. These shocks appear as discontinuities in the trajectory of an affected asset. To be robust against such shocks, typical cost functions like

$$E_t = \left| y(x_t, w) - y_t^d \right| \qquad (17.18)$$

are used which do not overweight large errors. A smoother version is given by

$$E_t = \frac{1}{a} \ln \cosh \left(a \left(y(x_t, w) - y_t^d \right) \right) , \qquad (17.19)$$

with parameter $a > 1$. We typically use $a \in [3, 4]$. This function approximates the parabola of the squared errors for small differences (Gaussian noise model), and is proportional to the absolute value of the difference for larger values of the difference (Laplacian noise model). The assumed noise model, the

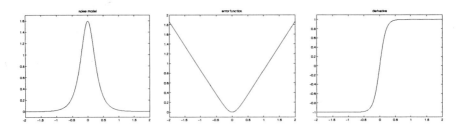

Fig. 17.9. The Laplacian-like noise model, left, the ln cosh error function 17.19, middle, and its corresponding derivative, right.

cost function 17.19 and its derivative for the case $a = 5$ is shown in fig. 17.9. The function ln cosh is motivated by the observation that the derivative of $|x|$ is $\text{sign}(x)$ and $\tanh(ax)$ is a smooth approximation of this step function with the integral $\int \tanh(az) dz = \frac{1}{a} \ln \cosh(az)$.

17.4.2 Robust Estimation with CDEN

This section describes a more general framework for robust estimation which is based on the theory of density estimation. The advantages are twofold. First, any parameters of the cost function (e. g. the a for eq. 17.19) can be determined by learning which avoids an artificial bias by setting those parameters to predefined values. Second, the proposed methods allow the modeling of heteroscedastic time series whose variance changes over time.

Probability density estimating neural networks have recently gained major attention in the neural network community as a more adequate tool to describe probabilistic relations than common feed-forward networks (see [35], [22] and [20]). Interest has hereby focussed on exploiting the additional information which is inherent to the conditional density, for example the conditional variance as a measure of prediction reliability [28], the representation of multi-valued mappings in the form of multi-modal densities to approach inverse problems [3], or the use of the conditional densities for optimal portfolio construction [18]. In this paper, we will use the *Conditional Density Estimation Network (CDEN)*, which is, among various other density estimation methods, extensively discussed in [22] and [20].

A possible architecture of the CDEN is shown in fig. 17.10. It is assumed that $p(y|x)$, the conditional density to be identified, may be formulated as the parametric density $p(y|\phi(x))$. The condition on x is realized by the parameter vector ϕ which determines the form of the probability distribution $p(y|\phi(x))$. Both $\phi(x)$ and $p(y|\cdot)$ may be implemented as neural networks with the output of the first determining the weights of the latter. Denoting the weights contained in the parameter prediction network $\phi(x)$ as w, we may thus write $p(y|x, w) = p(y|\phi_w(x))$. Assuming independence such that $p(y_1, \cdots, y_T|\cdot) = p(y_1|\cdot) \cdots p(y_T|\cdot)$ we minimize the negative Log-Likelihood error function

$$E = -\log p(y_1, \cdots, y_T|\cdot)$$

$$= -\log \prod_{t=1}^{T} p(y_t|x_t, w)$$

$$= -\sum_{t=1}^{T} \log p(y_t|x_t, w) \qquad (17.20)$$

by performing gradient descent using a variant of the common backpropagation algorithm, we give way to a maximum likelihood estimate of the weights in the parameter prediction network [22, 20].

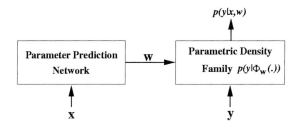

Fig. 17.10. Conditional Density Estimation Network, CDEN.

Specific problems with the discussed model are first approached by determining an appropriate density family $p(y|\phi(x))$. A powerful choice is a Gaussian mixture

$$p(y|\phi(x)) = \sum_{i=1}^{n} P_i(x) \, p(y|\mu_i(x), \sigma_i(x)), \quad P_i(x) \geq 0, \; \sum_{i}^{n} P_i(x) = 1, \qquad (17.21)$$

because they cover a wide range of probability models. For the univariate case (one output), the $p(y|\mu_i(x), \sigma_i(x))$ are normal density functions:

$$p(y|\mu_i(x), \sigma_i(x)) = \frac{1}{\sqrt{2\pi}\sigma_i(x)} e^{-\frac{1}{2}\left(\frac{y-\mu_i(x)}{\sigma_i(x)}\right)^2} . \qquad (17.22)$$

There are several ways to determine the individual density parameters contained in $(P_i, \mu_i, \sigma_i)_{i=1}^n$. Either they are set as the output of the parameter prediction network, or they are trained as x-independent, adaptable weights of the density network, or some of them may be given by prior knowledge (e. g. clustering, neuro-fuzzy).

A probability model $p(y|\cdot)$ based on the CDEN architecture in fig. 17.11 which perceives the presence of outliers and thus leads to robust estimators is illustrated in the left part of figure 17.12. The CDEN consists of two Gaussians with identical estimation for their mean $\mu(x)$. While its narrow Gaussian represents the distribution of the non-outliers part of the data, the wider one expresses the assumption that some data are located at larger distances from the prediction. The fact that outliers are basically exceptions may be reflected by an appropriate choice of the mixture weightings P_i, which may be regarded as prior probabilities for each Gaussian distribution. Using this probability model yields the following norm for the maximum-likelihood minimization:

$$E_t = -\log \left[\frac{P_1}{\sigma_1} e^{-\frac{1}{2}\left(\frac{y(x_t,w)-y_t^d}{\sigma_1}\right)^2} + \frac{P_2}{\sigma_2} e^{-\frac{1}{2}\left(\frac{y(x_t,w)-y_t^d}{\sigma_2}\right)^2} \right]. \quad (17.23)$$

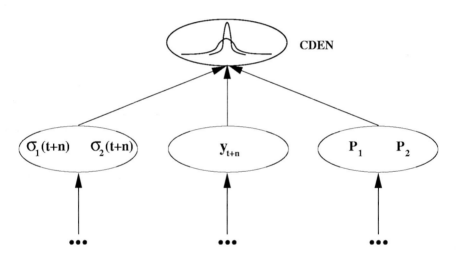

Fig. 17.11. A CDEN for limiting the influence of outliers.

The qualitative behavior of E_t in the one-dimensional case is illustrated in the middle and left part of figure 17.12 (compare also fig. 17.9). One may clearly see how the mixture limits the influence of outliers. Norms with influence-limiting properties are called M-estimators in regression [5].

The difficulty involved in using M-estimators is that they usually possess a set of parameters which have to be properly determined. In our case, the parameters are P_1, P_2, σ_1 and σ_2. In the framework of the CDEN architecture, they are considered as adaptable weights of the density network. The advantage is that unlike classical M-estimators, the parameters are determined by the data

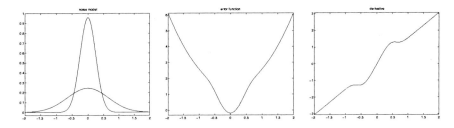

Fig. 17.12. The Gaussian mixture noise model, left, the error function 17.23, middle, and its corresponding derivative, right.

during training and thus do not bias the solution. One method which has proven to be successful in regression is to substitute eq. 17.23 by a mixture of a quadratic error function and a linear function of the absolute error to limit the influence of the outliers. An adaptive version of such an error measure may easily be implemented in the CDEN framework by substituting a mixture of Gaussian and Laplace distributions in equation 17.21. Other limiting error functions may be constructed alike.

Heteroscedasticity may arise due to changes in the risk associated with an investment. In the stock market, for example, the variance of a stock's return is commonly related to the company's debt-equity ratio due to the well-known leverage effect of a varying income on its return on equity. Heteroscedasticity has been extensively studied in time series analysis and is commonly approached using the (G)ARCH methodology. While the latter explains the conditional variances based on past residuals, the CDEN particularly accounts for nonlinear dependencies of the variances on past observations.

If we assume a normally distributed noise model with zero mean and variance σ^2, an appropriate representation is a single Gaussian with variable scale and location parameter. An implementation in the CDEN is straightforward. We use a parameter prediction network with two outputs, one for the conditional expectation and another one for the conditional variance. This special case of the CDEN has also been extensively investigated by Nix and Weigend [21]. During the training of the network, the weights are optimized with respect to

$$E_t = \left[\log\sqrt{2\pi}\sigma_t(x_t,w) + \frac{1}{2\sigma_t(x_t,w)^2}\left(y(x_t,w) - y_t^d\right)^2\right]. \quad (17.24)$$

Minimization according to eq. 17.17 and eq. 17.24 obviously only differs in that the individual errors $(y(x_t,w) - y_t^d)$ are weighted by estimates of the inverse variances $1/\sigma_t^2(x_t,w)$ in the CDEN. Training the CDEN thus corresponds to deriving a *generalized least square estimator (GLSE)*, except that we use an estimate $\hat{\sigma}_t(x_t,w)$ instead of the unknown σ_t.[3]

---

[3] Definition and properties of the GLSE are extensively discussed in [5]. The case where an estimate of σ is used during optimization is commonly denoted as a *two-stage estimation*.

17.5 Error Bar Estimation with CDEN

In addition to a more robust learning, the CDEN approach can be used to estimate the uncertainty associated with the prediction of the expected value $y_{t+n} := y(x_t, w)$. Here we assume a normally distributed error and have to optimize a likelihood function of the form

$$E = \frac{1}{T}\sum_{t=1}^{T}\left[\log\left(\sqrt{2\pi}\sigma(x_t, w_\sigma)\right) + \frac{(y(x_t, w) - y_t^d)^2}{2\sigma^2(x_t, w_\sigma)}\right]. \qquad (17.25)$$

If we assume a Laplacian noise model, the cost function is

$$E = \frac{1}{T}\sum_{t=1}^{T}\left[\log\left(2\sigma(x_t, w_\sigma)\right) + \frac{|y(x_t, w) - y_t^d|}{\sigma(x_t, w_\sigma)}\right]. \qquad (17.26)$$

The log cosh-approximation of the Laplacian model has the form

$$E = \frac{1}{T}\sum_{t=1}^{T}\left[\log\left(\pi\sigma(x_t, w_\sigma)\right) + \log\cosh\left(\frac{y(x_t, w) - y_t^d}{\sigma(x_t, w_\sigma)}\right)\right]. \qquad (17.27)$$

In fig. 17.13 we show a network architecture which we found useful to apply such error bar estimations. The architecture combines net internal preprocessing with the squared input approach and estimates in two branches the expected value $y(x_t, w)$ and the standard derivation $\sigma(x_t, w_\sigma)$. The CDEN cluster combines these pieces of information and computes the flow of the error.

We have found that the mixture of local and global analysis as shown in fig. 17.13 harmonizes well with the aim of solving a forecasting problem including error bars. To assure positive values for $\sigma(x)$ a suitable activation function $e^{(\cdot)}$ is used in the σ cluster. By using a positive offset (bias) one can avoid the singularities in the likelihood target function which are caused by very small $\sigma(x)$.

In fig. 17.14 the CDEN with one Gauss-function is combined with the architecture of section 17.3.7. Here we assume that the estimated "forces" contain the necessary information to approximate the uncertainty of our averaged forcecast:

$$\sigma^2(y_{t+n}|x_t) = \sigma^2(y_{t+n}|\text{forces}(x_t)) \qquad (17.28)$$

or, more specifically, using the acceleration and mean reverting forces of fig. 17.14:

$$\sigma^2(y_{t+n}|x_t) = \sum_{i=1}^{2m} w_i \cdot \text{force}_i^2(x_t). \qquad (17.29)$$

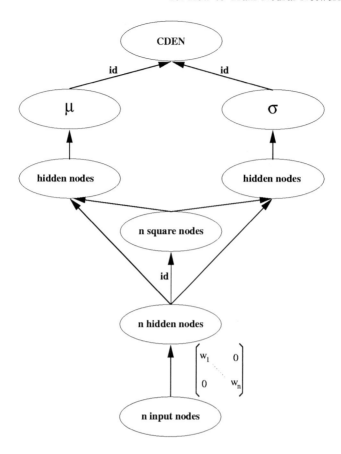

Fig. 17.13. Error bar estimation with CDEN.

By a linear combination of the squared values of the forces, this architecture learns the input dependent variance $\sigma(y_{t+n}|x_t)$ (see fig. 17.14 and eq. 17.13 for a description of forces). Thus, we are able to achieve a forecast as well as an error bar.

The interesting point in this combination is not only to be seen in the possibility of using the general frame as the means to identify dynamical systems. In this environment we can even analyze the responsibility of the forces, long range or short range, with respect to the uncertainty of the forecast. In several monthly forecast models we have found a monotonic increase of importance from the short to the long range forces.

Error bars and variance estimates are an essential piece of information for the typically used mean-variance approach in portfolio management [7]. The aim is to compute efficient portfolios which allocate the investor's capital to several assets in order to maximize the return of the investment for a certain level of

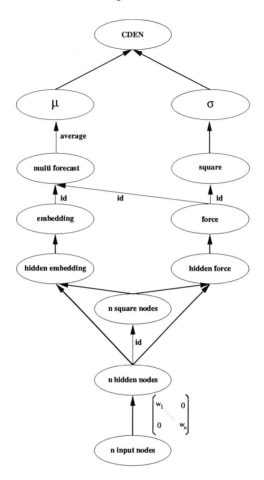

Fig. 17.14. Error bar estimation with CDEN using the architecture of section 17.3. The control-embedding and control-force clusters are suppressed for visual clarity.

risk. The variance is often estimated by linear models or is given a priori by an expert. In contrast, the CDEN approximates $\sigma(y_{t+n}|x_t)$ as a function of the current state of the financial market and of the forecast of the neural network.

Covariances can also be estimated with CDEN using a multivariate Gaussian in eq. 17.22. Implementation details can be found in [17, 19]. If one approximates the conditional density $p(y|\phi(x))$ with several Gaussians, the estimated expected mean, variance and covariance may be computed using typical moment generating transformations [23, 19].

17.6 Data meets Structure

17.6.1 The Observer-Observation Dilemma

Human beings believe that they are able to solve a psychological version of the *Observer-Observation Dilemma*. On the one hand, they use their observations to constitute an understanding of the laws of the world, on the other hand, they use this understanding to evaluate the correctness of the incoming pieces of information. Of course, as everybody knows, human beings are not free from making mistakes in this psychological dilemma. We encounter a similar situation when we try to build a mathematical model using data. Learning relationships from the data is only one part of the model building process. Overrating this part often leads to the phenomenon of overfitting in many applications (especially in economic forecasting). In practice, evaluation of the data is often done by external knowledge, i. e. by optimizing the model under constraints of smoothness and regularization [16]. If we assume that our model summarizes the best knowledge of the system to be identified, why shouldn't we use the model itself to evaluate the correctness of the data? One approach to do this is called Clearning [33]. In this paper, we present a unified approach of the interaction between the data and a neural network (see also [38]). It includes a new symmetric view on the optimization algorithms, here learning and cleaning, and their control by parameter and data noise.

17.6.2 Learning reviewed

We are especially interested in using the output of a neural network $y(x, w)$, given the input pattern, x, and the weight vector, w, as a forecast of financial time series. In the context of neural networks learning normally means the minimization of an error function E by changing the weight vector w in order to achieve good generalization performance. Again, we assume that the error function can be written as a sum of individual terms over all T training patterns, $E = \frac{1}{T} \sum_{t=1}^{T} E_t$. The often used sum-of-square error can be derived from the maximum-likelihood principle and a Gaussian noise model:

$$E_t = \frac{1}{2} \left(y(x, w) - y_t^d \right)^2, \qquad (17.30)$$

with y_t^d as the given target pattern. If the error function is a nonlinear function of the parameters, learning has to be done iteratively by a search through the weight space, changing the weights from step τ to $\tau + 1$ according to:

$$w^{(\tau+1)} = w^{(\tau)} + \Delta w^{(\tau)}. \qquad (17.31)$$

There are several algorithms for choosing the weight increment $\Delta w^{(\tau)}$, the easiest being *gradient descent*. After each presentation of an input pattern, the gradient $g_t := \nabla E_t|_w$ of the error function with respect to the weights is computed. In the batch version of gradient descent the increments are based on all training patterns

$$\Delta w^{(\tau)} = -\eta g = -\eta \frac{1}{T} \sum_{t=1}^{T} g_t, \tag{17.32}$$

whereas the pattern-by-pattern version changes the weights after each presentation of a pattern x_t (often randomly chosen from the training set):

$$\Delta w^{(\tau)} = -\eta g_t. \tag{17.33}$$

The learning rate η is typically held constant or follows an annealing procedure during training to assure convergence.

Our experiments have shown that small batches are most useful, especially in combination with Vario-Eta, a stochastic approximation of a Quasi-Newton method [9]:

$$\Delta w^{(\tau)} = -\frac{\eta}{\sqrt{\frac{1}{T}\sum(g_t-g)^2}} \cdot \frac{1}{N}\sum_{t=1}^{N} g_t, \tag{17.34}$$

with $N \leq 20$.

Let us assume, that the error function of a specific problem is characterized by a minimum in a narrow valley whose boundaries are parallel to the axes of the weight space. For the two dimensional case, such a situation is shown in fig. 17.15.

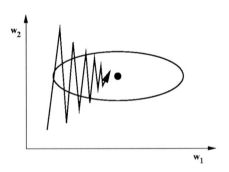

Fig. 17.15. Vario-Eta, a stochastic approximation of a Quasi-Newton method.

A gradient approach would follow a "zigzagging" track and would approximate the minimum very slowly. With Vario-Eta, the zigzagging along w_2 is damped and the drift along w_1 is accelerated. This behavior is similar to the weight trajectories classical Newton methods show in such a situation. The actual implementation uses stochastic approximations to compute the standard deviation.

The use of the standard deviation instead of the variance in Vario-Eta means an additional advantage in training large networks. Passing a long sequence of layers in a neural network, the error signals contain less and less information. The normalization in eq. 17.34 rescales the learning information for every weight. If one designed Vario-Eta as close as possible to second order methods, it would be appropriate to use the variance instead the standard deviation in the denominator, but then we would also lose the scaling property. In section 17.6.3, we will provide a further advantage of using the standard deviation.

Learning pattern-by-pattern or with small batches can be viewed as a stochastic search process because we can write the weight increments as:

$$\Delta w^{(\tau)} = -\eta \left[g + \left(\frac{1}{N} \sum_{t=1}^{N} g_t - g \right) \right]. \tag{17.35}$$

These increments consist of the terms g with a drift to a local minimum and of noise terms $(\frac{1}{N} \sum_{t=1}^{N} g_t - g)$ disturbing this drift.

17.6.3 Parameter Noise as an Implicit Penalty Function

Consider the Taylor expansion of $E(w)$ around some point w in the weight space

$$E(w + \Delta w) = E(w) + \sum_i \frac{\partial E}{\partial w_i} \Delta w_i + \frac{1}{2} \sum_{i,j} \frac{\partial^2 E}{\partial w_i \partial w_j} \Delta w_i \Delta w_j + \ldots . \tag{17.36}$$

Assume a given sequence of T disturbance vectors Δw_t, whose elements are uncorrelated over t with zero mean and variance (row-)vector var(Δw_i). The expected value $\langle E(w) \rangle$ can then be approximated by

$$\langle E(w) \rangle \approx \frac{1}{T} \sum_t E(w + \Delta w_t) = E(w) + \frac{1}{2} \sum_i \text{var}(\Delta w_i) \frac{\partial^2 E}{\partial w_i^2} \tag{17.37}$$

assuming that the first and second derivatives of E are stable if we are close to a local minimum. In eq. 17.37, noise on the weights acts implicitly as a penalty term to the error function given by the second derivatives $\frac{\partial^2 E}{\partial w_i^2}$. The noise variances var(Δw_i) operate as penalty parameters. As a consequence, flat minima solutions which may be important for achieving good generalization performance are favored [13].

Learning pattern-by-pattern introduces automatically such noise in the training procedure i.e., $\Delta w_t = -\eta \cdot g_t$. Close to convergence, we can assume that g_t is i.i.d. with zero mean and variance vector var(g_i) so that the expected value can be approximated by

$$\langle E(w) \rangle \approx E(w) + \frac{\eta^2}{2} \sum_i \text{var}(g_i) \frac{\partial^2 E}{\partial w_i^2} . \tag{17.38}$$

This type of learning introduces a local penalty parameter var(g_i), characterizing the stability of the weights $w = [w_i]_{i=1,\ldots,k}$. In a local minimum the sum of gradients for the weight w_i is $\sum g_{it} = 0$ whereas the variance var(g_i) may be large. In this case the solution is very sensitive against resampling of the data and therefore unstable. To improve generalization the curvature of the error function around such weights with high variance should be strongly penalized. This is automatically done by pattern-by-pattern learning.

The noise effects due to Vario-Eta learning $\Delta w_t(i) = -\frac{\eta}{\sqrt{\sigma_i^2}} \cdot g_{ti}$ leads to an expected value

$$\langle E(w) \rangle \approx E(w) + \frac{\eta^2}{2} \sum_i \frac{\partial^2 E}{\partial w_i^2}. \qquad (17.39)$$

By canceling the term var(g_i) in eq. 17.38, Vario-Eta achieves a simplified uniform penalty parameter, which depends only on the learning rate η. Whereas pattern-by-pattern learning is a slow algorithm with a locally adjusted penalty control, Vario-Eta is fast only at the cost of a simplified uniform penalty term.

local	learning rate (Vario-Eta)	→	global penalty
global	learning rate (pattern by pattern)	→	local penalty

Following these thoughts, we will now show that typical Newton methods should not be used in stochastic learning. Let us assume that we are close to a local minimum and that the Hessian H of the error function is not significantly changing anymore. On this supposition, the implied noise would be of the form $\Delta w_t = \eta H^{-1} g_t$. Because of the stable Hessian, the mean of the noise is zero so that the expected error function becomes

$$\langle E(w) \rangle \approx E(w) + \frac{\eta^2}{2} \sum_i \text{var}(g_i) \left(\frac{\partial^2 E}{\partial w_i^2} \right)^{-1}. \qquad (17.40)$$

In this case, we have again a local control of the penalty parameter through the variance of the gradients var(g_i), like in the pattern by pattern learning. But now, we are penalizing weights at points in the weight space where the inverse of the curvature is large. This means, we penalize flat minima solutions, which counters our goal of searching for stable solutions.

Typically Newton methods are used as cumulative learning methods so that the previous arguments do not apply. Therefore, we conclude, that second order methods should not be used in stochastic search algorithms. To support global

Table 17.1. Structure-Speed-Dilemma.

learning	structure	speed
pattern by pattern	+	−
VarioEta	−	+

learning and to exploit the bias to flat minima solution of such algorithms, we only use pattern-by-pattern learning or Vario-Eta in the following.

We summarize this section by giving some advice on how to achieve flat minima solutions (see also table 17.1):

- Train the network to a minimal training error solution with Vario-Eta, which is a stochastic approximation of a Newton method and therefore fast.
- Add a final phase of pattern-by-pattern learning with a uniform learning rate to fine tune the local curvature structure by the local penalty parameters (eq. 17.38). For networks with many layers, this step should be omitted because the gradients will vanish due to the long signal flows. Only Vario-Eta with its scaling capability can solve such optimization problems appropriately.
- Use a learning rate η as high as possible to keep the penalty effective. The training error may vary a bit, but the inclusion of the implicit penalty is more important.

We want to point out that the decision of which learning algorithm to use not only influences the speed and global behavior of the learning, but also, for fixed learning rates, leads to different structural solutions. This structural consequence has also to be taken into account if one analyzes and compares stochastic learning algorithms.

17.6.4 Cleaning reviewed

When training neural networks, one typically assumes that the input data is noise-free and one forces the network to fit the data exactly. Even the control procedures to minimize overfitting effects (i.e., pruning) consider the inputs as exact values. However, this assumption is often violated, especially in the field of financial analysis, and we are taught by the phenomenon of overfitting not to follow the data exactly. Clearning, as a combination of cleaning and learning, has been introduced in [33]. In the following, we focus on the cleaning aspects. The motivation was to minimize overfitting effects by considering the input data as being corrupted by noise whose distribution has to be learned also.

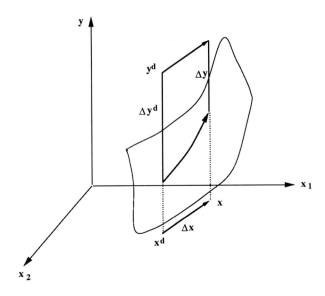

Fig. 17.16. If the slope of the modeled function is large, then a small shift in the input data decreases the output error dramatically.

The cleaning error function for the pattern t is given by the sum of two terms assuming same variance levels for input and output

$$E_t^{y,x} = \frac{1}{2}\left[(y_t - y_t^d)^2 + (x_t - x_t^d)^2\right] = E_t^y + E_t^x \quad (17.41)$$

with x_t^d, y_t^d as the observed data point. In the pattern-by-pattern learning, the network output $y(x_t, w)$ determines the weight adaptation as usual,

$$w^{(\tau+1)} = w^{(\tau)} - \eta \frac{\partial E^y}{\partial w}. \quad (17.42)$$

We also must memorize correction vectors Δx_t for all input data of the training set in order to present the cleaned input x_t to the network,

$$x_t = x_t^d + \Delta x_t. \quad (17.43)$$

The update rule for the corrections, initialized with $\Delta x_t^{(0)} = 0$ can be derived from typical adaptation sequences $x_t^{(\tau+1)} = x_t^{(\tau)} - \eta \frac{\partial E^{y,x}}{\partial x}$ leading to

$$\Delta x_t^{(\tau+1)} = (1-\eta)\Delta x_t^{(\tau)} - \eta(y_t - y_t^d)\frac{\partial y}{\partial x}. \quad (17.44)$$

This is a nonlinear version of the error-in-variables concept in statistics [27] (see also fig. 17.16 for the two dimensional case).

All of the necessary quantities, i. e. $(y_t - y_t^d)\frac{\partial y(x,w)}{\partial x}$ are computed by typical back-propagation algorithms anyway. We found that the algorithms work well if the same learning rate η is used for both the weight and cleaning updates. For regression, cleaning forces the acceptance of a small error in x, which can in turn decrease the error in y dramatically, especially in the case of outliers. Successful applications of cleaning are reported in [33] and [30].

Although the network may learn an optimal model for the cleaned input data, there is no easy way to work with cleaned data on the test set because for this data we do not know the output target difference for computing eq. 17.44. As a consequence, the model is evaluated on a test set with a different noise characteristic compared to the training set. We will later propose a combination of learning with noise and cleaning to work around this serious disadvantage.

17.6.5 Data Noise reviewed

Artificial noise on the input data is often used during training because it creates an infinite number of training examples and expands the data to empty parts of the input space. As a result, the tendency of learning by heart may be limited because smoother regression functions are produced.

Now, we consider again the Taylor expansion, this time applied to $E(x)$ around some point x in the input space. The expected value $\langle E(x) \rangle$ is approximated by

$$\langle E(x) \rangle \approx \frac{1}{T}\sum_t E(x + \Delta x_t) = E(x) + \frac{1}{2}\sum_j \text{var}(\Delta x_j)\frac{\partial^2 E}{\partial x_j^2} \qquad (17.45)$$

where $\frac{\partial^2 E}{\partial x_j^2}$ are the diagonal elements of the Hessian H_{xx} of the error function with respect to the inputs x. Again, in eq. 17.45, noise on the inputs acts implicitly as a penalty term to the error function with the noise variances $\text{var}(\Delta x_j)$ operating as penalty parameters (compare eq. 17.37). Noise on the input improves generalization behavior by favoring smooth models [3].

The noise levels can be set to a constant value, e. g. given by a priori knowledge, or adaptive as described now. We will concentrate on a uniform or normal noise distribution. Then, the adaptive noise level ξ_j is estimated for each input j individually. Suppressing pattern indices, we define the average residual errors ξ_j and ξ_j^2 as:

$$\text{uniform residual error: } \xi_j = \frac{1}{T}\sum_t \left|\frac{\partial E^y}{\partial x_j}\right|, \qquad (17.46)$$

$$\text{Gaussian residual error: } \xi_j^2 = \frac{1}{T}\sum_t \left(\frac{\partial E^y}{\partial x_j}\right)^2. \qquad (17.47)$$

Actual implementations use stochastic approximation, e. g. for the uniform residual error

$$\xi_j^{(\tau+1)} = (1 - \frac{1}{T})\xi_j^{(\tau)} + \frac{1}{T}\left|\frac{\partial E^y}{\partial x_j}\right|. \tag{17.48}$$

The different residual error levels can be interpreted as follows (table 17.2): A small level ξ_j may indicate an unimportant input j or a perfect fit of the network concerning this input j. In both cases, a small noise level is appropriate. On the other hand, a high value of ξ_j for an input j indicates an important but imperfectly fitted input. In this case high noise levels are advisable. High values of ξ_j lead to a stiffer regression model and may therefore increase the generalization performance of the network. Therefore, we use ξ_j or ξ_j^2 as parameter to control the level of noise for input j.

Table 17.2. The interpretation of different levels of the residual errors ξ.

observation of the residual error	interpretation	advice for the noise control
ξ small	perfect fit or unimportant input	use low noise
ξ large	imperfect fit but important input	use high noise

17.6.6 Cleaning with Noise

Typically, training with noisy inputs involves taking a data point and adding a random variable drawn from a fixed or adaptive distribution. This new data point x_t is used as an input to the network. If we assume, that the data is corrupted by outliers and other influences, it is preferable to add the noise term to the cleaned input. For the case of Gaussian noise the resulting new input is:

$$x_t = x_t^d + \Delta x_t + \xi\phi, \tag{17.49}$$

with ϕ drawn from the normal distribution. The cleaning of the data leads to a corrected mean of the data and therefore to a more symmetric noise distribution, which also covers the observed data x_t.

We propose a variant which allows more complicated and problem dependent noise distributions:

$$x_t = x_t^d + \Delta x_t - \Delta x_k, \qquad (17.50)$$

where k is a random number drawn from the indices of the memorized correction vectors $[\Delta x_t]_{t=1,\ldots,T}$. By this, we exploit the distribution properties of the correction vectors in order to generate a possibly asymmetric and/or dependent noise distribution, which still covers $x_t = x_t^d$ if $k = t$.

One might wonder why we want to disturb the cleaned input $x_t^d + \Delta x_t$ with an additional noisy term Δx_k. The reason for this is that we want to benefit from representing the whole input distribution to the network instead of only using one particular realization. This approach supplies a solution to the cleaning problem when switching from the training set to the test set as described in section 17.6.4.

17.6.7 A Unifying Approach: The Separation of Structure and Noise

In the previous sections we explained how the data can be separated into a cleaned part and an unexplainable noisy part. Analogously, the neural network is described as a time invariant structure (otherwise no forecasting is possible) and a noisy part.

| data | $\rightarrow$ cleaned data | + time invariant data noise |
| neural network | $\rightarrow$ time invariant parameters | + parameter noise |

We propose using cleaning and adaptive noise to separate the data and using learning and stochastic search to separate the structure of the neural network.

| data | $\leftarrow$ cleaning (neural network) | + adaptive noise (neural network) |
| neural network | $\leftarrow$ learning (data) | + stochastic search (data) |

The algorithms analyzing the data depend directly on the network whereas the methods searching for structure are directly related to the data. It should be clear that the model building process should combine both aspects in an alternating or simultaneous manner. We normally use learning and cleaning simultaneously. The interaction of the data analysis and network structure algorithms is a direct embodiment of the the concept of the Observer-Observation Dilemma.

The aim of the unified approach can be described, exemplary assuming here a Gaussian noise model, as the minimization of the error due to both, the structure and the data:

$$\frac{1}{2T}\sum_{t=1}^{T}\left[(y_t - y_t^d)^2 + (x_t - x_t^d)^2\right] \to \min_{x_t, w} \qquad (17.51)$$

Combining the algorithms and approximating the cumulative gradient g by $\tilde{g}$ using an exponential smoothing over patterns, we obtain

$$\begin{array}{c} \text{data} \\ \hline \Delta x_t^{(\tau+1)} = (1-\eta)\Delta x_t^{(\tau)} - \eta(y_t - y_t^d)\frac{\partial y}{\partial x} \\ x_t = x_t^d + \underbrace{\Delta x_t^{(\tau)}}_{\text{cleaning}} - \underbrace{\Delta x_k^{(\tau)}}_{\text{noise}} \end{array} \qquad (17.52)$$

$$\begin{array}{c} \text{structure} \\ \hline \tilde{g}^{(\tau+1)} = (1-\alpha)\tilde{g}^{(\tau)} + \alpha(y_t - y_t^d)\frac{\partial y}{\partial w} \\ w^{(\tau+1)} = w^{(\tau)} - \underbrace{\eta\tilde{g}^{(\tau)}}_{\text{learning}} - \underbrace{\eta(g_t - \tilde{g}^{(\tau)})}_{\text{noise}} \end{array}$$

The cleaning of the data by the network computes an individual correction term for each training pattern. The adaptive noise procedure according to eq. 17.50 generates a potentially asymmetric and dependent noise distribution which also covers the observed data. The implied curvature penalty, whose strength depends on the individual liability of the input variables, can improve the generalization performance of the neural network.

The learning of the structure involves a search for time invariant parameters characterized by $\frac{1}{T}\sum g_t = 0$. The parameter noise supports this exploration as a stochastic search to find better "global" minima. Additionally, the generalization performance may be further improved by the implied curvature penalty depending on the local liability of the parameters. Note that, although the description of the weight updates collapses to the simple form of eq. 17.33, we preferred the formula above to emphasize the analogy between the mechanism which handles the data and the structure.

In searching for an optimal combination of data and parameters, we have needed to model in both. This is not an indication of our failure to build a perfect model but rather it is an important element for controlling the interaction of data and structure.

17.7 Architectural Optimization

The initial network can only be a guess on the appropriate final architecture. One way to handle this inadequacy is to use a growing network. As a possibility in the opposite direction, pruning can be applied to shrink the network. From our experience the advantage of fast learning of growing networks is more than counterbalanced by the better learning properties of large architectures. At least in the first learning steps, large networks are not trapped as quickly in local minima. When applying the pruning methods according the *late stopping* concept, which trains the weights of the network until the error function converges (see section 17.8.1), the learning procedure had a sufficient number of adaptation steps to adjust the broad network structure. Another advantage of the pruning technique is the greater flexibility in generating sparse and possibly irregular network architectures.

17.7.1 Node-Pruning

We evaluate the importance of input- or hidden nodes by using as a test value

$$\text{test}_i = E(x_1, \ldots, x_i = \mu(x_i), \ldots, x_n) - E(x_1, \ldots, x_n), \qquad (17.53)$$

with $\mu(x_i)$ describing the mean value of the i-th input in the time series for the training set. This test value is a measure of the increase of the error if we omit one of the input series.

The creation of the bottleneck structure in the net internal preprocessing (section 17.2) is best performed by applying this test for the hidden neurons in the bottleneck on the training set. Thereby the disturbance of the learning is reduced to the minimum. In one of the steps of the training procedure (section 17.8) we will show that the pruning of the input neurons should be done on the validation set to improve the time stability of the forecasting.

In addition to the deletion of nodes the ranking of the inputs can give us a deeper insight into the task (which inputs are the most relevant and so on).

17.7.2 Weight-Pruning

The neural network topology represents only a hypothesis of the true underlying class of functions. Due to possible mis-specification, we may have defects of the parameter noise distribution. Pruning algorithms not only limit the memory of the network, but they also appear to be useful for correcting the noise distribution in different ways.

Stochastic-Pruning

Stochastic-Pruning [8] is basically a t-test on the weights w,

$$\text{test}_w = \frac{|w + g|}{\sqrt{\frac{1}{T} \sum (g_t - g)^2}}, \qquad (17.54)$$

with $g = \frac{1}{T}\sum_t g_t$. Weights with low test$_w$ values constitute candidates for pruning. From the viewpoint of our approach, this pruning algorithm is equivalent to the cancellation of weights with low *liability* as measured by the size of the weight divided by the standard deviation of its fluctuations. By this, we get a stabilization of the learning against resampling of the training data.

This easy to compute method worked well in combination with the early stopping concept in contrast to insufficient results with the late stopping concept. For early stopping Stochastic-Pruning acts like a t-test, whereas for late stopping the gradients of larger weights do not fluctuate enough to give useful test values. Thus, the pruning procedure includes an artificial bias to nonlinear models. Furthermore, Stochastic-Pruning is also able to revive already pruned weights.

Early-Brain-Damage

A further weight pruning method is EBD, *Early-Brain-Damage* [31], which is based on the often cited OBD pruning method [15]. In contrast to OBD, EBD allows its application before the training has reached a local minimum.

For every weight, EBD computes as a test value an approximation of the difference between the error function for $w = 0$ versus the value of the error function for the best situation this weight can have

$$\text{test}_w = E(0) - E(w_{\min}) = -gw + \frac{1}{2}w'Hw + \frac{1}{2}gH^{-1}g' \ . \tag{17.55}$$

The above approximation is motivated by a Taylor expansion of the error function. From

$$E(\tilde{w}) = E(w) + g(\tilde{w} - w) + \frac{1}{2}(\tilde{w} - w)'H(\tilde{w} - w) \tag{17.56}$$

we get

$$E(0) = E(w) - gw + \frac{1}{2}w'Hw \tag{17.57}$$

and as a solution to the minimum problem $E(\tilde{w}) \to \min$ we have

$$w_{\min} = w - H^{-1}g' \quad \text{together with} \quad E(w_{\min}) = E(w) - \frac{1}{2}gH^{-1}g' \ . \tag{17.58}$$

The difference of these two error values is the proposed EBD test. The Hessian H in this approach is computed in the same way as in the original OBD calculus [15].

One of the advantages of EBD over OBD is the possibility of performing the testing while being slightly away from a local minimum. In our training procedure we propose using noise even in the final part of learning so that we are only near a local minimum. Furthermore, EBD is also able to revive already pruned weights.

Similar to Stochastic Pruning, EBD favors weights with a low rate of fluctuations. If a weight is pushed around by a high noise, the implicit curvature

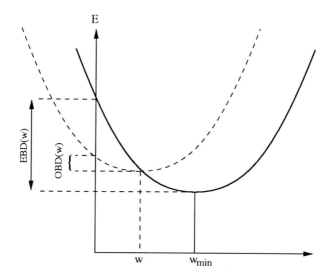

Fig. 17.17. EBD versus ODB weight pruning.

penalty would favor a flat minimum around this weight leading to its elimination by EBD.

Our practical experience shows that EBD pruning allows the creation of extremely sparse networks. We had examples where we could prune an initial network with 3000 weights down to a structure with around 50 weights. The first iterations of EBD pruning would typically give no improvement in generalization. This is due to the fact that EBD is testing the importance of a weight regarding the error function, i. e. the same information which is used by the backpropagation algorithm. To say it in another way: EBD cancels out only weights which are not disturbing the learning. Only at the end of the training procedure does the network have to give up a part of the coded structure which leads to an improvement in generalization.

Inverse-Kurtosis

A third pruning method we want to discuss is a method which we call *Inverse-Kurtosis*. The motivation follows an analysis of the following examples of possible distributions of gradient impulses forcing a weight shown in fig. 17.18.

If the network is trained to a local minimum the mean of all gradients by definition is equal to zero. Nevertheless the distribution of the gradients may differ. Now we have to analyze the difference of a peaked or very broad distribution versus a normal distribution. It is our understanding that the peaked distribution indicates a weight which reacts only to a small number of training patterns. A broader distribution, on the other hand, is a sign that many training patterns

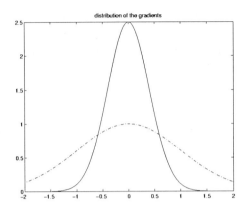

Fig. 17.18. Possible distributions of the gradients g_t, if weights are fixed.

focus on the optimization of this weight. In other words, a weight with a peaked distribution has learned by heart special events of the time series but it has not modeled a general underlying structure of the data. Therefore, Inverse-Kurtosis pruning harmonizes well with our network architecture which can encapsulate local structures with its squared layer. Furthermore, a weight with a very broad distribution of its gradients is pushed around by random events because it reacts to almost every pattern with a similar strength. A straight forward way to distinguish between the above distributions is based on the kurtosis to measure the difference to a normal distribution:

$$\text{distance}_w = \left(\frac{\frac{1}{T}\sum_{t=1}^{T}(g_t - g)^4}{\left(\frac{1}{T}\sum_{t=1}^{T}(g_t - g)^2\right)^2} - 3 \right)^2 . \quad (17.59)$$

To rank importance of the network weights based on this difference to a normal distribution we define the test values as

$$\text{test}_w = \frac{1}{\varepsilon + |\text{distance}_w|}, \quad (17.60)$$

with the small term $\varepsilon \approx 0.001$ to avoid numerical problems.

Similar to the previously mentioned pruning techniques we have now large test values for weights we do not want to eliminate. Note that in this test, we have neglected the size of the gradients and only taken into account the form of the distribution. The weights themselves are not a part of the evaluation, and we have not observed that this method has much effect on the distribution of the weight sizes found by the learning. In contrast, Stochastic Pruning and EBD have a tendency to prune out small weights because they explicitly refer to the size of the weights. Basically, Inverse-Kurtosis computes low test values

(i. e. probable pruning candidates) for weights which are encapsulating only local artifacts or random events in the data instead of modeling a general structure.

In our experience we found that this technique can give a strong improvement of generalization in the first several iterations of pruning.

Instability-Pruning

The local consequence of our learning until the minimal training error is that the cumulative gradient g for every weight is zero on the training set,

$$g = \frac{1}{T} \sum_{t \in T} g_t = 0. \tag{17.61}$$

If this condition was valid for the validation set our model would be perfectly stable:

$$g_V = \frac{1}{V} \sum_{t \in V} g_t = 0. \tag{17.62}$$

Since g_V typically is different from zero we have to check if the measured difference is significant enough to indicate instability. In the language of statistics this is a two sample test for the equality of the mean of two distributions. The following test value (Welch-test) measures this difference:

$$\text{distance}_w = \frac{g_V - g}{\sqrt{\frac{\sigma_V^2}{V} + \frac{\sigma_T^2}{T}}}, \tag{17.63}$$

which is approximately normally distributed with zero mean and unit variance. As a pruning test value we define

$$\text{test}_w = \frac{1}{\varepsilon + |\text{distance}_w|}. \tag{17.64}$$

This test value can be used to construct a ranking of the weights in the following way: Train the network until weights are near a local optimum and compute the test values. Then, take out 5% of the most unstable weights as measured by this test criterion and redo the learning step. Eliminating more than 5% of the weights is not recommandable because Instability-Pruning is not referring to the cost function used for learning, and thus may have a large impact on the error. This test of stability allows the definition of an interesting criterion when to stop the model building process. After pruning weights until a given stability level, we can check if the model is still approximating well enough. This final stopping criterion is possible due to the asymptotic normal distribution of the test. For example, if we define weights with $|\text{distance}_w| < 1$ as stable, then we prune weights with $\text{test}_w < 1$.

One may argue that the usage of the validation set in this approach is an implicit modeling of the validation set and therefore the validation set allows no error estimation on the generalization set. On the other hand, the typical need for an validation set is the estimation of a final stopping point in the pruning procedure. We have tried an alternative approach: Instead of using the validation set in terms of g_V and σ_V^2 in our comparison let us take a weighted measurement of the training set such that most recent data is given more importance:

$$\tilde{g} = \frac{2}{T(T+1)} \sum_{t \in T} t g_t, \qquad (17.65)$$

$$\tilde{\sigma}^2 = \frac{2}{T(T+1)} \sum_{t \in T} t(g_t - \tilde{g})^2. \qquad (17.66)$$

Checking the stability over the training set by substituting g_v resp. σ_V^2 with eq. 17.65 resp. 17.66 in eq. 17.63 avoids using the validation set. In our first experiences we observed that this version works as well as the previous one although the evaluation still needs more testing.

In comparison to EBD, which eliminates weights with only a small effect on the error function, Instability-Pruning introduces a new feature in the process of model building: stability over time. Additional information about when to choose which pruning method will be given in the following section 17.8.4.

17.8 The Training Procedure

The authors believe that complex real world problems should be attacked by a combination of various methods in order to handle the different types of difficulties which may arise during the optimization of neural networks. Thus, the aim of this section is to link all the previously described features to a complete and consistent training procedure.

17.8.1 Training Paradigms: Early vs. Late Stopping

One of the most well known techniques for attacking the overfitting problem is the early stopping procedure. In this procedure, we start the learning with a network initialized by small weights. In its most simple version one uses a validation set to estimate the beginning of overfitting, at which point the network learning is stopped. A more sophisticated variant of the procedure is to learn up to the start of overfitting and then to prune a part of the network structure, by say 10% of the weights. Then, one restarts the smaller network and repeats the steps. This sequence, which is schematically shown in in fig. 17.19, has to be iterated until a stable structure with no signs of overfitting is achieved.

In principle this procedure may work and it will generate a model with good generalization performance, but in many cases it will fail to do so, as we have observed in our experiments during the last several years. Difficulties with early stopping arise because the stopping point turns up after a few learning epochs through the training data. The authors have worked on examples where the

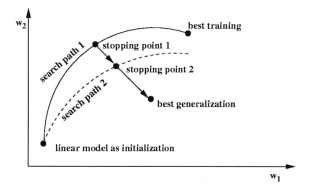

Fig. 17.19. Early stopping: After initialization with small weights the network is trained until the error on the validation set starts to increase (search path 1 to stopping point 1). Then, some weights are pruned, the remaining weights are reinitialized and training starts again until stopping point 2. This procedure has to be iterated until a stable structure with no signs of overfitting is achieved.

stopping point appeared as early as after the second epoch of learning. In this case, the solution is restricted to linear models, since the network has not been offered any chance to learn a complex nonlinearity from the data. A decreased learning rate does not mean a reduction of this bias, because it only slows down the movement away from the linear models. Using initial weights with larger values is also problematic for two reasons. The random initial (probably incorrect) specification of the network may lead to decreasing error curves due to shrinking weight values, but after a while overfitting will probably start again. Another important critique of this initialization is the intrinsic dependency of the solution on the starting point.

The same argument is true for some of the penalty term approaches for regularizing the networks. Weight decay, for example, is a regularizer with a bias towards linear models. Most of the other penalty functions (like data independent smoothness) set additional constraints on the model building process. These restrictions are not necessarily related to the task to be solved (see [16] for smoothing regularizers for neural networks with sigmoidal activation function in the hidden layer).

Following these thoughts, in this section we present a sequence of steps that follow a late stopping concept. We start with small weights to lower the influence of the random initialization. Then, we learn until the minimal error on the training data is achieved. Typically, the network shows clear overfitting at this point but we can be sure that the maximum nonlinearity which can be represented by the chosen network, has been learned from the data. At this point we must describe the way back to a solution with good generalization performance, as indicated in fig. 17.20.

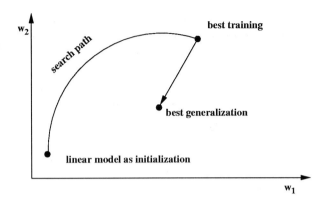

Fig. 17.20. Late stopping: After initialization with small weights the network is trained until the minimal training error. Subsequent optimization of the network structure by pruning increases generalization performance. Then, the network is trained again until the minimal training error.

Basically the two parts of the training, learning to the minimal training error and extracting an appropriate generalizing model, can be understood as a generation of a structural hypothesis followed by a falsification of the generated structure. These steps will be explained in detail in the next paragraphs.

17.8.2 Setup Steps

Before starting the learning process we have to specify the network architecture. Let us assume that our aim is the identification of a dynamical system. Then, we propose using the network architecture (fig. 17.8) of section 17.3, may be with the CDEN extension (fig. 17.14) of section 17.5 in order to additionally estimate error bars.

A further setup decision is the separation of the data set into a training set, a validation set and a generalization set. Possible separations are shown in fig. 17.21.

Considering forecasting models of time series (at least in economic applications), the sets should be explicitly separated in the order indicated in fig. 17.21.1. Otherwise there is no chance to test the model stability over time. If we randomly took out the validation set of the training set (fig. 17.21.2), we would have no chance to test this stability because the validation patterns are always embedded in training patterns.

In the proposed sequence we do not use the most recent data before the test set in the learning. As a consequence, one might tend to choose the validation patterns from the oldest part while using the most recent data in the training set (fig. 17.21.3). This leads to a better basis for the generation of our structural hypothesis by learning. Basically, the pruning methods are a falsification of the generated structure. Using the separation in fig. 17.21.3 with the validation set

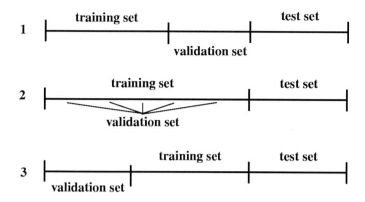

Fig. 17.21. Separation of the data.

including the oldest data can make this falsification misleading. In a fast changing world, model stability over time is an important performance characteristic of a good network model. Thus, we prefer the separation in fig. 17.21.1.

The several preprocessed time series which serve as inputs have to be checked for correlation because highly correlated inputs only serve to increase the amount of numerical instability in our model. To avoid this, we can introduce a bottleneck substructure to perform a principle component analysis. For our typical application with about 50 inputs, we use net internal preprocessing by the diagonal matrix. For models with a very large number of inputs (some hundreds) the bottleneck approach may be superior.

If using the net internal preprocessing by the diagonal matrix, it is necessary to check the pairwise correlation of the inputs. Following the common guideline of using lots of chart indicators in financial modeling it is typical that many of these indicators have a high correlation. Keeping them in the training set will cause the solutions of different optimization runs to give us different answers regarding the importance of the input factors.

Since we often start with a number of weights which is large relative to the number of training patterns, we also propose using a small weight decay during learning. The penalty parameter should be in the range of 10% of the learning rate because we are not interested in reducing the number of effective parameters. With this small value of the decay parameter, only those weights which simply learn nothing are pulled to zero. By this, we still achieve the minimal training error, but eliminate all the unnecessary weights which have an unpredictable effect on the test set.

17.8.3 Learning: Generation of Structural Hypothesis

If confronted with a large data set, such as in the task of forecasting daily returns, we propose training the weights with the VarioEta adaptation algorithm. As a stochastic approximation to a Newton method this is a fast learning technique.

Obeying the arguments about the implicit penalty of section 17.6 we should let the VarioEta training be followed by a simple pattern by pattern learning with a constant learning rate to achieve structurally correct learning. As mentioned, it is valuable to hold the learning rate as high as possible to benefit from the implied penalty term. However, if interested in monthly forecasting models, one may use only the pattern-by-pattern learning because learning speed is not relevant due to the small data set. On the other hand, the implied curvature penalty is more important to generate good models (see section 17.6).

We propose to start the cleaning and the cleaning noise procedure of section 17.6 from the beginning. In this way, one can observe the following interesting relationship and interaction between the stochastic learning and the cleaning noise which improves the learning behavior. Since the initial noise variance is set to zero, the noise to the input variables will start with a low level and will then increase rapidly during the first learning epochs. After several epochs, when the network has captured some of the structure in the data, the noise decays in parallel with the residual input error of the network. As a consequence, in the beginning of the training, the network can learn only the global structure in the data. Later in the training process, more and more detailed features of the data are extracted which leads simultaneously to lower cleaning noise levels (fig. 17.22). Cleaning[4] improves the learning process by sequencing the data structure from global to increasingly specialized features. Furthermore, noise improves the stochastic search in the weight space and therefore reduces the problem of local minima.

Fig. 17.22. Due to the large noise level in the beginning of the training phase, the network first learns global features. At later training steps, it is able to extract more and more details.

Depending on the problem, we observed that the implicitly controlled noise level can go down close to zero or it can converge to an intermediate level. We finish the step of learning to a minimal training error when we observe a stable

---

[4] Even in an early stopping concept Cleaning may improve the learning.

behavior of the error function on the training and validation set simultaneously with the stable behavior of the mean noise level for all inputs on the training set. The final model of this step is a structural hypothesis about the dynamical system we are analyzing.

17.8.4 Pruning: Falsification of the generated Structure

Next, we have to test the stability of our model over time. Especially in economic modeling, it is not evident that the structure of the capital markets we are extracting from the seventies or eighties are still valid now. We can check this stability by testing the network behavior on the validation set which follows the training set on the time axis.

By using input pruning on the validation set (see section 17.7), we are able to identify input series that may be of high relevance in the training set but may also have no or even a counterproductive effect on the validation data. Note that, the pruning of non-relevant time series is not possible before achieving the minimal training error because there exists no well defined relationship between all the inputs and the target variables before. On the other hand, unimportant or inconsistent inputs with respect to the validation set should not be eliminated later in order to facilitate the construction of models with high generalization performance by subsequent learning.

An alternative way to check the stability of the model is to perform weight pruning on the validation set using the Instability-Pruning algorithm. If the data set is very large, weight pruning has a significant advantage because only one pass through the data set is necessary, in comparison to n passes to get the ranking of the input pruning with n as the number of inputs. This becomes very important in the field of data-mining where we have to deal with hundreds or thousands of megabytes of training data.

Each time after pruning of weights or inputs, the network is trained again to achieve a stable behavior of the error curves on the training / validation set (fig. 17.24) and of the noise level (fig. 17.23). Here, by stable behavior of the error curve we mean that there is no significant downward trend (i. e. improving of the model) or upward trend (i. e. restructuring of the model). In the diagram shown in fig. 17.25, these stop conditions are marked with (∗).

To eliminate local artifacts in the network structure at this stage, we can include some iterations of Inverse-Kurtosis pruning and subsequent retraining. The process of iteratively pruning and training ends if a *drastic* decay, i. e. 5%, of this criterion is no longer observable. For several iterations of training and pruning, this strategy leads to a rapidly decreasing error on the validation set.

Due to the fact that the shrinking number of network parameters leads to increasing gradients, pattern by pattern learning with a constant learning rate is most useful because the implicit local curvature penalty becomes increasingly important.

One may criticize the proposed pruning policy because it is indeed an optimization on the validation set. Thus, the error on the validation set does not represent a good estimation of the generalization error anymore. We believe that there is no way to omit this pruning step since the test of the time stability is important for achieving models with high generalization performance.

In the subsequent part of the network learning, we use "Occam's Razor" by weight pruning based on test values computed on the training set. This may be EBD or Instability-Pruning (see section 17.7.2). The performance of both methods based on hundreds of experiments is comparable. After pruning about 10% resp. 5% of the active weights, we train again until stable error curves and noise levels are obtained.

On can combine these pruning methods by first eliminating weights with Instability-Pruning and generating very sparse networks using EBD afterwards. Instability-Pruning favors model stability over time whereas EBD allows sparse models which still approximate well.

If using the eleven layer architecture of fig. 17.8, weight pruning should only be applied to the weights connecting the preprocessing or square layer to the hidden layers for two reasons. First, these connectors represent the majority of the weights in the neural network. Second, it is important to apply the pruning techniques only to such weights which have the same distance to the output neurons. Thus, the test values of the gradient based pruning algorithms are comparable which leads to a reliable ranking of the weights.[5]

There is an interesting interaction between the pruning and the adaptive noise procedure (see fig. 17.23). At each pruning step, we eliminate some pa-

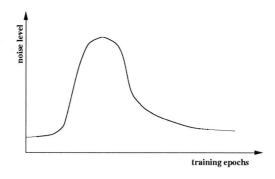

Fig. 17.23. Interaction between pruning and the adaptive noise level.

rameters which have memorized some of the structure in the data. Thereby, the

---

[5] If one decides not to use the eleven layer architecture but a typical neural network extended by random targets, one has to consider the following consequences for input and weight pruning. The test values have to be computed only with respect to the correct targets. By this, weights which are modeling random targets will loose their importance. EBD pruning works very well in identifying such flat minima weights (section 17.7.2), and thus, is our favorite pruning method for such a modeling approach.

residual input errors will increase, which leads to increased noise levels on the inputs. Consequently, after each pruning step the learning has to focus on more global structures of the data. Owing to the fact that by pruning the network memory gradually decreases, the network cannot rely on any particular features of the data but rather is forced to concentrate more and more on the general underlying structure. Interestingly, this part of the procedure can be viewed as being the opposite of learning until the minimal training error. There the network is forced by the cleaning noise to learn the global structures first before trying to extract specific characteristics from the data.

17.8.5 Final Stopping Criteria of the Training

The last question we have to answer is the definition of the final stopping point. Obviously, after many weight pruning and training iterations, the error will increase for the training and the validation set. Thus, the minimal error on the validation set should give us an estimate of when to stop the optimization. In practice, the error curves during learning are not smooth and monotonic curves with only one minimum. The most simple advice is to prune until all weights are eliminated while keeping a trace to store the best intermediate solution on the validation set. Fig. 17.24 displays the error behavior in the different phases of the training procedure assuming the worst situation from the viewpoint of an early stopping method due to the immediately increasing error on the validation set.

Instability-Pruning offers an alternative definition of a final stopping point. If we substitute EBD by this pruning method for applying Occam's Razor, each pruning / retraining iteration will increase the stability of the model because Instability-Pruning deletes unstable weights (see section 17.7.2).

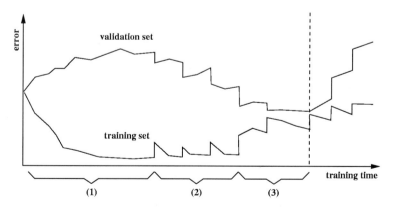

Fig. 17.24. The error behavior during the training. Region (1) describes the way to a minimal training error, (2) is the typical behavior during the input pruning on the validation set and (3) shows the consequence of Occam's razor in form of the pruning.

17.8.6 Diagram of the Training Procedure

The following diagram in fig. 17.25 shows the training steps combined into a unified training procedure.

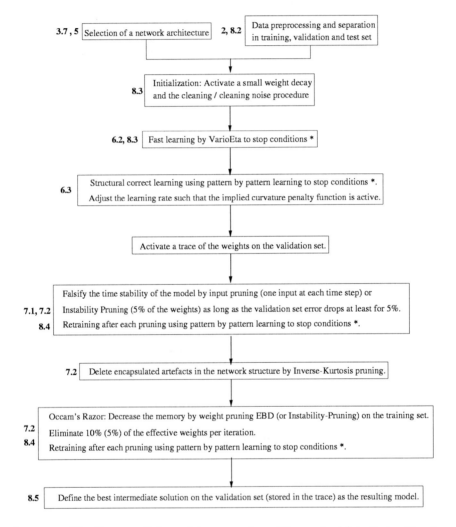

Fig. 17.25. The diagram of the training procedure. The marks (∗) indicates the stopping points (section 17.8.3 and 17.8.4). Box numbers correspond to sections.

17.9 Experiments

In a research project sponsored by the European Community we are using the proposed approach to estimate the returns of 3 financial markets for each of the G7 countries [12]. These estimations will be subsequently used in an asset allocation scheme to create a Markowitz-optimal portfolio. In this paper, we only report the results of the estimation of the German bond rate, which is one of the more difficult tasks due to the reunification of Germany and GDR. Here, we have to predict the return of the bond rate 6 months ahead. The inputs consist of 39 variables obtained by preprocessing 16 relevant financial time series[6]. The training set covers the time from April 1974 to December 1991, the validation set from January 1992 to May 1994. The results are collected using the out-of-sample data which runs from June 1994 to May 1996. To demonstrate the behavior of the algorithms, we compare our approach with a standard neural network with one hidden layer (20 neurons, tanh transfer function) and one linear output (eq. 17.30 as error function). First, we trained the neural network until convergence with pattern-by-pattern learning using a small batch size of 20 patterns. We refer to this network as a classical approach. Then, we trained the 11-layer network (fig. 17.8) using the unified approach as described in section 4.1. Due to small data sets we used pattern-by-pattern learning without VarioEta. The data were manipulated by the cleaning and noise method of eq. 17.50. We compare the resulting predictions of the networks on the basis of three performance measures (see tab. 17.3). First, the hit rate counts how often the sign of the return of the bond has been correctly predicted. As for the other measures, the step from the forecast model to a trading system is here kept very simple. If the output is positive, we buy shares of the bond, otherwise we sell them. The potential realized is the ratio of the return to the maximum possible return over the training (test) set. The annualized return is the average yearly profit of the trading systems. Our approach turns out to be superior. For example, we almost doubled the annualized return from 4.5% to 8.5% on the test set. In fig. 17.26, we compare the accumulated return of the two approaches on the test set. The unified approach not only shows a higher profitability, but also has by far a less maximal draw down.

Table 17.3. Comparison of the hit rate, the realized potential and the annualized return of the two networks for the training (test) data.

network	hit rate	realized potential	annualized return
our approach	96% (81%)	100% (75%)	11.22% (8.5%)
classical approach	93% (66%)	96% (44%)	10.08% (4.5%)

---

[6] At the moment, we are not allowed to be more specific on the actual data we used.

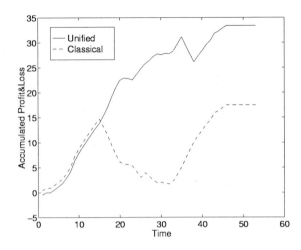

Fig. 17.26. Comparison of the accumulated profit&loss curve of the two approaches.

17.10 Conclusion

A typical regression analysis by neural networks begins with the statement: "In principal a neural network can model everything". If the data is not very reliable due to noise or missing factors, learning is only one part of the model building procedure. Often, task-independent regularization is used as weapon to reduce the uncertainty introduced by the data. More generally, the model building process demands more information (prior knowledge) about the specific task.

We have shown that there are a large number of techniques for including additional constraints which depend on the problem. Consequently, the resulting model is not only based on the training data but also on the additional constraints and also on the exact steps of the training procedure. From a Bayesian viewpoint, this can be described as an integration of priors in the model building process. Thus, this article can also be interpreted as a discussion of valuable priors and of how to combine them in order to achieve a maximal synergy.

The authors believe that the additional features are an outstanding advantage of neural networks and will lead to more robust and successful models. This paper gives an overview about some of the features we experienced as useful in our applications.

Finally, the unified training procedure is a recipe for building a model with neural networks by way of a relatively simple sequence of steps. An important aspect of this paper has been the study of the interactions between the different techniques. The described algorithms are integrated in the *Simulation Environment for Neural Networks*, **SENN**, a product of Siemens AG. More information can be found on the web page http://www.senn.sni.de.

References

1. D. H. Ackley, G. E. Hinton, and T. J. Sejnowski. A learning algorithm for Boltzmann machines. *Cognitive Science*, 9:147–169, 1985. [Reprinted in [2]].
2. J. A. Anderson and E. Rosenfeld, editors. *Neurocomputing: Foundations of Research*. The MIT Press, Cambridge, MA, 1988.
3. C. M. Bishop. *Neural Networks for Pattern Recognition*. Clarendon Press, Oxford, 1995.
4. L. Breiman. Bagging predictors. Technical Report TR No. 421, Department of Statistics, University of California, 1994.
5. H. Bunke and O. Bunke. *Nonlinear Regression, Functional Analysis and Robust Methods*, volume 2. John Wiley and Sons, 1989.
6. R. Caruana. Multitask learning. *Machine Learning*, 28:41, 1997.
7. E. J. Elton and M. J. Gruber. *Modern Portfolio Theory and Investment Analysis*. John Wiley & Sons, 1995.
8. W. Finnoff, F. Hergert, and H. G. Zimmermann. Improving generalization performance by nonconvergent model selection methods. In I. Aleksander and J Taylor, editors, *Neural Networks 2: Proc. of the Inter. Conference on Artificial Neural Networks, ICANN-92*, pages 233–236, 1992.
9. W. Finnoff, F. Hergert, and H. G. Zimmermann. Neuronale Lernverfahren mit variabler Schrittweite. 1993. Tech. report, Siemens AG.
10. G. Flake. Square unit augmented, radially extended multilayer perceptrons. Technical report, Siemens Corporate Research Center, Princeton, and this volume. 1997.
11. N. A. Gershenfeld. An experimentalist's introduction to the observation of dynamical systems. In B. L. Hao, editor, *Directions in Chaos*, volume 2, pages 310–384. World Scientific, Singapore, 1989.
12. P. Herve, P. Naim, and H. G. Zimmermann. Advanced Adaptive Architectures for Asset Allocation: A Trial Application. In *Forecasting Financial Markets*, 1996.
13. S. Hochreiter and J. Schmidhuber. Flat minima. *Neural Computation*, 9(1):1–42, 1997.
14. K. Hornik. Approximation Capabilities of Multilayer Feedforward Networks. *Neural Networks*, 4:251 – 257, 1991.
15. Y. le Cun, J. S. Denker, and S. A. Solla. Optimal brain damage. In D. S. Touretzky, editor, *Advances in Neural Information Processing Systems 2 (NIPS*89)*, pages 598–605, San Mateo, CA, 1990. Morgan Kaufmann.
16. J. E. Moody and T. S. Rögnvaldsson. Smoothing regularizers for projective basis function networks. In Michael C. Mozer, Michael I. Jordan, and Thomas Petsche, editors, *Advances in Neural Information Processing Systems*, volume 9, page 585. The MIT Press, 1997.
17. P. M. Williams. Using Neural Networks to Model Conditional Multivariate Densities. Technical Report CSRP 371, School of Cognitive and Computing Sciences, Univ. of Sussex, February 1995.
18. R. Neuneier. Optimal asset allocation using adaptive dynamic programming. In *Advances in Neural Information Processing Systems*, volume 8. MIT Press, 1996.
19. R. Neuneier. *Optimale Investitionsentscheidungen mit Neuronalen Netzen*. PhD thesis, Universität Kaiserslautern, Institut für Informatik, 1998.
20. R. Neuneier, W. Finnoff, F. Hergert, and D. Ormoneit. Estimation of Conditional Densities: A Comparison of Neural Network Approaches. In *Intern. Conf. on Artificial Neural Networks, ICANN*, volume 1, pages 689–692. Springer Verlag, 1994.

21. D.A. Nix and A.S. Weigend. Estimating the mean and variance of the target probability distribution. In *World Congress of Neural Networks*. Lawrence Erlbaum Associates, 1994.
22. D. Ormoneit. Estimation of Probability Densities using Neural Networks. Master's thesis, Fakultät für Informatik, Technische Universität München, 1993.
23. A. Papoulis. *Probability, Random Variables, and Stochastic Processes*. McGraw Hill, Inc., 3 edition, 1991.
24. M. P. Perrone. *Improving Regression Estimates: Averaging Methods for Variance Reduction with Extensions to General Convex Measure Optimization*. PhD thesis, Brown University, 1993.
25. A. P. Refenes, editor. *Neural Networks in the Capital Market*. Wiley & Sons, 1994.
26. T. D. Sanger. Optimal unsupervised learning in a single-layer linear feedforward network. *Neural Networks*, 2:459–473, 1989.
27. G. A. F. Seber and C. J. Wild. *Nonlinear Regression*. John Wiley & Sons, New York, 1989.
28. A. N. Srivastava and A. S. Weigend. Computing the probability density in connectionist regression. In M. Marinaro and P. G. Morasso, editors, *Proceedings of the International Conference on Artificial Neural Networks, Sorrento, Italy (ICANN 94)*, pages 685–688. Springer-Verlag, 1994. Also in Proceedings of the IEEE International Conference on Neural Networks, Orlando, FL (IEEE–ICNN'94), p. 3786–3789, IEEE-Press.
29. F. Takens. Detecting strange attractors in turbulence. In D. A. Rand and L. S. Young, editors, *Dynamical Systems and Turbulence*, volume 898 of *Lecture Notes in Mathematics*, pages 366–381. Springer, 1981.
30. B. Tang, W. Hsieh, and F. Tangang. Clearning neural networks with continuity constraints for prediction of noisy time series. In *Progres in Neural Information Processing (ICONIP'96)*, pages 722–725, Berlin, 1996. Springer.
31. V. Tresp, R. Neuneier, and H. G. Zimmermann. Early brain damage. In *Advances in Neural Information Processing Systems*, volume 9. MIT Press, 1997.
32. A. S. Weigend and H. G. Zimmermann. Exploiting local relations as soft constraints to improve forecasting. *Computational Intelligence in Finance*, 6(1), January 1998.
33. A. S. Weigend, H. G. Zimmermann, and R. Neuneier. The observer-observation dilemma in neuro-forecasting: Reliable models from unreliable data through clearning. In R. Freedman, editor, *AI Applications on Wall Street, June 1995, New York.*, pages 308–317. Software Engineering Press. (Tel. 301 948-5390), 1995.
34. A. S. Weigend, D. E. Rumelhart, and B. A. Huberman. Generalization by weight-elimination with application to forecasting. In Richard P. Lippmann, John E. Moody, and David S. Touretzky, editors, *Advances in Neural Information Processing Systems*, volume 3, pages 875–882. Morgan Kaufmann, San Mateo, 1991.
35. H. White. Parametrical statistical estimation with artificial neural networks. Technical report, University of California, San Diego, 1991.
36. H. G. Zimmermann and A. S. Weigend. Representing dynamical systems in feedforward networks: A six layer architecture. In A. S. Weigend, Y. Abu-Mostafa, and A.-P. N. Refenes, editors, *Decision Technologies for Financial Engineering: Proceedings of the Fourth International Conference on Neural Networks in the Capital Markets (NNCM-96)*, Singapore, 1997. World Scientific.
37. H. G. Zimmermann. Neuronale Netze als Entscheidungskalkül. In H. Rehkugler and H. G. Zimmermann, editor, *Neuronale Netze in der Ökonomie*. Verlag Franz Vahlen, 1994.

38. H. G. Zimmermann and R. Neuneier. The observer-observation dilemma in neuro-forecasting. In *Advances in Neural Information Processing Systems*, volume 10. MIT Press, 1998.

Author Index

Andersen, L.N., 113

Back, A., 299
Bottou, L., 9
Burns, I., 299

Caruana, R., 165

Denker, J.S., 239

Finke, M., 315
Flake, G.W., 145
Fritsch, J., 315

Giles, C.L., 299

Hansen, L.K., 113
Hirzinger, G., 193
Horn, D., 133

Intrator, N., 133

Larsen, J., 113
Lawrence, S., 299
LeCun, Y., 9, 239
Lyon, R.F., 275

Müller, K.-R., 9
Moody, J., 347

Naftaly, U., 133
Neuneier, R., 373

Orr, G.B., 9

Plate, T., 93, 227
Prechelt, L., 55

Rögnvaldsson, T.S., 71

Schraudolph, N.N., 207
Simard, P.Y., 239
Smagt, P. van der, 193
Svarer, C., 113

Tsoi, A.C., 299

Victorri, B., 239

Webb, B.J., 275

Yaeger, L.S., 275

Zimmermann, H.G., 373

Subject Index

ACID 328
ACID/HNN 328
acoustic modeling 322, 328
– connectionist 326
Apple Computer
– eMate 275
– MessagePad 275
– Newton 275
architectural optimization 405–410
Arrhythmia database 304
asset allocation 419
autoregressive model 305
averaging 381

backpropagation 11, 305
– flatness 198
Bayes rule 316, 326
Bayesian
– *a posteriori* probabilities 299, 301
– model prior 74
bias/variance 73, 360
– decomposition 136
– ensembles 136
box-whiskers plots 305

CART 325
CDEN 388
centering
– activity 209
– error 209
– slope 210
class
– equalizing class membership 303
– frequencies 300
– probabilities 299
classification 122, 315
– ECG 303
– error 320
– figure-of-merit 301
– handwritten characters 278
– k-nearest neighbor 240
– LVQ 240
– medical 310
– memory based 240
– modular 316, 331

– trees 319
cleaning 399
clearning 399
clustering 321
– agglomerative 327
conditional density estimation 388
conditioning 193
confusion matrix 303, 321
conjugate directions 34
context
– geometric 291
– lexical 288
convergence 308
covariance 27
covariance estimation 394
credit assignment 224
cross-validation 115, 338, 361
– generalized (GCV) 360
– nonlinear 361

data
– artificial training data 308
– autonomous vehicle navigation 179
– bilinear problem 80
– combustion engine 81
– data noise 401
– Deterding's vowel 152, 217
– distribution 182
– ECG 303
– hill-plateau 148
– Index of Industrial Production 348, 365
– Mackey-Glass 125
– MIT-BIH Arrhythmia database 304
– NETtalk 184
– NIST1 251
– NIST2 251
– overlapping distributions 309
– Peterson and Barney 122
– pneumonia 173
– power load 80
– Proben1 61
– riverflow 80
– split of data 412
– sunspots 81, 134

428   Subject Index

- Switchboard   333, 338
- test set   61
- training set   61
- two-spirals   150
- US Postal Service   251
- validation set   61
- waste of   61

decimation
- of d.c. errors   209
- of linear errors   210

decision boundaries   308
decision tree   317, 325
- phonetic   330
deformable prototype   241
dendrogram   330
density estimation   388
distance
- Euclidean   240
- invariant   243
- tangent   242, 246
divide & conquer   316
Doonesbury Effect   275, 293

early stopping   75, 410
- in MTL nets   184
effectice number of parameters   361
efficiency   58
- of stopping criteria   60, 64
Eigenvalue spread of hessian   39
ensemble
- averaging   133
- forecast   353
error   387, 391
- approximation   194
- cross-entropy   300
- cross-validation   115
- error-in-variables   400
- estimate   57
- evolution   57
-- idealized   56
-- realistic   57
- generalization   55
- generalized least square estimation 391
- increase   59
- Kullback-Leibler   329
- Likelihood
-- Log   388
-- scaled   326
- LnCosh   387

- local minima   57
- mean squared sensitivity error   304
- metrics   178
- prediction error   359
-- final (FPE)   360
-- generalized (GPE)   361
-- squared (PSE)   360
- residual pattern   194
- training   55
- validation   55
error bar estimation   392–394
estimation
- conditional density   388
- covariance   394
- density   388
- error bar   392–394
- generalization error   57
- robust   387–391
-- with CDEN   388
-- with LnCosh   387
- weight decay parameter of   76

factoring posteriors   327
false positive rate   303
feature selection   182
feed-forward network
- Hessian   196
- Jacobian   196
- linearly augmented   193–205
final stopping   417
focus of attention   179
Fusion   305

GARCH   391
Gaussian mixture   389
generalization   9, 115, 333, 338
generalization loss   58
German bond   419
GL   58
GLSE   391
gradient ascent
- stochastic   336
gradient descent
- regularization parameters   118

handwriting recognition   250, 315
- architecture   279
- online   275
Hessian   25, 196
- backpropagating diagonal hessian   37

- conditioning 223
- Eigenvalues 39
- maximum Eigenvalue 28, 44
-- online computation 44
-- power method 44
-- Taylor expansion 44
- minimum Eigenvalue 22
- product of hessian and vector 38
- shape 41
- singular 196–198
- square Jacobian approximation 35, 36
heteroscedasticity 388, 391
hidden Markov models 323
hidden nodes
- number of 183, 306
hierarchical
- classifier 337, 340
- structure 317
hierarchy of neural networks 327, 328, 331, 334
hyperparameters 351

ill-conditioned 193
ill-posed problems 72
information divergence 328
information shocks 387
input
- normalization 16, 31
-- decorrelation 17
-- equalization 17
-- zero mean 16
- representation 278
inputs
- force 374
- momentum 374
interaction layer 379
invariance transformations 267
- diagonal hyperbolic transformation 269
- parallel hyperbolic transformation 268
- rotation 268
- scaling 268
- thickening 269
- X-translation 268
- Y-translation 268

Jacobian 35, 36, 196
JanusRTk 334

Kullback-Leibler 329

late stopping 411
learning 395
- batch 13
-- advantages 14
- BFGS 34
- bold driver 215
- condition 193–205
- conjugate gradient 33, 150, 195, 308
-- Fletcher-Reeves 33
-- Polak-Ribiere 34
- Gauss-Newton 35
- gradient descent 300
-- convergence 23
-- divergence 24
- hierarchical 182
- hill-climbing 303
- hints 181
- Levenberg-Marquardt 35, 195
- life-long 165
- line search 33
- multitask 166
- Newton 32, 150
- online 13
- Quasi-Newton 34, 195, 396
- rankprop 174
- rate 20, 187, 336
-- adaptation 21, 337
-- annealing 14, 286
-- individual 41
-- maximal 25
-- momentum 21
-- optimal 25
- RPROP 61
- second order 31, 195–196
- single task 166
- stochastic 13
-- advantages 13
- stochastic diagonal Levenberg Marquardt 41
- sub-optimal 307
- tangent propagation 259
- variable metric 195
- vario-η 214, 396
Lie group 263
linearly augmented feed-forward network 193–205
- equivalence with feed-forward network 200

Subject Index

- universal approximation 200
local minimum 193–205, 307
local versus global features 145, 157, 161
LVCSR 322, 323, 326

M-estimators 390
macroeconomic forecasting 347
mean-variance approach 393
medical
- diagnosis 173
- risk prediction 173
MIT-BIH Arrhythmia database 304
mixture
- densities 324, 328
- of Gaussians 327
model
- complexity 333
- interpretation 353
- selection 319, 352, 357
-- architecture 358
-- input 358
MTL 166
multinomial pdf 335

network information criterion (NIC) 361
neural network
- architecture 375–386
- bottleneck network 376
- capacity 183
- expert 327
- feed-forward 123, 125
- internal preprocessing 375
- linear 28
- multi-layer perceptron 28, 317, 333
- size 183
- vanilla 350
noise
- Gaussian 387, 395
- Laplacian 387
noise reduction 223
noise/nonstationarity tradeoff 349
nonlinear calibration 311

Observer-Observation Dilemma 395, 403
outliers 376, 387, 390
output representation 178
overfitting 55, 57

parameter noise 397
penalty

- implicit 397
- implied curvature 404
performance
- criteria 303
- measures 303
polyphones 325
portfolio management 393
positive predictivity 303
power method 44
PQ 59
predicate
- stopping 58
predicting
- posteriors 317
Premature Ventricular Contraction 305
preprocessing
- by bottleneck network 376
- by diagonal connector 375
- by squared inputs 378
- of data 374–375
price/performance ratio 58
prior 74
- bias reduction 294
- knowledge 246, 320
- probabilities 299
- scaling 300
probabilistic sampling 302
pronunciation graph 323
pruning 363, 405–410
- Early-Brain-Damage 406
- input variables 351, 363
- Instability-Pruning 409
- Inverse-Kurtosis 407
- of nodes 405
- of weights 405
- sensitivity-based (SBP) 363
- Stochastic-Pruning 405

radial basis function network 148, 157, 378
recurrent network 317
regularization 56, 73, 113, 116, 184, 245, 352
- adaptive 117
- choice of 117
- for recurrent networks 355
- smoothing 354
- weight decay 120, 123, 125
representational capacity 308
risk 394

Subject Index   431

robust estimation   387–391
– with CDEN   388
– with LnCosh   387

saddle point   193
scaling   300
search
– context driven   288
segmentation
– character   277
– word   292
SENN   420
sensitivity   303
– average absolute gradient   364
– average gradient   364
– delta output   366
– evolution   366
– output gradient   366
– pruning   363
– RMS gradient   364
separation of structure and noise   403
setup steps   412
shortcut connections   210
shuffle test   169
simulation
– weight decay   82
softmax   335
speech recognition   300, 315, 322
– connectionist   316
– statistical   317, 322
– system   334
sphering   32
squared inputs   145, 378
state tying   326
STL   166
stopping criterion   58
– early stopping   410
– efficiency   60
– final stopping   417
– late stopping   411
– robustness   60, 64
– rules for selecting   60
– threshold
– – for stopping training   58
– tradeoff
– – time versus error   58, 60, 65
structural instability   386
Structure-Speed-Dilemma   399
Supraventricular Contraction   305

tangent propagation   257
targets
– embeddings   384
– forces   384
– random   382
test set   61
time series
– linear models   349
– noise   349
– nonlinearity   349
– nonstationarity   349
– prediction   125, 135, 178
training
– algorithm   336
– – RPROP   61
– procedure   410–418
– set   61
transfer   165
– inductive   165
– sequential vs. parallel   181
tricks
– backpropagating diagonal hessian   37
– backpropagating second derivatives   37
– choice of targets   19
– computing hessian information   36
– data warping   283
– ensemble with different initial weights   133
– error emphasis   285
– frequency balancing   302
– Hessian
– – finite difference   36
– – maximum Eigenvalue   44
– – minimum Eigenvalue   22
– – square Jacobian approximation   36
– initializing the weights   20
– multitask learning   166
– negative training   282
– nonlinear cross-validation   361
– normalizing output errors   281
– post scaling   302
– prior frequency balancing   284
– prior scaling   300
– sensitivity-based pruning   363
– shuffling examples   15
– sigmoid   18
– squared input units   145
– tangent distance   252
– – elastic tangent distance   253

– – hierarchy of distances   255
– – smoothing   253
– – speeding   255
– tangent propagation   263
– training big nets   316
– variance reduction with ensembles   133
– weight decay parameter estimate   77, 79
triphone models   325

unequal misclassification costs   299
unit
– radial basis function   22
– sigmoid   18
– squared input   145
UP   59
user adaptation   296

validation set   55, 61
variable metric   195
variance   62
VC dimension   245
visualization   353, 365

weight
– decay   75, 307
– – parameter estimate   76, 77
– elimination   307
– quantized (low precision)   287
– updates
– – magnitude of   302
whitening   32
word error rate   338

Lecture Notes in Computer Science

For information about Vols. 1–1447

please contact your bookseller or Springer-Verlag

Vol. 1448: M. Farach-Colton (Ed.), Combinatorial Pattern Matching. Proceedings, 1998. VIII, 251 pages. 1998.

Vol. 1449: W.-L. Hsu, M.-Y. Kao (Eds.), Computing and Combinatorics. Proceedings, 1998. XII, 372 pages. 1998.

Vol. 1450: L. Brim, F. Gruska, J. Zlatuška (Eds.), Mathematical Foundations of Computer Science 1998. Proceedings, 1998. XVII, 846 pages. 1998.

Vol. 1451: A. Amin, D. Dori, P. Pudil, H. Freeman (Eds.), Advances in Pattern Recognition. Proceedings, 1998. XXI, 1048 pages. 1998.

Vol. 1452: B.P. Goettl, H.M. Halff, C.L. Redfield, V.J. Shute (Eds.), Intelligent Tutoring Systems. Proceedings, 1998. XIX, 629 pages. 1998.

Vol. 1453: M.-L. Mugnier, M. Chein (Eds.), Conceptual Structures: Theory, Tools and Applications. Proceedings, 1998. XIII, 439 pages. (Subseries LNAI).

Vol. 1454: I. Smith (Ed.), Artificial Intelligence in Structural Engineering. XI, 497 pages. 1998. (Subseries LNAI).

Vol. 1455: A. Hunter, S. Parsons (Eds.), Applications of Uncertainty Formalisms. VIII, 474 pages. 1998. (Subseries LNAI).

Vol. 1456: A. Drogoul, M. Tambe, T. Fukuda (Eds.), Collective Robotics. Proceedings, 1998. VII, 161 pages. 1998. (Subseries LNAI).

Vol. 1457: A. Ferreira, J. Rolim, H. Simon, S.-H. Teng (Eds.), Solving Irregularly Structured Problems in Prallel. Proceedings, 1998. X, 408 pages. 1998.

Vol. 1458: V.O. Mittal, H.A. Yanco, J. Aronis, R-. Simpson (Eds.), Assistive Technology in Artificial Intelligence. X, 273 pages. 1998. (Subseries LNAI).

Vol. 1459: D.G. Feitelson, L. Rudolph (Eds.), Job Scheduling Strategies for Parallel Processing. Proceedings, 1998. VII, 257 pages. 1998.

Vol. 1460: G. Quirchmayr, E. Schweighofer, T.J.M. Bench-Capon (Eds.), Database and Expert Systems Applications. Proceedings, 1998. XVI, 905 pages. 1998.

Vol. 1461: G. Bilardi, G.F. Italiano, A. Pietracaprina, G. Pucci (Eds.), Algorithms – ESA'98. Proceedings, 1998. XII, 516 pages. 1998.

Vol. 1462: H. Krawczyk (Ed.), Advances in Cryptology - CRYPTO '98. Proceedings, 1998. XII, 519 pages. 1998.

Vol. 1463: N.E. Fuchs (Ed.), Logic Program Synthesis and Transformation. Proceedings, 1997. X, 343 pages. 1998.

Vol. 1464: H.H.S. Ip, A.W.M. Smeulders (Eds.), Multimedia Information Analysis and Retrieval. Proceedings, 1998. VIII, 264 pages. 1998.

Vol. 1465: R. Hirschfeld (Ed.), Financial Cryptography. Proceedings, 1998. VIII, 311 pages. 1998.

Vol. 1466: D. Sangiorgi, R. de Simone (Eds.), CONCUR'98: Concurrency Theory. Proceedings, 1998. XI, 657 pages. 1998.

Vol. 1467: C. Clack, K. Hammond, T. Davie (Eds.), Implementation of Functional Languages. Proceedings, 1997. X, 375 pages. 1998.

Vol. 1468: P. Husbands, J.-A. Meyer (Eds.), Evolutionary Robotics. Proceedings, 1998. VIII, 247 pages. 1998.

Vol. 1469: R. Puigjaner, N.N. Savino, B. Serra (Eds.), Computer Performance Evaluation. Proceedings, 1998. XIII, 376 pages. 1998.

Vol. 1470: D. Pritchard, J. Reeve (Eds.), Euro-Par'98: Parallel Processing. Proceedings, 1998. XXII, 1157 pages. 1998.

Vol. 1471: J. Dix, L. Moniz Pereira, T.C. Przymusinski (Eds.), Logic Programming and Knowledge Representation. Proceedings, 1997. IX, 246 pages. 1998. (Subseries LNAI).

Vol. 1472: B. Freitag, H. Decker, M. Kifer, A. Voronkov (Eds.), Transactions and Change in Logic Databases. Proceedings, 1996, 1997. X, 396 pages. 1998.

Vol. 1473: X. Leroy, A. Ohori (Eds.), Types in Compilation. Proceedings, 1998. VIII, 299 pages. 1998.

Vol. 1474: F. Mueller, A. Bestavros (Eds.), Languages, Compilers, and Tools for Embedded Systems. Proceedings, 1998. XIV, 261 pages. 1998.

Vol. 1475: W. Litwin, T. Morzy, G. Vossen (Eds.), Advances in Databases and Information Systems. Proceedings, 1998. XIV, 369 pages. 1998.

Vol. 1476: J. Calmet, J. Plaza (Eds.), Artificial Intelligence and Symbolic Computation. Proceedings, 1998. XI, 309 pages. 1998. (Subseries LNAI).

Vol. 1477: K. Rothermel, F. Hohl (Eds.), Mobile Agents. Proceedings, 1998. VIII, 285 pages. 1998.

Vol. 1478: M. Sipper, D. Mange, A. Pérez-Uribe (Eds.), Evolvable Systems: From Biology to Hardware. Proceedings, 1998. IX, 382 pages. 1998.

Vol. 1479: J. Grundy, M. Newey (Eds.), Theorem Proving in Higher Order Logics. Proceedings, 1998. VIII, 497 pages. 1998.

Vol. 1480: F. Giunchiglia (Ed.), Artificial Intelligence: Methodology, Systems, and Applications. Proceedings, 1998. IX, 502 pages. 1998. (Subseries LNAI).

Vol. 1481: E.V. Munson, C. Nicholas, D. Wood (Eds.), Principles of Digital Document Processing. Proceedings, 1998. VII, 152 pages. 1998.

Vol. 1482: R.W. Hartenstein, A. Keevallik (Eds.), Field-Programmable Logic and Applications. Proceedings, 1998. XI, 533 pages. 1998.

Vol. 1483: T. Plagemann, V. Goebel (Eds.), Interactive Distributed Multimedia Systems and Telecommunication Services. Proceedings, 1998. XV, 326 pages. 1998.

Vol. 1484: H. Coelho (Ed.), Progress in Artificial Intelligence – IBERAMIA 98. Proceedings, 1998. XIII, 421 pages. 1998. (Subseries LNAI).

Vol. 1485: J.-J. Quisquater, Y. Deswarte, C. Meadows, D. Gollmann (Eds.), Computer Security – ESORICS 98. Proceedings, 1998. X, 377 pages. 1998.

Vol. 1486: A.P. Ravn, H. Rischel (Eds.), Formal Techniques in Real-Time and Fault-Tolerant Systems. Proceedings, 1998. VIII, 339 pages. 1998.

Vol. 1487: V. Gruhn (Ed.), Software Process Technology. Proceedings, 1998. VIII, 157 pages. 1998.

Vol. 1488: B. Smyth, P. Cunningham (Eds.), Advances in Case-Based Reasoning. Proceedings, 1998. XI, 482 pages. 1998. (Subseries LNAI).

Vol. 1489: J. Dix, L. Fariñas del Cerro, U. Furbach (Eds.), Logics in Artificial Intelligence. Proceedings, 1998. X, 391 pages. 1998. (Subseries LNAI).

Vol. 1490: C. Palamidessi, H. Glaser, K. Meinke (Eds.), Principles of Declarative Programming. Proceedings, 1998. XI, 497 pages. 1998.

Vol. 1491: W. Reisig, G. Rozenberg (Eds.), Lectures on Petri Nets I: Basic Models. XII, 683 pages. 1998.

Vol. 1492: W. Reisig, G. Rozenberg (Eds.), Lectures on Petri Nets II: Applications. XII, 479 pages. 1998.

Vol. 1493: J.P. Bowen, A. Fett, M.G. Hinchey (Eds.), ZUM '98: The Z Formal Specification Notation. Proceedings, 1998. XV, 417 pages. 1998.

Vol. 1494: G. Rozenberg, F. Vaandrager (Eds.), Lectures on Embedded Systems. Proceedings, 1996. VIII, 423 pages. 1998.

Vol. 1495: T. Andreasen, H. Christiansen, H.L. Larsen (Eds.), Flexible Query Answering Systems. IX, 393 pages. 1998. (Subseries LNAI).

Vol. 1496: W.M. Wells, A. Colchester, S. Delp (Eds.), Medical Image Computing and Computer-Assisted Intervention – MICCAI'98. Proceedings, 1998. XXII, 1256 pages. 1998.

Vol. 1497: V. Alexandrov, J. Dongarra (Eds.), Recent Advances in Parallel Virtual Machine and Message Passing Interface. Proceedings, 1998. XII, 412 pages. 1998.

Vol. 1498: A.E. Eiben, T. Bäck, M. Schoenauer, H.-P. Schwefel (Eds.), Parallel Problem Solving from Nature – PPSN V. Proceedings, 1998. XXIII, 1041 pages. 1998.

Vol. 1499: S. Kutten (Ed.), Distributed Computing. Proceedings, 1998. XII, 419 pages. 1998.

Vol. 1501: M.M. Richter, C.H. Smith, R. Wiehagen, T. Zeugmann (Eds.), Algorithmic Learning Theory. Proceedings, 1998. XI, 439 pages. 1998. (Subseries LNAI).

Vol. 1502: G. Antoniou, J. Slaney (Eds.), Advanced Topics in Artificial Intelligence. Proceedings, 1998. XI, 333 pages. 1998. (Subseries LNAI).

Vol. 1503: G. Levi (Ed.), Static Analysis. Proceedings, 1998. IX, 383 pages. 1998.

Vol. 1504: O. Herzog, A. Günter (Eds.), KI-98: Advances in Artificial Intelligence. Proceedings, 1998. XI, 355 pages. 1998. (Subseries LNAI).

Vol. 1506: R. Koch, L. Van Gool (Eds.), 3D Structure from Multiple Images of Large-Scale Environments. Proceedings, 1998. VIII, 347 pages. 1998.

Vol. 1507: T.W. Ling, S. Ram, M.L. Lee (Eds.), Conceptual Modeling – ER '98. Proceedings, 1998. XVI, 482 pages. 1998.

Vol. 1508: S. Jajodia, M.T. Özsu, A. Dogac (Eds.), Advances in Multimedia Information Systems. Proceedings, 1998. VIII, 207 pages. 1998.

Vol. 1510: J.M. Zytkow, M. Quafafou (Eds.), Principles of Data Mining and Knowledge Discovery. Proceedings, 1998. XI, 482 pages. 1998. (Subseries LNAI).

Vol. 1511: D. O'Hallaron (Ed.), Languages, Compilers, and Run-Time Systems for Scalable Computers. Proceedings, 1998. IX, 412 pages. 1998.

Vol. 1512: E. Giménez, C. Paulin-Mohring (Eds.), Types for Proofs and Programs. Proceedings, 1996. VIII, 373 pages. 1998.

Vol. 1513: C. Nikolaou, C. Stephanidis (Eds.), Research and Advanced Technology for Digital Libraries. Proceedings, 1998. XV, 912 pages. 1998.

Vol. 1514: K. Ohta,, D. Pei (Eds.), Advances in Cryptology – ASIACRYPT'98. Proceedings, 1998. XII, 436 pages. 1998.

Vol. 1515: F. Moreira de Oliveira (Ed.), Advances in Artificial Intelligence. Proceedings, 1998. X, 259 pages. 1998. (Subseries LNAI).

Vol. 1516: W. Ehrenberger (Ed.), Computer Safety, Reliability and Security. Proceedings, 1998. XVI, 392 pages. 1998.

Vol. 1517: J. Hromkovič, O. Sýkora (Eds.), Graph-Theoretic Concepts in Computer Science. Proceedings, 1998. X, 385 pages. 1998.

Vol. 1518: M. Luby, J. Rolim, M. Serna (Eds.), Randomization and Approximation Techniques in Computer Science. Proceedings, 1998. IX, 385 pages. 1998.

Vol. 1520: M. Maher, J.-F. Puget (Eds.), Principles and Practice of Constraint Programming - CP98. Proceedings, 1998. XI, 482 pages. 1998.

Vol. 1521: B. Rovan (Ed.), SOFSEM'98: Theory and Practice of Informatics. Proceedings, 1998. XI, 453 pages. 1998.

Vol. 1522: G. Gopalakrishnan, P. Windley (Eds.), Formal Methods in Computer-Aided Design. Proceedings, 1998. IX, 529 pages. 1998.

Vol. 1524: G.B. Orr, K.-R. Müller (Eds.), Neural Networks: Tricks of the Trade. VI, 432 pages. 1998.

Vol. 1526: M. Broy, B. Rumpe (Eds.), Requirements Targeting Software and Systems Engineering. Proceedings, 1997. VIII, 357 pages. 1998.

Vol. 1529: D. Farwell, L. Gerber, E. Hovy (Eds.), Machine Translation and the Information Soup. Proceedings, 1998. XIX, 532 pages. 1998. (Subseries LNAI).

Vol. 1531: H.-Y. Lee, H. Motoda (Eds.), PRICAI'98: Topics in Artificial Intelligence. XIX, 646 pages. 1998. (Subseries LNAI).

Vol. 1096: T. Schael, Workflow Management Systems for Process Organisations. Second Edition. XII, 229 pages. 1998.